# Polyhedral and Semidefinite Programming Methods in Combinatorial Optimization

# FIELDS INSTITUTE MONOGRAPHS

THE FIELDS INSTITUTE FOR RESEARCH IN MATHEMATICAL SCIENCES

# Polyhedral and Semidefinite Programming Methods in Combinatorial Optimization

Levent Tunçel

**American Mathematical Society**
Providence, Rhode Island

**The Fields Institute**
**for Research in Mathematical Sciences**
Toronto, Ontario

FIELDS

2010 *Mathematics Subject Classification.* Primary 90C22, 90C27, 15A39, 52A41, 65Y20, 05C69, 90C05, 90C25, 90C51, 68Q25.

For additional information and updates on this book, visit
**www.ams.org/bookpages/fim-27**

**Library of Congress Cataloging-in-Publication Data**

Tunçel, Levent, 1965–
Polyhedral and semidefinite programming methods in combinatorial optimization / Levent Tunçel.
p. cm. — (Fields Institute monographs ; 27)
Includes bibliographical references and index.
ISBN 978-0-8218-3352-0 (alk. paper)
1. Semidefinite programming. 2. Combinatorial optimization. 3. Convex Geometry. I. Title.

QA402.5.T86 2010
519.7—dc22

2010031316

AMS softcover ISBN: 978-1-4704-2811-2



∞ The paper used in this book is acid-free and falls within the guidelines
established to ensure permanence and durability.
This publication was prepared by the Fields Institute.
`http://www.fields.utoronto.ca`
Visit the AMS home page at `http://www.ams.org/`

10 9 8 7 6 5 4 3 2 1 21 20 19 18 17 16

# Contents

# Preface

Since the early 1960's, polyhedral methods have had a central role to play in both the theory and practice of combinatorial optimization. Since the early 1990's, a new technique, semidefinite programming, has been increasingly applied to some combinatorial optimization problems. The semidefinite programming problem is the problem of optimizing a linear function of matrix variables, subject to finitely many linear inequalities and the positive semidefiniteness condition on some of the matrix variables. On certain problems, such as maximum cut, maximum satisfiability, maximum stable set and geometric representations of graphs, semidefinite programming techniques yield important, new results. In this monograph, we provide the necessary background to work with semidefinite optimization techniques, usually by drawing parallels to the development of polyhedral techniques and with a special focus on combinatorial optimization, graph theory and lift-and-project methods.

The core of this monograph is based on ten lectures given at the Fields Institute during the academic term Fall-1999. This activity was a part of a special year of activities at the Fields Institute under the heading *Graph Theory and Combinatorial Optimization*. During the terms Fall-2001, Fall-2003, Spring-2005, Spring-2006, Spring-2007 as well as Fall-2008, I gave a course entitled *Semidefinite Optimization* at the Department of Combinatorics and Optimization, Faculty of Mathematics, University of Waterloo. During the course, I used and expanded some of the material from my Fields Institute lectures. These lecture notes and the handouts that I prepared evolved to the current monograph.

As prerequisites for this monograph, a solid background in mathematics at the undergraduate level and some exposure to linear optimization are required. Some familiarity with computational complexity theory and the analysis of algorithms would be helpful.

The chapters are exactly in the same order as the lectures. The first chapter familiarizes the audience with the basic concepts, notation, and lays down some theory to motivate the focus of the monograph, sometimes by way of analogy to the mainstream polyhedral approaches. Duality theory is paramount. As a result, instead of continuing with the material in Chapter 12 (which covers some examples of convex sets that can be represented as the feasible regions of Semidefinite Optimization problems) which would be the right way to go for an application-oriented audience, I took a risk and chose to cover duality theory as early as possible (Chapter 2). Then comes the theory of algorithms for convex optimization (Chapters 3 and 4). In Chapter 3, I give a quick overview of the Ellipsoid Method and in Chapter 4, I go through the theory of interior-point methods, with a focus on symmetric, primal-dual algorithms. This portion of the monograph (Chapters 1–4) aims to establish rigorously most of the fundamental tools needed for Semidefinite Optimization. Chapters 5 and 6 cover various impressive results in Combinatorial Optimization and Graph Theory involving Semidefinite Optimization in a central way. Chapter 7 starts moving towards more abstract approaches in combinatorial optimization which use Semidefinite Optimization (Lift-and-Project Operators). Chapter 8 covers some of the basic techniques to analyze Lift-and-Project procedures with a special emphasis on the stable set problem. Chapter 9 considers yet further abstraction and generalization of these methods and prepares the audience for Chapter 10. The latter chapter is a collection of pointers to various wonderful results, some from other areas of mathematics and some establishing connections to such areas. Chapter 11 is a quick application to a cute theorem in number theory. Chapter 12 brings the lectures to a close in a nice, straightforward way with some obviously interesting open questions. Open problems of seemingly varying difficulty have been sprinkled throughout the text.

I thank Joseph Cheriyan for providing many interesting references over the years and Steven Karp, Graeme Kemkes, Lingchen Kong, Cristiane Sato, and Marcel Silva for many very useful remarks on earlier versions of the monograph. I also thank six anonymous referees for their very useful remarks and suggestions. I thank the editor Carl Riehm for his work and patience.

My research efforts were supported in part by Natural Sciences and Engineering Research Council of Canada, and by a Premier's Research Excellence Award from Ontario, Canada. The support is gratefully acknowledged.

Levent Tunçel
Waterloo, Canada 2010

CHAPTER 1

# Introduction

We will investigate various problems in several branches of mathematics from the point of view of linear optimization with symmetric positive semidefinite matrices; that is, the problem of minimizing or maximizing a linear function of a finite dimensional symmetric matrix (variable) subject to finitely many linear equations and inequalities on this matrix variable and the important restriction that the matrix variable be positive semidefinite. Special attention will be given to those relevant discrete mathematics problems arising in combinatorial optimization and graph theory.

## 1.1 Linear Programming

We denote $n$-dimensional Euclidean space over the real numbers by $\mathbb{R}^n$. Given $c \in \mathbb{R}^n \setminus \{0\}$ and $\gamma \in \mathbb{R}$, the set

$$\left\{x \in \mathbb{R}^n : \; c^T x \leq \gamma\right\}$$

defines a *closed half-space* in $\mathbb{R}^n$. We begin with the most important geometric object in linear programming, a *polyhedron*. $P \subseteq \mathbb{R}^n$ is called a *polyhedron* if it can be expressed as the intersection of finitely many closed half-spaces in $\mathbb{R}^n$. Then the *Linear Programming (LP) problem* is the problem of optimizing (computing the minimum or the maximum value of) a linear function of $n$ variables (called the *objective function*) over a polyhedron in $\mathbb{R}^n$. We usually want an *optimal solution* as well (not just the optimum value of the linear function over the polyhedron, but an element of the polyhedron—*a feasible solution*—which attains the *optimum objective value*).

Let $A \in \mathbb{R}^{m \times n}$, $b \in \mathbb{R}^m$ and $c \in \mathbb{R}^n$ be given. Then, we have the LP problem

$$\begin{array}{lllcl} (LP) & \text{minimize} & c^T x & & \\ & \text{subject to:} & Ax & = & b, \\ & & x & \geq & 0 \end{array}$$

called the *primal* problem. Here, $x \in \mathbb{R}^n$ and we can assume $\text{rank}(A) = m$. (Thus, $m \leq n$ also.) Since otherwise, upon application of Gaussian Elimination to $Ax = b$, we either prove that the problem $(LP)$ has no solution (*infeasible*) or we find all redundant equations in $Ax = b$; we eliminate the redundant equations and redefine $A$ and $b$ accordingly. Gaussian Elimination can be performed efficiently. (We will discuss the notion of efficiency a little bit more rigorously when we discuss computational complexity.)

The *dual* of $(LP)$ is defined as

$$\begin{array}{llllcl} (LD) & \text{maximize} & b^T y & & & \\ & \text{subject to:} & A^T y & + & s = & c, \\ & & & & s \geq & 0. \end{array}$$

When referring to a solution of $A^T y + s = c$, we sometimes ignore $s$ or $y$. (One determines the other uniquely, since $A$ is of full row rank.)

The polyhedron defining the feasible region of $(LP)$ is

$$\{x \in \mathbb{R}^n : Ax = b, x \geq 0\}.$$

The polyhedra corresponding to $(LD)$ are

$$\{y \in \mathbb{R}^m : A^T y \leq c\}, \ \ \{(y, s) \in \mathbb{R}^m \oplus \mathbb{R}^n_+ : A^T y + s = c\}$$
$$\text{and } \{s \in \mathbb{R}^n : Fs = Fc, s \geq 0\},$$

where the rows of $F \in \mathbb{R}^{(n-m)\times n}$ form a basis for the null space of $A$.

Sometimes it is convenient to express polyhedra as

$$\{x \in \mathbb{R}^n : \bar{A}x = \bar{b}, \hat{A}x \leq \hat{b}\},$$

where $\bar{A} \in \mathbb{R}^{r\times n}$, $\text{rank}(\bar{A}) = r$ and $\hat{A} \in \mathbb{R}^{(m-r)\times n}$ for some $m \geq r \geq 0$.

We denote by $\bar{e}$, the vector of all ones. The size of $\bar{e}$ should be clear from the context. A polyhedron $P \subset \mathbb{R}^n$ is called a *polytope* if $P$ is bounded; that is, there exists a positive constant $M$ such that

$$P \subseteq \{x \in \mathbb{R}^n : -M\bar{e} \leq x \leq M\bar{e}\}.$$

A polyhedron $P$ is called *rational* if

$$P = \{x \in \mathbb{R}^n : Ax \leq b\},$$

where $A \in \mathbb{Q}^{m\times n}$, $b \in \mathbb{Q}^m$. In words, a polyhedron is rational if it admits a linear algebraic inequality representation with all rational coefficients.

In solving and manipulating system of linear equations, Gaussian Elimination is a very useful tool. For systems of linear inequalities an analogous role is served by Fourier-Motzkin Elimination. The following theorem of the alternative can be proven using Fourier-Motzkin Elimination (see for instance, Chvátal [85]).

**Theorem 1.1** *Let $A \in \mathbb{R}^{m\times n}$, $b \in \mathbb{R}^n$ be given. Then exactly one of the following systems has a solution:*

(I) $Ax \leq b$;
(II) $A^T u = 0, u \geq 0, b^T u < 0$.

Proving that (I) and (II) above cannot both have solutions is as easy as proving the next fact. We denote by $\mathbb{R}^n_+$, the nonnegative orthant in $\mathbb{R}^n$.

**Proposition 1.2** *(Weak Duality Relation) Let $A, b, c$ be given as above, defining $(LP)$ and $(LD)$. Then, for every $\bar{x}$ feasible in $(LP)$, for every $\bar{y}$ feasible in $(LD)$,*

$$c^T\bar{x} \geq b^T\bar{y}.$$

**Proof** Let $\bar{x}$, $\bar{y}$ be feasible in $(LP)$ and $(LD)$ respectively. Then

$$c^T\bar{x} = (A^T\bar{y} + \bar{s})^T\bar{x} = b^T\bar{y} + \bar{x}^T\bar{s} \geq b^T\bar{y},$$

where the last inequality used $x, s \in \mathbb{R}^n_+$. □

Note that any LP problem can be transformed into the form of $(LP)$ or $(LD)$; however, these transformations do not necessarily preserve all geometric properties of the set of feasible solutions. So, the theory developed for one form of LP can be used for the other after following such a transformation through and translating the conclusions of results to the original form. For example, consider the primal-dual pair of LP problems

$$\min\left\{c^T x : Ax \le b\right\}$$

and

$$\max\left\{b^T y : A^T y = c, y \le 0\right\}.$$

The weak duality relation (Proposition 1.2) applies exactly the same to this pair.

Similarly, using Theorem 1.1, we can prove many other theorems of the alternative for various forms of systems of linear equations and inequalities. The following basic, but fundamental fact is known as Farkas' Lemma.

**Lemma 1.3** *Let $A \in \mathbb{R}^{m\times n}$, $b \in \mathbb{R}^m$ be given. Then exactly one of the following systems has a solution:*

(I) $Ax = b, x \ge 0$;
(II) $A^T y \le 0, b^T y > 0$.

**Proof** Let

$$\tilde{A} := \begin{bmatrix} A \\ -A \\ -I \end{bmatrix}, \quad \tilde{b} := \begin{bmatrix} b \\ -b \\ 0 \end{bmatrix}.$$

Then there exists $v \in \mathbb{R}^n$ such that $\tilde{A}v \le \tilde{b}$ iff there exists $x \in \mathbb{R}^n_+$ such that $Ax = b$. Applying Theorem 1.1 to the system $\tilde{A}v \le \tilde{b}$, we conclude that exactly one of the following systems has a solution:

(I) $\tilde{A}v \le \tilde{b}$ (or equivalently, $Ax = b, x \ge 0$);
(II) $A^T u - A^T w - z = 0, u \ge 0, w \ge 0, z \ge 0, b^T(u - w) < 0$
(or equivalently, $A^T y \le 0, b^T y > 0$).

□

We say that an *LP problem is unbounded* if for every given real number $M$, there exists a feasible solution of the LP with objective value strictly better than $M$. That is, there exists a sequence $\left\{x^{(k)}\right\}$ of vectors in the feasible region such that $c^T x^{(k)} \to -\infty$, for a minimization problem. Now, we are ready to prove the Fundamental Theorem of Linear Programming.

**Theorem 1.4** *(Fundamental Theorem of LP) Every LP problem has exactly one of the following properties:*

(a) *LP is infeasible;*
(b) *LP is unbounded;*
(c) *LP has an optimal solution.*

**Proof** Let $P \subseteq \mathbb{R}^n$ be a polyhedron. Without loss of generality, we may assume that our LP problem is

$$\min\left\{c^T x : x \in P\right\}$$

and

$$P = \{x \in \mathbb{R}^n : Ax \le b\}.$$

First of all, no LP can have more than one of the three properties listed. Secondly, if $P = \emptyset$, then (a) holds. So, we may assume that $P \ne \emptyset$. If the LP has an optimal solution, then (c) holds. Hence, we may assume $P \ne \emptyset$ and the LP problem has no optimal solution. Then

- either there exists $\left\{x^{(k)}\right\} \subseteq P$ such that $c^T x^{(k)} \to -\infty$,

- or, there exists $\{x^{(k)}\} \subseteq P$ such that $c^T x^{(k)} \to v$ and $v < c^T x$ for every $x \in P$.

In the former case, the LP problem satisfies (b). Thus, we may assume we are in the latter case. Then, the system

$$Ax \leq b, c^T x \leq v$$

has no solution. By Theorem 1.1, there exist $u \geq 0$, $\alpha \geq 0$ such that

$$A^T u = -\alpha c, b^T u < -\alpha v.$$

If $\alpha > 0$, then upon setting $\bar{y} := -\frac{u}{\alpha}$, we obtain

$$\bar{y} \leq 0, A^T \bar{y} = c, b^T \bar{y} > v,$$

which contradicts Proposition 1.2 for the LP $\min \{c^T x : x \in P\}$ and its dual. So, $\alpha$ must be zero. Then, we have $u \geq 0$ such that $A^T u = 0$ and $b^T u < 0$. Thus by Theorem 1.1, the system $Ax \leq b$ has no solution, a contradiction to $P \neq \emptyset$. Therefore, for every LP problem exactly one of (a), (b), (c) must hold. □

Every polyhedron is a *convex set*. ($G \subseteq \mathbb{R}^n$ is *convex* if for every pair $u, v \in G$ the line segment $\{\lambda u + (1 - \lambda)v : \lambda \in [0, 1]\}$ lies in $G$.) Let $G \subseteq \mathbb{R}^n$ be a convex set. Then $\bar{x} \in G$ is an *extreme point* of $G$ if there do not exist $u, v \in G$ such that $u \neq v$ and $\bar{x} = \frac{1}{2}u + \frac{1}{2}v$. Note that if a polyhedron $P$ contains a line then $P$ cannot have any extreme points. This brings us to the next definition.

We say that a polyhedron $P \subseteq \mathbb{R}^n$ is *pointed* if $P$ contains no line. Also, in general, a convex set $G \subseteq \mathbb{R}^n$ is *pointed*, if $G$ contains no line.

**Theorem 1.5** *If the feasible region of an LP problem is a pointed polyhedron, then:*

(a) *whenever the LP problem is feasible, it has a feasible solution that is an extreme point of the feasible region;*
(b) *whenever the LP problem has optimal solution(s), it has an optimal solution that is an extreme point of the feasible region.*

**Proof** (a) Let $P \subseteq \mathbb{R}^n$ denote the feasible region of LP. Since $P$ is a polyhedron,

$$P = \{x \in \mathbb{R}^n : Ax \leq b\},$$

for some $A \in \mathbb{R}^{m \times n}$ and $b \in \mathbb{R}^m$. Suppose LP has a feasible solution, and $P$ is pointed. Let $\bar{x} \in P$ be such a feasible solution that among all feasible solutions, $\bar{x}$ satisfies the largest number of inequalities from $Ax \leq b$ as equality. Partition the rows of $A$ and $b$ as follows (after reordering if necessary):

$$\begin{bmatrix} A^= \\ A^< \end{bmatrix} := A, \quad \begin{bmatrix} b^= \\ b^< \end{bmatrix} := b$$

such that

$$A^= \bar{x} = b^= \text{ and } A^< \bar{x} < b^<.$$

If $\bar{x}$ is an extreme point of $P$ then we are done. So, we may assume, $\bar{x}$ is not extreme. Then, there exists $h \in \mathbb{R}^n \setminus \{0\}$ such that

$$(\bar{x} - h), (\bar{x} + h) \in P.$$

Consider the largest $\lambda_1$ (possibly $+\infty$) such that $(\bar{x} - \lambda_1 h) \in P$ and the largest $\lambda_2$ (again, possibly $+\infty$) such that $(\bar{x} + \lambda_2 h) \in P$. Since $P$ does not contain any lines,

$$\lambda^* := \min\{\lambda_1, \lambda_2\} \in [1, +\infty).$$

Consider

$$\hat{x} := \begin{cases} \bar{x} - \lambda^* h, & \text{if } \lambda^* = \lambda_1, \\ \bar{x} + \lambda^* h, & \text{if } \lambda^* = \lambda_2. \end{cases}$$

Since $A^= \bar{x} = b$ and $(\bar{x} - h), (\bar{x} + h) \in P$, we have $A^= h = 0$. Thus, $A^= \hat{x} = b^=$. Moreover, $A^< \hat{x} \leq b^<$; however, $\hat{x}$ satisfies at least one more inequality of $Ax \leq b$ as equality than $\bar{x}$ did, a contradiction to the choice of $\bar{x}$.

(b) Let $v^*$ be the optimum objective value of the LP. Consider the polyhedron

$$\bar{P} := P \cap \{x \in \mathbb{R}^n : c^T x = v^*\}.$$

Since the LP has an optimal solution, $\bar{P} \neq \emptyset$. Since $P$ is pointed and $\bar{P} \subseteq P$, $\bar{P}$ is pointed. Therefore, part (a) applies and yields the desired result.

□

**Theorem 1.6** *Let $A, b, c$ be given as above, defining $(LP)$ and $(LD)$. Then*

(a) *(Weak Duality Relation) For every $\bar{x}$ feasible in $(LP)$, for every $\bar{y}$ feasible in $(LD)$, $c^T \bar{x} \geq b^T \bar{y}$; moreover, if in addition, $c^T \bar{x} = b^T \bar{y}$ then $\bar{x}$ is optimal in $(LP)$ and $\bar{y}$ is optimal in $(LD)$;*

(b) *(Duality Theorem, type I) if $(LP)$ has an optimal solution, then so does $(LD)$ and their optimum values coincide;*

(c) *(Duality Theorem, type II) if $(LP)$ and $(LD)$ both have feasible solutions, then they both have optimal solutions and their optimum values are the same.*

**Proof** (a) We already proved the first portion of part (a) in Proposition 1.2. The second portion follows from the first.

(c) Suppose both $(LP)$ and $(LD)$ have feasible solutions. If there exist feasible solutions $\bar{x}, \bar{y}$ such that $c^T \bar{x} = b^T \bar{y}$ then we are done by part (a). Suppose such feasible solutions do not exist (we are seeking a contradiction). Then the system

$$\begin{bmatrix} A & 0 \\ -A & 0 \\ -I & 0 \\ 0 & A^T \\ c^T & -b^T \end{bmatrix} \begin{bmatrix} x \\ y \end{bmatrix} \leq \begin{bmatrix} b \\ -b \\ 0 \\ c \\ 0 \end{bmatrix}$$

has no solution. By Theorem 1.1, there exist $u, v, z, w, \alpha \geq 0$ such that

$$A^T(u - v) - z + \alpha c = 0, \quad Aw - \alpha b = 0, \quad b^T(u - v) + c^T w < 0.$$

If $\alpha > 0$, then

$$\bar{y} := \frac{v - u}{\alpha}, \quad \bar{x} := \frac{w}{\alpha}$$

satisfy

$$A^T \bar{y} \leq c, A\bar{x} = b, \bar{x} \geq 0, \text{ and } c^T \bar{x} < b^T \bar{y},$$

a contradiction to the Weak Duality Relation. Thus, we may assume $\alpha = 0$. Then there exist $u, v, w \geq 0$ such that

$$A^T(v - u) \leq 0, Aw = 0, b^T(v - u) > c^T w.$$

Since $(LP)$ and $(LD)$ are feasible, there exist $\hat{x}$, $\hat{y}$ such that

$$A\hat{x} = b, \hat{x} \geq 0; \quad A^T \hat{y} \leq c.$$

We deduce, on the one hand

$$c^T w < b^T (v - u) = \hat{x}^T A^T (v - u) \leq 0,$$

on the other hand

$$c^T w \geq \hat{y}^T Aw = 0,$$

a contradiction.

(b) Suppose $(LP)$ has an optimal solution. Then by the Weak Duality Relation, $(LD)$ cannot be unbounded. By Theorem 1.4, $(LD)$ is either infeasible or has an optimal solution. If $(LD)$ is infeasible, that is

$$A^T y \leq c$$

has no solution, then by Theorem 1.1, there exists $u \geq 0$ such that $Au = 0$, $c^T u < 0$. Let $\hat{x}$ be an optimal solution of $(LP)$. Then $(\hat{x} + \alpha u)$ is feasible in $(LP)$ for every $\alpha \geq 0$, and

$$c^T (\hat{x} + \alpha u) \to -\infty \text{ as } \alpha \to \infty,$$

a contradiction. Therefore, $(LD)$ has an optimal solution. The rest of the claim now follows from part (c).

□

Note that in the above theorem, parts (b) and (c) are essentially equivalent. We already proved (c) ⇒ (b). Now, let us show the converse. Assume (b). If $(LP)$ and $(LD)$ both have feasible solutions, then neither is infeasible. Moreover, by the Weak Duality Relation, neither is unbounded. Therefore, by the Fundamental Theorem of LP, both have optimal solutions. Finally, by part (b), these optimal values must coincide.

**1.1.1 A Bit of Computational Complexity.** Now let us describe what we mean by $L$, the *input size* of an LP problem. Given $a \in \mathbb{R}$, $\lceil a \rceil$ is the smallest integer $k$ such that $a \leq k$. Similarly, $\lfloor a \rfloor$ is the largest integer $k$ such that $k \leq a$. Thus, $\lfloor a \rfloor$ and $\lceil a \rceil$ are the *floor* and *ceiling* of $a$ respectively.

Given $\alpha \in \mathbb{Z}$, $\text{size}(\alpha) := \lceil \log_2 (|\alpha| + 1) \rceil + 1$. Then for $A \in \mathbb{Z}^{m \times n}$,

$$\text{size}(A) := \sum_{i=1}^{m} \sum_{j=1}^{n} \text{size}(a_{ij}).$$

Finally, the size of the input of an LP problem described by $A, b, c$ (all with integer entries) is

$$L := \text{size}(A) + \text{size}(b) + \text{size}(c).$$

That is, the size of the input is the number of bits (in base 2) required to express the data. It will become clear shortly that the base of the logarithm here is irrelevant for our purposes (any base, larger than 1, will do—as it would result in a constant multiplicative factor). Also note that the notion of size is extendible to rational numbers (represent each rational number by the ratio of two relatively prime integers and store each integer as above).

Let $h, l : \mathbb{Z}_+ \to \mathbb{Z}_+$. We say $h = O(l)$ (*$h$ is order $l$*) if there exist constants $c_1, c_2 \in \mathbb{Z}$ such that

$$h(n) \leq c_1 l(n), \ \forall n \geq c_2.$$

We say $h = \Omega(l)$ ($h$ *is omega of* $l$) if there exist constants $c_1, c_2 \in \mathbb{Z}$ such that

$$h(n) \geq c_1\, l(n), \ \forall\, n \geq c_2.$$

Finally, we say $h = \Theta(l)$ ($h$ *is theta of* $l$) if

$$h = O(l) \text{ and } h = \Omega(l).$$

Consider the algorithms which are based on the *elementary arithmetic operations* addition, subtraction, multiplication, division and comparison applied to pairs of rational numbers. If an algorithm performs $O\left(L^k\right)$ elementary arithmetic operations for some $k = O(1)$ independent of the input (and the sizes of all the numbers arising during the execution of the algorithm do not exceed a polynomial function of $L$) then the algorithm is called a *polynomial time algorithm*. A proper computational model to use for this part of our discussion is the Turing Machine Model; but covering the Turing Machine Model in detail would take us far afield. (Instead, see the books [137], [280] and the references therein.)

*Does there exist a polynomial time algorithm for LP?* This question intrigued researchers in Optimization, Combinatorics and Theoretical Computer Science during the 1970s. Many believed that such algorithms must exist. The definitive answer came in 1979. Leonid Khachiyan proved [193], using a variant of the Ellipsoid Method (proposed in the 1970's by Shor [332] and Yudin and Nemirovskii [374, 267]) that Linear Programming problems can be solved in polynomial time. We will see part of the arguments. The bound on the number of arithmetic operations required by the algorithm was $O(n^4\, L)$. A few years later, Narendra Karmarkar [192] proposed a polynomial-time interior-point algorithm with an $O(n^{3.5}\, L)$ bound on the number of arithmetic operations. This algorithm sparked great interest and activity and created the whole field of modern interior-point methods. In 1988, Renegar [303] improved the complexity bound to $O(n^3\, L)$. Since then many interior-point algorithms matched this bound (see Gonzaga [149]), but in terms of these two measures $n$ and $L$, $O(n^3\, L)$ is still the best bound at the time of this writing. It turned out that certain primal-dual algorithms can be very efficient in practice as well. (These are related to theoretical algorithms proposed by Kojima-Mizuno-Yoshise [205], Monteiro-Adler [258, 259], and Monteiro-Adler-Resende [260].)

Currently, on many large scale LP problems interior-point algorithms provide the most effective, that is, fast and robust (in the sense that the number of iterations required seems to be almost constant) solution technique. However, for many other classes of LP problems, variants of the simplex method are still the algorithms of choice. Currently, no variant of simplex method is known to be a polynomial-time algorithm and many of the favorite variants have corresponding examples (usually derived from the Klee-Minty cube [201]) proving that they require exponentially many pivots in the dimension, in the worst case. A related, long outstanding question in the area is the so-called *Hirsch Conjecture*. Let $P \subseteq \mathbb{R}^d$ be a polytope defined by $n$ linear inequalities (or as the intersection of $n$ closed half-spaces in $\mathbb{R}^d$). Construct a graph $G(P)$ as follows: for each extreme point of $P$ make a node in $G(P)$, and put an edge between the nodes $u$ and $v$ in $G(P)$ if the extreme points $u$ and $v$ lie on an edge of the polytope $P$. The *Hirsch Conjecture* states that for every polytope $P$, the length of the shortest path from $u$ to $v$ is at most $(n-d)$ for every pair of nodes $u, v$ in $G$. So, the conjecture suggests that no matter which extreme point of $P$ we start from, following the edges of the polytope, we can reach any other extreme point of $P$ in $(n-d)$ steps. Each step corresponds to traversing an edge in $G(P)$. Since each (nondegenerate) iteration of the Simplex Method and its variants corresponds to traversing an edge in $G(P)$, the Hirsch Conjecture is tied into understanding the performance of these algorithms. In this regard, the main question is whether the length of the longest shortest path in $G(P)$ can be bounded by a polynomial function of $n$ for every $P$ with $n$ facets. For a survey of the work in this area, see Kalai's article [187] and the references therein.

When the data $(A, b, c)$ are rational, we can either express a rational number by the ratio of two integers (and just store the integers) or scale the data (multiply by the lowest common multiple of the denominators) so that the new data has only integer components. As we mentioned above, we have

**Theorem 1.7** *LP problems with rational data ($A \in \mathbb{Q}^{m\times n}$, $b \in \mathbb{Q}^m$ and $c \in \mathbb{Q}^n$) can be solved in polynomial time.*

We had already mentioned that Gaussian Elimination can be implemented efficiently. In fact, for $A \in \mathbb{Z}^{m\times n}$, the number of elementary arithmetic operations is $O\left(\max\{m, n\}^3\right)$. However, the implementation must be carefully designed. If we choose the pivot as the entry with the largest absolute value (among all valid pivot candidates) we can guarantee that the sizes of all the numbers arising throughout the execution of the algorithm stays bounded by a polynomial function of the input size (otherwise, even if the initial data have only the entries 0,1,2, the growth in the intermediate numbers can be exponential). See, Edmonds [110]. For related complexity issues, also see Lovász [242] and Schrijver [319].

Note that the complexity bound above on the number of elementary arithmetic operations does not involve the sizes of the integers in the data. Algorithms with such strong complexity bounds are called *strongly polynomial time algorithms*.

Under certain restrictions on the coefficient matrix $A$, a strongly polynomial bound (depending only on $n$) is possible (see Tardos [341, 342], Vavasis and Ye [360] and Ho and Tunçel [176]). However, the general problem is open:

**Open Problem 1:** Does there exist a strongly polynomial time algorithm for linear programming? (I.e., the number of elementary arithmetic operations is $O(n^k)$, where $k$ is an absolute constant independent of the input.)

A "yes" answer to the above open problem would correspond to a polynomial time algorithm in the real-number machine model of Blum, Shub and Smale [48]. In this real number machine model, we are concerned with bounding the number of elementary arithmetic operations on the original data (*given* real numbers). From our current viewpoint, this model gives a higher level view of the intrinsic complexity of the algorithms (by moving away from the bit-complexity aspect of the Turing Machine Model) and is more suitable for continuous optimization algorithms. In fact, just like our discussion above about the Hirsch Conjecture, it is usually very convenient to take even a higher level view of complexity and study *iteration complexity* of algorithms. Recall that Hirsch Conjecture only involves moves (going from one extreme point to an adjacent extreme point) which represent an iteration (strictly speaking, a nondegenerate one) of the Simplex Method. So, this kind of view of the Simplex Method and its variants would only care about the number of iterations and not worry about how many arithmetic operations were performed, etc. Many of the algorithms that we discuss, solve in each iteration, a linear system of equations and do certain matrix multiplications and decompositions. To get this higher level view, we can simply step back from the bit operations as well as elementary arithmetic operations and consider the bound on the number of iterations of the algorithm where each iteration involves an absolute constant number of matrix multiplications/decompositions and linear equation solving involving $O(n)$-by-$O(n)$ matrices.

## 1.2 Semidefinite Programming

For $X \in \mathbb{R}^{n\times n}$, the *trace* of $X$ is the sum of its diagonal entries. We denote the trace of $X$ by $\mathrm{Tr}(X)$. Thus, for $X \in \mathbb{R}^{n\times n}$,

$$\mathrm{Tr}(X) = \sum_{i=1}^{n} X_{ii}.$$

$\Sigma^n$ denotes the space of $n \times n$ symmetric matrices over the reals. An inner-product on $\mathbb{R}^{n\times n}$ and $\Sigma^n$ is

$$\langle X, S\rangle := \mathrm{Tr}(X^T S),$$

where

$$\mathrm{Tr}(X^T S) = \sum_{i=1}^{n}\sum_{j=1}^{n} X_{ij}S_{ij} = \mathrm{Tr}(SX^T).$$

The latter also implies that for every nonsingular $P \in \mathbb{R}^{n\times n}$,

$$\mathrm{Tr}(PXP^{-1}) = \mathrm{Tr}(XP^{-1}P) = \mathrm{Tr}(X).$$

Note however that $\mathrm{Tr}(XYZ) \neq \mathrm{Tr}(XZY)$.

Let $X \in \mathbb{R}^{n\times n}$ be given. Then the roots $\lambda_1, \lambda_2, \ldots, \lambda_n$, of the polynomial equation

$$\det(X - \lambda I) = 0$$

are the *eigenvalues* of $X$. If $X$ is a symmetric matrix over the reals, then all the eigenvalues of $X$ are real. We usually order them:

$$\lambda_1(X) \geq \lambda_2(X) \geq \cdots \geq \lambda_n(X).$$

Sometimes we write them as a vector $\lambda(X)$, $\lambda : \Sigma^n \to \mathbb{R}^n$.

Let $\mathrm{Diag} : \mathbb{R}^n \to \Sigma^n$ denote the linear transformation which maps a given $x \in \mathbb{R}^n$ onto the diagonal matrix $\begin{pmatrix} x_1 & 0 & \ldots & 0 \\ 0 & x_2 & \ldots & 0 \\ \vdots & \vdots & \ddots & \vdots \\ 0 & 0 & \ldots & x_n \end{pmatrix}$. Its *adjoint* is $\mathrm{diag} : \Sigma^n \to \mathbb{R}^n$ which extracts the diagonal of a given symmetric matrix. In general, let $\mathcal{A} : \Sigma^n \to \mathbb{R}^m$ be a linear transformation. Then its *adjoint* $\mathcal{A}^* : \mathbb{R}^m \to \Sigma^n$ is a linear transformation. We can define it by describing the image of a basis of $\mathbb{R}^m$ under $\mathcal{A}^*$. Below, we give a definition which describes for each $y \in \mathbb{R}^m$ what $\mathcal{A}^*(y)$ is in terms of the inner product $\langle\cdot,\cdot\rangle_{(\Sigma^n)}$ chosen on $\Sigma^n$ and the inner product $\langle\cdot,\cdot\rangle_{(\mathbb{R}^m)}$ chosen on $\mathbb{R}^m$:

$$\langle \mathcal{A}^*(y), X\rangle_{(\Sigma^n)} = \langle y, \mathcal{A}(X)\rangle_{(\mathbb{R}^m)}, \forall y \in \mathbb{R}^m, X \in \Sigma^n.$$

For instance, we can define $\mathcal{A}$ by

$$[\mathcal{A}(X)]_i := \langle A_i, X\rangle, \forall i \in \{1, 2, \ldots, m\},$$

where $A_i \in \Sigma^n$, for $i \in \{1, 2, \ldots, m\}$ are given. Then

$$\mathcal{A}^*(y) = \sum_{i=1}^{m} y_i A_i.$$

**Theorem 1.8** *For every $X \in \Sigma^n$, there exists $Q \in \mathbb{R}^{n \times n}$, orthogonal ($Q^TQ = I$) such that*

$$X = Q\mathrm{Diag}(\lambda(X))Q^T.$$

Let $X \in \Sigma^n$ and $Q \in \mathbb{R}^{n\times n}$ be the orthogonal matrix in the statement of Theorem 1.8. Let us denote by $q^{(1)}, q^{(2)}, \ldots, q^{(n)}$ the columns of $Q$ and denote by $e_j$ the $j$th unit vector. Then, note that

$$Xq^{(j)} = Q\mathrm{Diag}(\lambda(X))e_j = \lambda_j(X)Qe_j = \lambda_j(X)q^{(j)},$$

that is, $q^{(j)}$ is an eigenvector of $X$ associated with the eigenvalue $\lambda_j(X)$.

By the above observations about trace and Theorem 1.8,

$$\mathrm{Tr}(X) = \sum_{j=1}^{n} \lambda_j(X).$$

We are ready to talk about norms on $\Sigma^n$. Recall that for $h \in \mathbb{R}^n$, $\|h\|_p := \left(\sum_{j=1}^n |h_j|^p\right)^{\frac{1}{p}}$. We have

$$\text{Frobenius norm:} \quad \|X\|_F := \langle X, X\rangle^{1/2} = \|\lambda(X)\|_2 = \sqrt{\sum_{j=1}^{n} (\lambda_j(X))^2};$$

$$\text{operator } p\text{-norm:} \quad \|X\|_p := \max\left\{\|Xh\|_p : h \in \mathbb{R}^n, \|h\|_p = 1\right\}.$$

Note that $\|X\|_2 = \max_j\{|\lambda_j(X)|\}$.

We say that $X \in \Sigma^n$ is *positive semidefinite* if $\forall h \in \mathbb{R}^n$, $h^TXh \geq 0$. $X$ is *positive definite* if $\forall h \in \mathbb{R}^n\backslash\{0\}$, $h^TXh > 0$. We denote this by $X \succeq 0$ and $X \succ 0$, respectively. For $U, V \in \Sigma^n$, we write $U \succeq V$ ($U \succ V$) to mean $(U - V) \succeq 0$ ($(U - V) \succ 0$).

We denote the closed convex cone of $n$-by-$n$ symmetric positive semidefinite matrices by $\Sigma^n_+$. $\Sigma^n_{++}$ denotes the cone of $n$-by-$n$ symmetric positive definite matrices.

For $X \in \Sigma^n_{++}$, we can define its unique symmetric positive semidefinite square root as

$$X^{1/2} := Q\left[\mathrm{Diag}(\lambda(X))\right]^{1/2} Q^T.$$

We extend this definition to $X \in \Sigma^n_+$.

Another important tool (from both the theoretical and computational viewpoints) in dealing with symmetric positive semidefinite matrices is the *Cholesky Decomposition*. See, for instance, Golub and Van Loan [148].

**Theorem 1.9** *Let $X \in \Sigma^n$. Then $X \in \Sigma^n_+$ iff there exists $B \in \mathbb{R}^{n\times n}$ lower-triangular such that $X = BB^T$.*

We give a constructive proof which describes an algorithm to compute $B$. The algorithm can be applied to any symmetric matrix. If the algorithm fails to construct $B$ as above then it delivers a certificate proving that $X$ is not positive semidefinite. One kind of certificate is $h \in \mathbb{R}^n$ such that $h^TXh < 0$. Another kind is $i, j$ such that $X_{ii} = 0$ and $X_{ij} \neq 0$ (every positive semidefinite matrix has the property that $X_{ii} = 0$ implies the whole row, and the column, $i$ is zero—also see part (e) of the next proposition). To prove this little claim, suppose that for some $j \neq i$, $X_{ij} = \alpha \neq 0$. Consider $h := \beta e_i + e_j$ ($\beta \in \mathbb{R}$ to be chosen later). Then

$$h^TXh = 2\alpha\beta + X_{jj}.$$

The latter is negative iff $\beta \text{sign}(\alpha) < -\frac{X_{jj}}{2|\alpha|}$. The last inequality can be satisfied by a proper choice of $\beta$. Thus, the little claim is proved. Next, we prove the theorem.

**Proof** First, we do the easy direction. If there exists $B \in \mathbb{R}^{n \times n}$ lower-triangular such that $X = BB^T$, then for every $h \in \mathbb{R}^n$, $h^T X h = h^T B B^T h = \|B^T h\|_2^2 \geq 0$. Thus, $X \in \Sigma_+^n$. For the other direction, we proceed by induction on $n$. Let $n := 1$. Then, if $X_{11} \geq 0$, let $B_{11} := \sqrt{X_{11}}$ and we are done. If $X_{11} < 0$, then let $h := 1$. We have $h^T X h = X_{11} < 0$ (hence $X \notin \Sigma_+^1$).

Suppose the claim is true for all $n \leq (k-1)$. Let $X \in \Sigma^k$. Write

$$X =: \begin{pmatrix} X_{11} & x^T \\ x & \bar{X} \end{pmatrix}.$$

If $X_{11} < 0$ then let $h := e_1$. We obtain $h^T X h = X_{11} < 0$ (hence $X \notin \Sigma_+^k$). If $X_{11} = 0$ then for $X$ to be positive semidefinite, the $(k-1)$-vector $x$ must be the zero vector. If $X_{11} = 0$ and $x = 0$, then we are done by the induction hypothesis, since $\bar{X} \in \Sigma^{k-1}$. Now, we can assume $X_{11} > 0$. We let $b := \frac{1}{\sqrt{X_{11}}} x$ and

$$\tilde{X} := \bar{X} - \frac{1}{X_{11}} x x^T.$$

If $\tilde{X} \succeq 0$, then there exists $\tilde{B} \in \mathbb{R}^{(k-1)\times(k-1)}$ lower triangular, such that $\tilde{X} = \tilde{B}\tilde{B}^T$. Then

$$\begin{pmatrix} \sqrt{X_{11}} & 0 \\ b & \tilde{B} \end{pmatrix} \begin{pmatrix} \sqrt{X_{11}} & b^T \\ 0 & \tilde{B}^T \end{pmatrix} = X,$$

as desired. Otherwise ($\tilde{X} \notin \Sigma_+^{k-1}$), there exists $\tilde{h} \in \mathbb{R}^{k-1}$ such that $\tilde{h}^T \tilde{X} \tilde{h} < 0$. Let $h_0 := -\frac{x^T \tilde{h}}{X_{11}}$. Then

$$\begin{pmatrix} h_0, & \tilde{h}^T \end{pmatrix} \begin{pmatrix} X_{11} & x^T \\ x & \bar{X} \end{pmatrix} \begin{pmatrix} h_0 \\ \tilde{h} \end{pmatrix} = -\frac{1}{X_{11}} \left(x^T \tilde{h}\right)^2 + \tilde{h}^T \bar{X} \tilde{h} = \tilde{h}^T \tilde{X} \tilde{h} < 0.$$

□

**Proposition 1.10** *Let $X \in \Sigma^n$. Then, TFAE*[1]

(a) *$X$ is positive semidefinite;*
(b) *$\lambda_j(X) \geq 0$, for every $j \in \{1, 2, \ldots, n\}$;*
(c) *there exist $\mu \in \mathbb{R}_+^n$ and $h^{(i)} \in \mathbb{R}^n$, for every $i \in \{1, 2, \ldots, n\}$ such that*

$$X = \sum_{i=1}^n \mu_i h^{(i)} {h^{(i)}}^T;$$

(d) *there exists $B \in \mathbb{R}^{n \times n}$ such that $X = BB^T$ (here, $B$ can be chosen as a lower triangular matrix—the Cholesky decomposition of $X$);*
(e) *for every nonempty $J \subseteq \{1, 2, \ldots, n\}$, $\det(X_J) \geq 0$, where $X_J := \{[X_{ij}] : i, j \in J\}$;*
(f) *for every $S \in \Sigma_+^n$, $\langle X, S \rangle \geq 0$.*

Note that the number of nonzero eigenvalues of $X \in \Sigma_+^n$ is equal to the rank of $X$. In part (d), if $\text{rank}(X) = r$, then we can choose $B \in \mathbb{R}^{n \times r}$. In part (e), we call $X_J$ the *symmetric minors of $X$*.

[1]TFAE: the following are equivalent.

We are ready to define the semidefinite programming problem in standard form and its dual. Suppose $C \in \Sigma^n$, $b \in \mathbb{R}^m$ and a linear transformation $\mathcal{A} : \Sigma^n \to \mathbb{R}^m$ are given. Then we define

$$\begin{array}{llrcl} (P) & \inf & \langle C, X\rangle & & \\ & \text{subject to:} & \mathcal{A}(X) & = & b, \\ & & X & \succeq & 0, \end{array}$$

$$\begin{array}{llrclcl} (D) & \sup & b^T y & & & & \\ & \text{subject to:} & \mathcal{A}^*(y) & + & S & = & C, \\ & & & & S & \succeq & 0. \end{array}$$

As we noted, $\mathcal{A}$ can be represented in a more explicit form. So, let $A_1, A_2, \ldots, A_m \in \Sigma^n$ be given such that for every $X \in \Sigma^n$, $[\mathcal{A}(X)]_i = \langle A_i, X\rangle$. Then we can write $(P)$ and $(D)$ as

$$\begin{array}{llrcll} (P) & \inf & \langle C, X\rangle & & & \\ & \text{subject to:} & \langle A_i, X\rangle & = & b_i, & \forall i \in \{1, 2, \ldots, m\} \\ & & X & \succeq & 0, & \end{array}$$

$$\begin{array}{llrclcl} (D) & \sup & b^T y & & & & \\ & \text{subject to:} & \sum_{i=1}^m y_i A_i & + & S & = & C, \\ & & & & S & \succeq & 0. \end{array}$$

A good way to write the SDP problems from both theoretical and practical viewpoints is to treat the constraints $X \in \Sigma^n_+$, $S \in \Sigma^n_+$ as $X \in \Sigma^{n_1}_+ \oplus \Sigma^{n_2}_+ \oplus \cdots \oplus \Sigma^{n_r}_+$ and $S \in \Sigma^{n_1}_+ \oplus \Sigma^{n_2}_+ \oplus \cdots \oplus \Sigma^{n_r}_+$ respectively. For example,

$$X = \begin{bmatrix} (n_1 \times n_1) & 0 & \cdots & 0 \\ 0 & (n_2 \times n_2) & \cdots & 0 \\ \vdots & \vdots & \ddots & \vdots \\ 0 & 0 & \cdots & (n_r \times n_r) \end{bmatrix} \succeq 0.$$

LP is a special case with $r := n$, $n_1 := n_2 := \cdots := n_n := 1$. The interior of the nonnegative orthant $\mathbb{R}^n_+$ is denoted by $\mathbb{R}^n_{++}$.

**Proposition 1.11** (1) $\Sigma^n_{++} = \operatorname{int}\left(\Sigma^n_+\right)$.

(2) *Let $X \in \Sigma^n$. Then, TFAE*

(a) *$X$ is positive definite;*

(b) *$\lambda_j(X) > 0$, for every $j \in \{1, 2, \ldots, n\}$;*

(c) *there exist $\mu \in \mathbb{R}^n_{++}$ and $h^{(i)} \in \mathbb{R}^n$, for every $i \in \{1, 2, \ldots, n\}$ linearly independent such that*

$$X = \sum_{i=1}^n \mu_i h^{(i)} {h^{(i)}}^T;$$

(d) *there exists $B \in \mathbb{R}^{n \times n}$ nonsingular such that $X = BB^T$ (here, $B$ can be chosen as a lower triangular matrix—the Cholesky decomposition of $X$);*

(e) *for every $J_k := \{1, 2, \ldots, k\}$, $k \in \{1, 2, \ldots, n\}$, $\det(X_{J_k}) > 0$;*

(f) *for every $S \in \Sigma^n_+ \setminus \{0\}$, $\langle X, S\rangle > 0$;*

(g) *$X \succeq 0$ and $\operatorname{rank}(X) = n$.*

In part (e), we call $X_{J_k}$ the *leading principle minors of* $X$. Note that Proposition 1.11(2)(e) cannot be relaxed to $\det(X_{J_k}) \geq 0$, to give a characterization of positive semidefiniteness of $X$ as the example

$$X := \begin{bmatrix} 1 & 1 & 0 & 0 \\ 1 & 1 & 0 & 0 \\ 0 & 0 & -1 & 1 \\ 0 & 0 & 1 & -1 \end{bmatrix}$$

shows. For this example, we have $\det(X_{J_1}) = 1$, $\det(X_{J_2}) = \det(X_{J_3}) = \det(X_{J_4}) = 0$. However, $X \notin \Sigma_+^4$ since $e_3^T X e_3 = -1$ (so, we do need to consider every symmetric minor of $X$). However, note that if $X$ is not positive semidefinite then there exists a permutation matrix $P$ such that $\det\left((PXP^T)_{J_k}\right) < 0$ for some $k$.

As we already observed, if $X \succeq 0$, then $\operatorname{diag}(X) \geq 0$; moreover,

$$X_{ii} = 0 \implies X_{ij} = X_{ji} = 0, \forall j.$$

Also, if $X \succ 0$ then $\operatorname{diag}(X) > 0$. There are sharper, more useful conditions ensuring positive semidefiniteness (or positive definiteness) of a symmetric matrix. For this, we step back and move to the larger space of matrices $\mathbb{R}^{n\times n}$. The Gershgorin Disk Theorem is:

**Theorem 1.12** *Let $M \in \mathbb{R}^{n\times n}$. Define for each $i \in \{1, 2, \ldots, n\}$*

$$B_i(M) := \left\{ \lambda \in \mathbb{C} : |\lambda - M_{ii}| \leq \sum_{j\neq i} |M_{ij}| \right\}.$$

*Then the union of these disks $B_i(M)$ contains all the eigenvalues of $M$.*

We would like to apply this theorem to symmetric matrices. Let $X \in \Sigma^n$. Then we say that $X$ is *diagonally dominant* if $X_{ii} \geq \sum_{j\neq i} |X_{ij}|$ for every $i$. Similarly, we say that $X$ is *strictly diagonal dominant* if the above inequalities are strict for all $i$. We have that for $X \in \Sigma^n$,

$$X \text{ is diagonally dominant} \implies X \succeq 0;$$

and

$$X \text{ is strictly diagonal dominant} \implies X \succ 0.$$

Now, let's go back to our discussion of the optimization problems. The above definitions of $(P)$ and $(D)$ also arise as a special case of general, convex optimization problems in conic form. Let $K \subseteq \mathbb{R}^d$ be a convex cone (i.e., $\forall x, v \in K$ and $\alpha \in \mathbb{R}_{++}$, we have $\alpha x \in K$ and $(x+v) \in K$). As before, let $\mathcal{A} : \mathbb{R}^d \to \mathbb{R}^m$ be a given linear transformation. We assume that $c \in \mathbb{R}^d$ and $b \in \mathbb{R}^m$ are also given. Then we define the following primal-dual pair of optimization problems:

$$\begin{array}{llrcl} (CP) & \inf & \langle c, x\rangle & & \\ & \text{subject to:} & \mathcal{A}(x) & = & b, \\ & & x & \in & K, \end{array}$$

$$\begin{array}{llrcrcl} (CD) & \sup & \langle b, y\rangle & & & & \\ & \text{subject to:} & \mathcal{A}^*(y) & + & s & = & c, \\ & & & & s & \in & K^*, \end{array}$$

where

$$K^* := \left\{ s \in \mathbb{R}^d : \langle x, s\rangle \geq 0, \forall x \in K \right\}$$

is called the *dual* of $K$.

**Definition 1.13** Let $K \subseteq \mathbb{R}^d$ be a pointed, closed convex cone. Then $K$ is *self-dual* if there exists an inner-product under which $K = K^*$.

The set of nonsingular, linear transformations on $\mathbb{R}^d$, keeping $K$ invariant is called the *automorphism group of* $K$ and is denoted by $\text{Aut}(K)$.

**Definition 1.14** Let $K \subseteq \mathbb{R}^d$ be a pointed, closed convex cone with nonempty interior. Then $K$ is *homogeneous* if for every pair $x, v \in \text{int}(K)$, there exists $T \in \text{Aut}(K)$ such that $T(v) = x$ (i.e., the automorphism group of $K$ acts transitively on the interior of $K$).

**Definition 1.15** Let $K \subseteq \mathbb{R}^d$ be a pointed, closed convex cone with nonempty interior. Then $K$ is *symmetric* if $K$ is homogeneous and self-dual.

Our cones, if nonempty and closed, must contain the origin by definition. We define the notion of an *extreme ray* next. $\{\alpha v : \alpha \in \mathbb{R}_+\}$ for $v \in K \setminus \{0\}$ defines a *ray* inside $K$. Such a ray $R \subseteq K$ is called an *extreme ray of* $K$ if for every pair of rays $R_1, R_2 \subseteq K$, such that $R_1 + R_2 \supseteq R$ implies either $R_1 = R$ or $R_2 = R$ possibly both. The union of all extreme rays of $K$ is denoted by $\text{Ext}(K)$. We also use $\text{ext}(K)$ to denote the set of normalized extreme rays of $K$ (where each ray is represented by a single nonzero element of $K$, e.g., with the same norm, or with the same value on a fixed linear function). For compact, convex sets, we use the notation $\text{ext}(\cdot)$ to denote the set of extreme points of the compact, convex set.

A classical theorem of Carathéodory implies that if $K$ is a compact convex set in $\mathbb{R}^d$, then every point in $K$ can be expressed as a convex combination of at most $(d+1)$ points from $\text{ext}(K)$. If $K$ is a pointed, closed convex cone in $\mathbb{R}^d$, then every point in $K$ can be expressed as a nonnegative linear combination of at most $d$ points from $\text{ext}(K)$.

**Theorem 1.16** *$\Sigma^n_+$ is a pointed, closed convex cone with nonempty interior. Also, $\Sigma^n_+$ is homogeneous and self-dual; hence, $\Sigma^n_+$ is a symmetric cone. Moreover,*

$$\text{ext}(\Sigma^n_+) = \left\{hh^T : h \in \mathbb{R}^n, \|h\|_2 = 1\right\}.$$

**Proof** We first prove $\left(\Sigma^n_+\right)^* = \Sigma^n_+$ under the trace inner-product. Let $\bar{S} \in \Sigma^n_+$. Then, for every $X \in \Sigma^n_+$, we have

$$\langle X, \bar{S}\rangle = \text{Tr}\left(\bar{S}^{1/2} X \bar{S}^{1/2}\right) \geq 0.$$

Thus, $\bar{S} \in \left(\Sigma^n_+\right)^*$ and $\left(\Sigma^n_+\right)^* \supseteq \Sigma^n_+$. Let $\bar{S} \in \left(\Sigma^n_+\right)^*$. Denote by $h^{(1)}, h^{(2)}, \ldots, h^{(n)}$ the eigenvectors of $\bar{S}$ (and $\left\|h^{(i)}\right\|_2 = 1, \forall i$). Then for every $i \in \{1, 2, \ldots, n\}$, $h^{(i)}\left(h^{(i)}\right)^T \in \Sigma^n_+$ and

$$0 \leq \langle \bar{S}, h^{(i)} {h^{(i)}}^T\rangle = {h^{(i)}}^T \bar{S} h^{(i)} = \lambda_i(\bar{S}).$$

Hence, $\bar{S} \in \Sigma^n_+$. Therefore, $\left(\Sigma^n_+\right)^* \subseteq \Sigma^n_+$. We proved that $\Sigma^n_+$ is self-dual. This also implies that $\Sigma^n_+$ is a closed convex cone. Note that for every $n$, $I \in \text{int}(\Sigma^n_+) = \Sigma^n_{++}$. Suppose $\Sigma^n_+$ is not pointed. Then there exists $\bar{X} \in \Sigma^n_+ \setminus \{0\}$ such that $-\bar{X} \in \Sigma^n_+$. Using Proposition 1.10 part (b), we conclude $\lambda(\bar{X}) = 0$, whence $\bar{X} = 0$, a contradiction. Therefore, $\Sigma^n_+$ is pointed. Let $\bar{X} \in \Sigma^n_{++}$. Then the linear map $T_{\bar{X}} : \Sigma^n \to \Sigma^n$ defined as

$$T_{\bar{X}}(\cdot) := \bar{X}^{1/2} \cdot \bar{X}^{1/2}$$

is nonsingular and satisfies $T_{\bar{X}}(\Sigma_+^n) = \Sigma_+^n$. Let $\bar{V} \in \Sigma_{++}^n$. Then

$$T_{\bar{X}}\left(T_{\bar{V}^{-1}}(\bar{V})\right) = \bar{X},$$

proving that $\Sigma_+^n$ is a homogeneous cone. The proof of the identity $\text{ext}(\Sigma_+^n) = \{hh^T : h \in \mathbb{R}^n, \|h\|_2 = 1\}$ is left for the reader. $\square$

Given two sets $K_1 \subseteq \mathbb{R}^{d_1}$ and $K_2 \subseteq \mathbb{R}^{d_2}$, their *direct sum* is denoted by $K_1 \oplus K_2$ and it is given by

$$K_1 \oplus K_2 := \left\{ \begin{pmatrix} u \\ v \end{pmatrix} \in \mathbb{R}^{d_1+d_2} : u \in K_1, v \in K_2 \right\}.$$

Note that $(K_1 \oplus K_2)^* = K_1^* \oplus K_2^*$ and the theorem above imply that any direct sum structure that the primal SDP has is also inherited by the dual SDP.

**Theorem 1.17** *(Weak Duality Relation—SDP) Let $\overline{X}$ be feasible in $(P)$ and $(\overline{y}, \overline{S})$ be feasible in $(D)$. Then*

$$\langle C, \overline{X}\rangle - b^T\overline{y} = \langle \overline{X}, \overline{S}\rangle \geq 0.$$

**Proof** Let $\overline{X}$ be feasible in $(P)$ and $(\overline{y}, \overline{S})$ be feasible in $(D)$. Then

$$\langle C, \overline{X}\rangle - b^T\overline{y} = \langle C, \overline{X}\rangle - \mathcal{A}(\overline{X})^T\overline{y} = \langle C, \overline{X}\rangle - \langle \overline{X}, \mathcal{A}^*(\overline{y})\rangle = \langle C - \mathcal{A}^*(\overline{y}), \overline{X}\rangle = \langle \overline{X}, \overline{S}\rangle.$$

The above equations do not use the full power of the feasibility of $(\overline{X}, \overline{y}, \overline{S})$, but only that $\mathcal{A}(\overline{X}) = b$ and $\mathcal{A}^*(\overline{y}) + \overline{S} = C$. Since $\overline{X}, \overline{S} \in \Sigma_+^n$, by Proposition 1.10 part (f) the result follows. $\square$

**Corollary 1.18** (a) *If $(P)$ is unbounded then $(D)$ is infeasible.*
(b) *If $(D)$ is unbounded then $(P)$ is infeasible.*
(c) *If for feasible solutions $\overline{X}$ of $(P)$ and $(\overline{y}, \overline{S})$ of $(D)$ we have $\langle \overline{X}, \overline{S}\rangle = 0$ then $\overline{X}$ is optimal in $(P)$ and $(\overline{y}, \overline{S})$ is optimal in $(D)$.*

Note: The dual of $(D)$ is $(P)$.

**Proof** (a) Suppose $(D)$ is feasible. Let $\overline{y} \in \mathbb{R}^m$ be a feasible solution of $(D)$. Then by Theorem 1.17, for every feasible solution $X$ of $(P)$, we have

$$\langle C, X\rangle \geq b^T\overline{y}.$$

Thus, $(P)$ cannot be unbounded.

(b) Suppose $(P)$ is feasible and let $\overline{X}$ be a feasible solution of $(P)$. Then by Theorem 1.17, for every feasible solution $y$ of $(D)$, we have

$$b^Ty \leq \langle C, \overline{X}\rangle.$$

Hence, $(D)$ cannot be unbounded.

(c) Let $\overline{X}$ be a feasible solution of $(P)$ and let $(\overline{y}, \overline{S})$ be a feasible solution of $(D)$. Then by Theorem 1.17,

$$\langle C, X\rangle \geq b^T\overline{y}, \text{ for every feasible } X \text{ in } (P).$$

Since $\langle \overline{X}, \overline{S}\rangle = 0$, we have $\langle C, \overline{X}\rangle = b^T\overline{y}$. Therefore, $\overline{X}$ is optimal in $(P)$. Similarly, $(\overline{y}, \overline{S})$ is optimal in $(D)$.

$\square$

If $\mathcal{A}$ is surjective (in matrix representation, this assumption is equivalent to $\{A_i : i \in \{1, 2, \ldots, m\}\}$ is linearly independent) then every $\overline{S} \in \Sigma^n$ satisfying $\mathcal{A}^*(y) + S = C$ for some $y \in \mathbb{R}^m$, uniquely identifies the corresponding $\overline{y}$.

The next fact is a generalization of complementary slackness conditions from LP to SDP.

**Proposition 1.19** *Let $X, S \in \Sigma_+^n$. Then*

$$\langle X, S\rangle = 0 \quad \textit{iff} \quad XS = 0.$$

**Proof** "$\Leftarrow$" is trivial since $\operatorname{Tr}(0) = 0$. For the converse, suppose $X, S \in \Sigma_+^n$ and $\langle X, S\rangle = 0$. Then $\operatorname{Tr}\left(X^{1/2}SX^{1/2}\right) = 0$. Since $S \succeq 0$ and $X^{1/2} \in \Sigma^n$, we have $X^{1/2}SX^{1/2} \succeq 0$. Therefore, $\lambda\left(X^{1/2}SX^{1/2}\right) \geq 0$. Since $\operatorname{Tr}\left(X^{1/2}SX^{1/2}\right) = 0$, we must have $\lambda\left(X^{1/2}SX^{1/2}\right) = 0$. This implies

$$0 = X^{1/2}SX^{1/2} = \left(X^{1/2}S^{1/2}\right)\left(X^{1/2}S^{1/2}\right)^T.$$

Therefore, $X^{1/2}S^{1/2} = 0$. Thus,

$$XS = X^{1/2}\left(X^{1/2}S^{1/2}\right)S^{1/2} = 0.$$

□

The following classical and useful fact is a variational characterization of eigenvalues.

**Theorem 1.20** *Let $X \in \Sigma^n$. Then*

$$\begin{aligned}
\lambda_k(X) &= \min_{\substack{L \subseteq \mathbb{R}^n, \text{ a subspace} \\ \dim(L) = n-k+1}} \ \max_{h \in L \setminus \{0\}} \frac{h^T X h}{h^T h} \\
&= \max_{\substack{L \subseteq \mathbb{R}^n, \text{ a subspace} \\ \dim(L) = k}} \ \min_{h \in L \setminus \{0\}} \frac{h^T X h}{h^T h},
\end{aligned}$$

*for every $k \in \{1, 2, \ldots, n\}$.*

The above theorem includes the well-known cases

$$\lambda_1(X) = \max_{h \in \mathbb{R}^n \setminus \{0\}} \frac{h^T X h}{h^T h},$$

and

$$\lambda_n(X) = \min_{h \in \mathbb{R}^n \setminus \{0\}} \frac{h^T X h}{h^T h}.$$

**Corollary 1.21** (a) *Let $X, S \in \Sigma^n$. Then,*

$$\lambda_k(X) + \lambda_n(S) \leq \lambda_k(X + S) \leq \lambda_k(X) + \lambda_1(S),$$

*for every $k \in \{1, 2, \ldots, n\}$.*

(b) *Let $X, \overline{X} \in \Sigma^n$ such that $X \succeq \overline{X}$. Then $\lambda_j(X) \geq \lambda_j(\overline{X})$, for every $j \in \{1, 2, \ldots, n\}$.*

(c) *(Interlacing eigenvalues) Let $X \in \Sigma^n$, and let $\overline{X}$ be a $r \times r$ symmetric minor of $X$. Then*

$$\lambda_{n-(r-k)}(X) \leq \lambda_k(\overline{X}) \leq \lambda_k(X),$$

*for every $k \in \{1, 2, \ldots, r\}$.*

**Lemma 1.22** *(Schur Complement) Let $X \in \Sigma^n$ and $T \in \Sigma^m_{++}$. Then*

$$M := \begin{pmatrix} T & U^T \\ U & X \end{pmatrix} \succeq 0 \iff X - UT^{-1}U^T \succeq 0.$$

*Moreover, $M \succ 0$ iff $X - UT^{-1}U^T \succ 0$.*

**Proof** Note the following decomposition of $M$:

$$\begin{pmatrix} I & 0 \\ UT^{-1} & I \end{pmatrix}\begin{pmatrix} T & 0 \\ 0 & X - UT^{-1}U^T \end{pmatrix}\begin{pmatrix} I & T^{-1}U^T \\ 0 & I \end{pmatrix} = \begin{pmatrix} T & U^T \\ U & X \end{pmatrix}.$$

Note that we wrote

$$M = L\begin{pmatrix} T & 0 \\ 0 & X - UT^{-1}U^T \end{pmatrix}L^T,$$

where $L$ is nonsingular (in our case $\det(L) = 1$). Therefore,

$$M \succeq 0 \iff X - UT^{-1}U^T \succeq 0.$$

Also,

$$M \succ 0 \iff X - UT^{-1}U^T \succ 0.$$

□

$(X - UT^{-1}U^T)$ is called the *Schur complement of $T$ in $M$*. One interesting special case (which already came up in the proof of Theorem 1.9) is

$$\begin{pmatrix} 1 & x^T \\ x & X \end{pmatrix} \succeq 0 \iff X - xx^T \succeq 0$$

and

$$\begin{pmatrix} 1 & x^T \\ x & X \end{pmatrix} \succ 0 \iff X - xx^T \succ 0.$$

Let $X, S \in \mathbb{R}^{n\times n}$. The *Hadamard product of* $X$ and $S$ is defined as

$$(X \odot S)_{ij} := X_{ij}S_{ij}, \forall i, j.$$

So, $(X \odot S)$ is an $n \times n$ matrix which is the pointwise product of $X$ and $S$. We can prove that

$$X, S \succeq 0 \Rightarrow (X \odot S) \succeq 0. \tag{1.1}$$

Also, it is easy to establish:

$$X, S \succ 0 \Rightarrow (X \odot S) \succ 0.$$

Now, let us apply the Schur Complement Lemma to conclude a nice result about the Hadamard product of two symmetric, positive definite matrices and the inverse of the Hadamard product of their inverses. Let $X, S \succ 0$. Then, $X^{-1}, S^{-1} \succ 0$ and by the Schur Complement Lemma,

$$\begin{bmatrix} X & I \\ I & X^{-1} \end{bmatrix} \succeq 0 \text{ and } \begin{bmatrix} S & I \\ I & S^{-1} \end{bmatrix} \succeq 0.$$

Applying (1.1) to the last two matrices, we deduce

$$\begin{bmatrix} X \odot S & I \\ I & X^{-1} \odot S^{-1} \end{bmatrix} \succeq 0.$$

Finally, the Schur Complement Lemma implies

$$X \odot S \succeq \left(X^{-1} \odot S^{-1}\right)^{-1} \text{ as well as } X^{-1} \odot S^{-1} \succeq (X \odot S)^{-1}.$$

Let $A \in \mathbb{R}^{m\times n}$ and $B \in \mathbb{R}^{p\times q}$. Then the *Kronecker product of* $A$ and $B$ is defined as

$$A \otimes B := \begin{pmatrix} A_{11}B & \cdots & A_{1n}B \\ \vdots & \ddots & \vdots \\ A_{m1}B & \cdots & A_{mn}B \end{pmatrix} \in \mathbb{R}^{mp\times nq}.$$

A better way of dealing with the Kronecker product is to treat it as a new linear transformation defined from the linear transformations described by $A$ and $B$. To this end, we can equivalently define

$$(A \otimes B)\mathrm{vec}\,(H) := \mathrm{vec}\,(BHA^T), \forall H \in \mathbb{R}^{q\times n}, \tag{1.2}$$

where $\mathrm{vec}\,(H)$ is the vector formed from the columns of $H$. That is,

$$\mathrm{vec}\,(H) = [H_{11}, H_{21}, \ldots, H_{q1}, H_{12}, \ldots, H_{q2}, \ldots, H_{qn}]^T.$$

For compatible matrices:

$$(A \otimes B)(C \otimes D) = (AC \otimes BD); \quad (A \otimes B)^T = (A^T \otimes B^T).$$

Note that the dimension of $\Sigma^n$ is $n(n+1)/2$. So, let $\mathrm{s2vec}\,(X) \in \mathbb{R}^{n(n+1)/2}$ denote the isometry between $\Sigma^n$ and $\mathbb{R}^{n(n+1)/2}$ that takes the lower triangular part of $X$ column-wise and multiplies the strict lower triangular part by $\sqrt{2}$. Then the trace inner-product in $\Sigma^n$ can be expressed as the vector inner-product $\langle X, S\rangle = \mathrm{s2vec}\,(X)^T\mathrm{s2vec}\,(S)$. For instance,

$$\mathrm{s2vec}\,(X) = \left[X_{11}, \sqrt{2}X_{21}, \ldots, \sqrt{2}X_{n1}, X_{22}, \sqrt{2}X_{32}, \ldots, \sqrt{2}X_{n2}, \ldots, X_{nn}\right]^T.$$

In addition, we can express the matrix representations of certain linear transformations on $\Sigma^n$ (identified with $\mathbb{R}^{n(n+1)/2}$) using the *symmetric Kronecker product*. For every $H \in \Sigma^n$ we require,

$$(A\overset{s}{\otimes}B)\mathrm{s2vec}\,(H) := \mathrm{s2vec}\left(\frac{1}{2}(BHA^T + AHB^T)\right).$$

Note that by definition, $A\overset{s}{\otimes}B = B\overset{s}{\otimes}A$.

Moreover, it is easy to check that $\left(A\overset{s}{\otimes}B\right)^T = A^T\overset{s}{\otimes}B^T$.

We can express $\overset{s}{\otimes}$ using $\otimes$. Let $Q \in \mathbb{R}^{(n(n+1)/2)\times n^2}$ be defined as follows:

$$Q_{(i,j),(k,l)} := \begin{cases} 1 & \text{if } i = j = k = l, \\ \frac{1}{\sqrt{2}} & \text{if } i = k \neq j = l \text{ or } i = l \neq j = k, \\ 0 & \text{otherwise.} \end{cases}$$

Then $QQ^T = I$, $Q^TQ$ is the matrix representation of the orthogonal projection of $\mathbb{R}^{n\times n}$ (as $\mathbb{R}^{n^2}$) onto $\Sigma^n$ (as a subspace of $\mathbb{R}^{n^2}$). It can be shown that

$$A\overset{s}{\otimes}B = \frac{1}{2}Q\,(A \otimes B + B \otimes A)\,Q^T = Q(A \otimes B)Q^T. \tag{1.3}$$

Finally, we also have

**Theorem 1.23** *Let $A, B \in \Sigma^n$. Then,*

$$A\overset{s}{\otimes}B \succeq 0 \iff A \otimes B \succeq 0;$$

$$A\overset{s}{\otimes}B \succ 0 \iff A \otimes B \succ 0.$$

Kronecker product has close connections to Hadamard product. For example, if $X, S \succeq 0$, then $X \otimes S \succeq 0$. Since $X \odot S$ is a symmetric minor of $X \otimes S$, this gives us another way (using Proposition 1.10) to conclude $X \odot S \succeq 0$.

## 1.3 Fundamentals of Polyhedral Theory

One of the first things we should understand about polyhedra in Euclidean spaces is their dimensions; or put differently: in which Euclidean space do they naturally belong? To that end, we start with the definition of *affine independence.* $v^{(1)}, v^{(2)}, \ldots, v^{(k)} \in \mathbb{R}^n$ are *affinely independent* if $\begin{pmatrix} v^{(1)} \\ 1 \end{pmatrix}, \begin{pmatrix} v^{(2)} \\ 1 \end{pmatrix}, \ldots, \begin{pmatrix} v^{(k)} \\ 1 \end{pmatrix} \in \mathbb{R}^{n+1}$ are linearly independent.

(Also, $v^{(1)}, v^{(2)}, \ldots, v^{(k)} \in \mathbb{R}^n$ are affinely independent iff $(v^{(2)} - v^{(1)}), (v^{(3)} - v^{(1)}), \ldots, (v^{(k)} - v^{(1)})$ are linearly independent.)

**Definition 1.24** Let $P \subseteq \mathbb{R}^n$ be a polyhedron. The *dimension of* $P$ is

$$\dim(P) := (\text{maximum number of affinely independent points in } P) - 1.$$

Since the maximum number of affinely independent points in $\mathbb{R}^n$ is $(n+1)$, $\dim(P) \leq n$.

**Definition 1.25** If $P = \{x \in \mathbb{R}^n : Ax \leq b\}$ and $A$ has no zero rows, then $\hat{x} \in P$ such that $A\hat{x} < b$ is called an *interior point* of $P$.

**Definition 1.26** $P$ is *full dimensional* if $\dim(P) = n$.

Note that $P$ is full dimensional iff $P$ has interior points. If $P = \{x \in \mathbb{R}^n : \bar{A}x = \bar{b},\ Ax \leq b\}$, such that the rank of $\bar{A}$ is $k$, and there exists $\hat{x} \in P$ such that $A\hat{x} < b$, then $\dim(P) = (n - k)$.

$\alpha \in \mathbb{R}^n$, $\beta \in \mathbb{R}$ such that

$$P \subseteq \{x \in \mathbb{R}^n : \alpha^T x \leq \beta\}$$

give rise to the *valid inequality* $(\alpha^T x \leq \beta)$ for $P$.

Let $P \subseteq \mathbb{R}^n$ be a polyhedron. Also let $(\alpha^T x \leq \beta)$ be a valid inequality for $P$. Then

$$P_f := \{x \in P : \alpha^T x = \beta\}$$

is a *face* of $P$. If $\emptyset \subset P_f \subset P$ then $P_f$ is called a *proper face* of $P$. Proper faces of $P$ that are of the highest possible dimension get a special name.

**Definition 1.27** A face $P_f$ of $P$ is a *facet* of $P$ if $\dim(P_f) = \dim(P) - 1$.

Facets are very important in describing polyhedra as the following fact illustrates.

**Proposition 1.28** *For each facet $P_f$ of $P$, at least one of the inequalities representing $P_f$ (i.e., a valid inequality $(\alpha^T x \leq \beta)$ for $P$, such that $P_f = \{x \in P : \alpha^T x = \beta\}$) is necessary in describing $P$.*

Once we include an inequality representation of each facet, the polyhedron at hand is completely described and any other inequality that represents a lower dimensional face (that is of dimension less than $[\dim(P) - 1]$) will be implied by the inequalities representing the facets of $P$ (in fact, these implications can be expressed as appropriate linear combinations of the facet representing inequalities of $P$).

**Proposition 1.29** *Every inequality that represents a face of dimension less than $[\dim(P) - 1]$ is irrelevant to the description of $P$.*

**Proof** For a contradiction, suppose $P_f$ is such a face and the corresponding inequality $(\alpha^T x \leq \beta)$ is relevant. Let

$$P =: \{x \in \mathbb{R}^n : \bar{A}x = \bar{b},\ Ax \leq b,\ \alpha^T x \leq \beta\}$$

such that $\bar{A}x = \bar{b}$ describes a maximal set of equations satisfied by every point in $P$. So, there exists $\hat{x} \in \mathbb{R}^n$ such that

$$\bar{A}\hat{x} = \bar{b},\ A\hat{x} \leq b,\ \alpha^T \hat{x} > \beta$$

and there exists $\tilde{x} \in P$ such that

$$\bar{A}\tilde{x} = \bar{b},\ A\tilde{x} < b,\ \alpha^T \tilde{x} < \beta.$$

Hence, there exists $\lambda \in (0,1)$ such that $z := \lambda\hat{x} + (1-\lambda)\tilde{x}$ has the property

$$\bar{A}z = \bar{b},\ Az < b,\ \alpha^T z = \beta.$$

However, this implies, $\dim\{x \in P : \alpha^T x = \beta\} = \dim(P) - 1$. That is, $P_f$ is a facet of $P$. This is a contradiction. □

Next, we summarize what we learned about the representation of polyhedra as the feasible regions of linear optimization problems.

**Theorem 1.30** *Every full dimensional polyhedron $P$ has a unique (up to scalar multiplication), minimal representation by a finite set of linear inequalities. In particular, for each facet $P_f$ of $P$, there exists a unique inequality $\alpha^T x \leq \beta$ representing it. If $\dim(P) = n - k$, $k \geq 1$, then the representation*

$$P = \{x \in \mathbb{R}^n : \bar{A}x = \bar{b},\ Ax \leq b\},$$

*such that $\bar{A}$ is $k \times n$, $\operatorname{rank}(\bar{A}) = k$ is unique up to multiplication of the inequalities in $Ax \leq b$ by positive scalars and translations of inequalities in $Ax \leq b$ by linear combinations of the equations in the system $\bar{A}x = \bar{b}$.*

Also note that there are many possible representations of the affine space $\{x \in \mathbb{R}^n : \bar{A}x = \bar{b}\}$ which are characterized by the bases of $\mathbb{R}^k$.

We have been discussing the representations of polyhedra as intersections of closed half-spaces. A useful concept extended to arbitrary subsets of $\mathbb{R}^n$ is that of the *convex hull*. Given $G \subseteq \mathbb{R}^n$, *convex hull* of $G$, denoted $\operatorname{conv}(G)$, is the intersection of all convex sets containing $G$. In the sense of containment, $\operatorname{conv}(G)$ is the smallest convex set containing $G$. Note that if $G \subset \mathbb{R}^n$ is a compact convex set, with nonempty interior, then defining

$$K_G := \left\{\begin{pmatrix} \lambda \\ \lambda x \end{pmatrix} \in \mathbb{R}^{n+1} : \lambda \geq 0,\ x \in G\right\},$$

we have

$$\begin{aligned} K_G^* &= \left\{\begin{pmatrix} \beta \\ -\alpha \end{pmatrix} \in \mathbb{R}^{n+1} : \lambda(\alpha^T x) \leq \lambda\beta, \text{ for every } \lambda \geq 0,\ x \in G\right\} \\ &= \left\{\begin{pmatrix} \beta \\ -\alpha \end{pmatrix} \in \mathbb{R}^{n+1} : \alpha^T x \leq \beta, \text{ for every } x \in G\right\} \\ &= \left\{\begin{pmatrix} \beta \\ -\alpha \end{pmatrix} \in \mathbb{R}^{n+1} : \alpha^T x \leq \beta, \text{ for every } x \in \operatorname{ext}(G)\right\}, \end{aligned}$$

where $\operatorname{ext}(G)$ denotes the set of extreme points of $G$.

Let us turn to $P \subseteq \mathbb{R}^n$ being a polyhedron, and define $P_I := \operatorname{conv}(P \cap \mathbb{Z}^n)$. Two of the central questions in Combinatorial Optimization are:

- When is $P = P_I$?
- How do we get to a representation of $P_I$ by a system of linear inequalities, starting from a description of $P$ by a system of linear inequalities ?

Let us go over some of the classical answers to such questions in combinatorial optimization. For the question of when $P = P_I$, let us assume $P$ is pointed and express $P$ as $\{x \in \mathbb{R}^n : Ax \le b, x \ge 0\}$. First, we look at the characterizations of $P = P_I$ when $b \in \mathbb{Z}^m$ is completely arbitrary. So, the question becomes "what property of $A$ guarantees that $P = P_I$ for every $b \in \mathbb{Z}^m$?"

**1.3.1 Total Unimodularity (TUM).** $A \in \mathbb{Z}^{m \times n}$ is *Totally Unimodular* (TUM), if every square submatrix of $A$ has determinant $-1, 0$, or 1.

**Theorem 1.31** *Let $A \in \{-1, 0, 1\}^{m \times n}$. Then TFAE*

(i) *$A$ is TUM;*
(ii) $\text{ext}\{x \in \mathbb{R}^n : \ Ax \le b, \ x \ge 0\} \subseteq \mathbb{Z}^n_+, \ \forall\, b \in \mathbb{Z}^m$;
(iii) $\text{ext}\{x \in \mathbb{R}^n : \ a \le Ax \le b, \ c \le x \le d\} \subseteq \mathbb{Z}^n, \ \forall\, a, b \in \mathbb{Z}^m, \ c, d \in \mathbb{Z}^n$;
(iv) *no square submatrix of $A$ has determinant $\pm 2$;*
(v) *for every $I \subseteq \{1, 2, \dots, n\}$, there exist $I_1, I_2$ making up a partition of $I$ such that*

$$\left| \sum_{i \in I_1} A_{ij} - \sum_{i \in I_2} A_{ij} \right| \le 1, \ \forall j.$$

The above theorem gives a tool for (re)proving many interesting theorems in combinatorial optimization. E.g, Maximum Flow-Minimum Cut theorem of Ford and Fulkerson, Chain-antichain theorem of Dilworth, Kőnig's (also known as Birkhoff-von Neumann) Theorem. It is a good exercise to reprove these using Theorem 1.31 and the duality theory of LP.

There is also a constructive description of TUM matrices.

1.3.1.1 *Network Matrices.* Let $G = (V, A)$, with the node set $V$ and arc set $A$, be a directed graph, $T = (V, A_0)$ be a directed tree on $V$. Construct $M \in \{-1, 0, 1\}^{A_0 \times A}$ as follows: For every arc $a = (v, w) \in A$ and for every arc $a' \in A_0$, the corresponding entry of $M$ is

$$M_{a',a} := \begin{cases} 1 & \text{if the unique } v \to w \text{ path in } T \text{ uses } a' \text{ positively;} \\ -1 & \text{if the unique } v \to w \text{ path in } T \text{ uses } a' \text{ negatively;} \\ 0 & \text{if the unique } v \to w \text{ path in } T \text{ does not use } a'. \end{cases}$$

Thus, a matrix $M$ with $-1, 0, 1$ entries is called a *network matrix* if it arises as above from a directed graph $G$ and a directed tree $T$. Tutte [352] proved that every network matrix is TUM.

Seymour [322] proved that all TUM matrices arise from some network matrices and two fixed $5 \times 5$ matrices by some simple operations. Moreover, such matrices can be recognized in polynomial time. Therefore, given an integer programming problem, we can decide in polynomial time whether its coefficient matrix is TUM and if so, we can solve the underlying LP in polynomial time and recover an extreme point solution of the LP that is optimal, thus, solving the original combinatorial optimization problem.

However, given a matrix $A \in \{-1, 0, 1\}^{m \times n}$, it is $\mathcal{NP}$-hard to find the largest submatrix of $A$ that is TUM.

Next in the hierarchy are the *balanced matrices.*

**1.3.2 Balanced Matrices.** The concept of Total Unimodularity answered the question of when $P = P_I$ independent of $b \in \mathbb{Z}^m$. Balanced matrices arise as an answer to the same question when $P$ models packing or covering problems with $A \in \{0,1\}^{m \times n}$. For *packing problems*

$$P = \{x \in \mathbb{R}^n : Ax \leq b, x \geq 0\} \text{ and } b \in \{1, +\infty\}^m.$$

For *covering problems*

$$P = \{x \in \mathbb{R}^n : Ax \geq b, x \geq 0\} \text{ and } b \in \{0,1\}^m.$$

In the packing problem problem, the feasible solutions try to *pack* the columns of $A$ into $b$ and $b_i = +\infty$ effectively means that the $i$th constraint does not exist. Similarly, in the covering problem, we are trying to *cover* the vector $b$ by the columns of $A$ and $b_i = 0$ effectively means that the $i$th constraint does not exist.

Let

$$\begin{aligned} \mathcal{B} \ := \ & \left\{B \in \{0,1\}^{k \times k} : \ k \geq 3, \ \ k \text{ is odd}, \ B\bar{e} = B^T\bar{e} = 2\bar{e}, \right. \\ & \left. \qquad B \text{ does not contain the submatrix } \begin{pmatrix} 1 & 1 \\ 1 & 1 \end{pmatrix} \right\}. \end{aligned}$$

$A \in \{0,1\}^{m \times n}$ is called *balanced* if it does not contain any submatrix that is in $\mathcal{B}$. E.g.,

$$A := \begin{pmatrix} 1 & 1 & 1 & 1 \\ 1 & 1 & 0 & 0 \\ 1 & 0 & 1 & 0 \\ 1 & 0 & 0 & 1 \end{pmatrix}$$

is balanced but not TUM.

**Theorem 1.32** *Let $A \in \{0,1\}^{m \times n}$ with no zero rows or columns. Then TFAE*

(i) *$A$ is balanced;*
(ii) *$\{x \in \mathbb{R}^n : \ Ax \leq b, \ x \geq 0\}$ is integral for every $b \in \{1, +\infty\}^m$;*
(iii) *$\{x \in \mathbb{R}^n : \ Ax \geq b, \ x \geq 0\}$ is integral for every $b \in \{0,1\}^m$.*

It is possible that $\{x \in \mathbb{R}^n : \ Ax \leq \bar{e}, \ x \geq 0\}$ is integral but $A$ is not balanced.

**1.3.3 Perfect Graphs.** We continue focusing on matrices $A$ with only 0,1 entries. The rows (or the columns) of $A$ can be used to encode some information about discrete substructures. Next, as an extreme instance of this, we focus on the case when the columns of $A$ represent the nodes in a given graph $G = (V, E)$ and the rows of $A$ represent *cliques* (subgraphs $H = (W, F)$ such that $W \subseteq V$, $F \subseteq E$ such that for every pair $u, v \in W$, $\{u, v\} \in F$) in $G$. Such matrices $A$ are called *clique-node incidence matrices*. This discussion brings us to the beautiful subject of perfect graphs.

Given an undirected graph $G = (V, E)$, a subset $\mathcal{S} \subseteq V$ is a *stable set of* $G$ if for every $\{i, j\} \in E$ at most one of $i$, $j$ is in $\mathcal{S}$. An undirected graph $G = (V, E)$ is *perfect* if for every node induced subgraph $H = (W, F)$ of $G$ (that is, $W \subseteq V$ and $F := \{\{u, v\} \in E : u, v \in W\}$) the size of the maximum cardinality stable set in $H$ is the same as the minimum number of cliques of $H$ required to cover all nodes of $H$.

Let $A_{\text{clq}}(G)$ denote the clique-node incidence matrix of $G$.

**Theorem 1.33** *$G$ is perfect iff $\{x : \ A_{\text{clq}}(G)x \leq \bar{e}, \ x \geq 0\}$ is integral.*

In fact, $\{x : Ax \leq \bar{e},\ x \geq 0\}$ is integral iff $A = A_{\text{clq}}(G)$ for some $G$. The corresponding matrices $A \in \{0,1\}^{m\times n}$ are called *perfect*.

$\overline{G}$ denotes the complement of $G$.

**Theorem 1.34** *(Perfect Graph Theorem) $G$ is perfect iff $\overline{G}$ is.*

An *odd-hole* is an odd cycle of length at least five such that the cycle has no chord. An *odd anti-hole* is the complement of an odd-hole.

**Theorem 1.35** *(Strong Perfect Graph Theorem) A graph is perfect iff it does not contain any odd-hole or odd-antihole.*

If we step back and look at the development of our discussion starting with the question of $P = P_I$ and reaching the concept of perfect graphs and perfect matrices, we see that an analogous development of the subject follows the route of covering problems.

**1.3.4 Ideal Matrices.** We saw that one can reach the concept of perfect matrices starting from the concept of balanced matrices, taking the characterization of Theorem 1.32 (ii) and replacing $b$ by $\bar{e}$. So, for a parallel development involving covering polyhedra, we start from the characterization of Theorem 1.32 (iii) and replace $b$ by $\bar{e}$. This leads to the concept of *ideal matrices*.

For this subsection, we assume $A \in \{0,1\}^{m\times n}$ such that $A$ does not have any zero rows.

**Definition 1.36** $A \in \{0,1\}^{m\times n}$ is *ideal* if the polyhedron $\{x \in \mathbb{R}^n : Ax \geq \bar{e}\}$ is integral.

**Theorem 1.37** *Let $A \in \{0,1\}^{m\times n}$. Then TFAE*

(a) *$A$ is ideal;*

(b) $\min\left\{c^T x : Ax \geq \bar{e}, x \geq 0\right\}$ *has an integral solution for every $c \in \{0,1,+\infty\}^n$.*

Although the ideal matrices did not attract the same level of attention that the perfect matrices commanded since the 1960's, there has been increasing interest and plenty of activity in the area.

**1.3.5 Extension to $\{-1,0,+1\}$ Matrices.** The notions of *balanced*, *perfect* and *ideal* matrices have been extended to the matrices with $\{-1,0,1\}$ entries.

**Definition 1.38** Let $A \in \{-1,0,1\}^{m\times n}$. Then $A$ is *balanced* if every square submatrix with exactly two nonzero entries in every row and column has the property that the sum of the entries of the submatrix is zero, mod 4.

Let $a_- \in \mathbb{Z}^m_+$ encode the number of $(-1)$s in the rows of $A$.

**Theorem 1.39** *Let $A \in \{-1,0,1\}^{m\times n}$ be a balanced matrix. Then the polytope $\{x \in \mathbb{R}^n : Ax = \bar{e} - a_-, 0 \leq x \leq \bar{e}\}$ is integral.*

**Definition 1.40** Let $A \in \{-1,0,1\}^{m\times n}$. Then $A$ is *perfect* if the polytope

$$\{x \in \mathbb{R}^n : Ax \leq \bar{e} - a_-, 0 \leq x \leq \bar{e}\}$$

is integral. $A$ is *ideal* if the polytope

$$\{x \in \mathbb{R}^n : Ax \geq \bar{e} - a_-, 0 \leq x \leq \bar{e}\}$$

is integral.

**Theorem 1.41** *Let $A \in \{-1,0,1\}^{m\times n}$. Then TFAE*

(a) *the matrix $A$ is balanced;*

(b) *every submatrix of $A$ is perfect;*
(c) *every submatrix of $A$ is ideal.*

Every TUM matrix is balanced and every balanced matrix is both perfect and ideal. There are perfect matrices that are not ideal and there are ideal matrices which are not perfect.

**1.3.6 A Theorem for the General Setting.** In the most general setting, we have $P := \{x \in \mathbb{R}^n : Ax \leq b\}$, where $A \in \mathbb{Q}^{m \times n}$ and $b \in \mathbb{Q}^m$. Recall $P_I := \text{conv}(P \cap \mathbb{Z}^n)$.

**Theorem 1.42** *TFAE*

(i) $P = P_I$;
(ii) *every non-empty face of $P$ contains an integer vector;*
(iii) *every minimal face of $P$ contains an integer vector;*
(iv) *for every $c \in \mathbb{R}^n$ such that $\max\{c^T x : Ax \leq b\}$ is attained, the optimum value is attained by an integer vector $x$;*
(v) *for every $c \in \mathbb{Z}^n$ such that $\max\{c^T x : Ax \leq b\}$ is attained, the optimum value is integer;*
(vi) *every rational supporting hyperplane of $P$ contains an integer vector.*

**1.3.7 Total Dual Integrality (TDI).** A system of inequalities $\{Ax \leq b\}$ (for $A \in \mathbb{Q}^{m \times n}$, $b \in \mathbb{Q}^m$), is *Totally Dual Integral* if for every $c \in \mathbb{Z}^n$ such that $\max\{c^T x : Ax \leq b\}$ is attained, the dual $\min\{b^T y : A^T y = c,\ y \geq 0\}$ has an optimal solution that is integer.

Using Theorem 1.42, we have

**Corollary 1.43** *Let $P := \{x \in \mathbb{R}^n : Ax \leq b\}$. If $b \in \mathbb{Z}^m$ and $\{Ax \leq b\}$ is TDI then $P = P_I$.*

**Theorem 1.44** *If $G$ is perfect, then*

$$\left\{ \begin{array}{c} A_{\text{clq}}(G)x \leq \bar{e} \\ x \geq 0 \end{array} \right\}$$

*is TDI.*

By establishing the TDIness of the "right system" and using the theory of LP duality one can obtain many interesting combinatorial theorems.

Note that any rational system can be made TDI by scaling or by adding redundant inequalities (these are also good exercises). Also note that $P = P_I$ does not imply that the minimal description (by facets) of $P$ is TDI.

Next, we consider a classical answer to the question "how do we get to a representation of $P_I$ by a system of linear inequalities, starting from a description of $P$ by a system of linear inequalities?"

**1.3.8 Gomory-Chvátal Cuts and Their Rank.** Let $A \in \mathbb{Q}^{m \times n}$ and $b \in \mathbb{R}^m$ and define $P := \{x \in \mathbb{R}^n : Ax \leq b\}$. Note that for every $u \in \mathbb{Q}^m_+$ such that $u^T A \in \mathbb{Z}^n$, the linear inequality

$$u^T Ax \leq \lfloor u^T b \rfloor$$

is valid for $P_I$. Moreover, if $P \neq P_I$, then there exists $\bar{u} \in \mathbb{Q}^m_+$ such that $\bar{u}^T A \in \mathbb{Z}^n$ and there exists $\bar{x} \in (P \setminus \mathbb{Z}^n)$ such that

$$\bar{u}^T A\bar{x} > \lfloor \bar{u}^T b \rfloor$$

that is, the valid inequality $\bar{u}^T Ax \leq \lfloor \bar{u}^T b \rfloor$ for $P_I$ *cuts* the fractional point $\bar{x}$ in $P$.

The set of all linear inequalities obtained this way are called Gomory-Chvátal Cuts.

Gomory proved in early 1960's that such cuts can be used in a simplex method based approach to solve combinatorial optimization problems and proved the finite termination of his method (with some restriction of the underlying techniques: Cut selection and anti-cycling rules in simplex method etc.).

An elegant theorem due to Chvátal shows us that repeating the above procedure finitely many times suffices to "compute" $P_I$ from $P$.

Formally, we define

$$GC(\{Ax \leq b\}) := \left\{Ax \leq b,\ u^T Ax \leq \lfloor u^T b \rfloor,\ \forall u \in \mathbb{Q}_+^m \text{ such that } u^T A \in \mathbb{Z}^n\right\},$$

and

$$GC^2(\cdot) := GC(GC(\cdot)).$$

I abuse (very slightly) the notation and use $GC(\cdot)$ on polyhedra as well (then the definition of $GC$ is applied to the facets of the polyhedron).

**Definition 1.45** The smallest integer $k$ for which $GC^k(P) = P_I$ is called *the Chvátal rank of $P_I$ relative to $P$*. Given a valid inequality $(\alpha^T x \leq \beta)$ for $P_I$, we call the smallest $k$ for which $(\alpha^T x \leq \beta)$ is valid for $GC^k(P)$, *the Chvátal rank of $(\alpha^T x \leq \beta)$ relative to $P$*.

The following is an extension of Chvátal's Theorem.

**Theorem 1.46** *Let $A$ be a rational matrix. For every $b \in \mathbb{R}^m$, define*

$$P(b) := \{x \in \mathbb{R}^n : Ax \leq b\}.$$

*Then there exists an integer $k(A)$ such that*

$$GC^{k(A)}(P(b)) = [P(b)]_I,$$

*for every $b \in \mathbb{R}^m$.*

$k(A)$ can be very large, it cannot be bounded in terms of $n$ alone, in general.

## 1.4 Further Bibliographical Notes

There are hundreds of books on LP (or Linear Optimization). The seminal book of *the Father of LP*, George B. Dantzig [1914–2005], is still a key reference [100] in the field. Together with Dantzig, two other main contributors, John von Neumann [1903–1957] and Leonid Kantarovich [1912–1986] stand out as the main founders of the theory of Linear Optimization during the first half of the 20th century. However, some of the relevant work on the theory of systems of linear inequalities and the theorems of the alternative go back at least to Fourier's paper from 1826, Gordan's Theorem (from 1873), Stiemke's Theorem (from 1915) and Farkas' Lemma (from 1902).

We already made some bibliographical comments in Section 1.1.1 about LP complexity. Also see, Chvátal [87], Goldfarb and Todd [147] as well as Schrijver [319] and Ye [373]. For further details on the fundamental material discussed in Section 1.3, see Schrijver's books [319, 317], the monograph by Cornuéjols [95], the book by Cook, Cunningham, Schrijver and Pulleyblank [91], and the book by Nemhauser and Wolsey [264].

Elementary but fundamental observation Theorem 1.20 is usually attributed to a subset of Courant-Fischer-Weyl. Theorem 1.23 is from [350]. For a nice introduction to Kronecker product, see Brewer [65, 66] and Van Loan [358]. One of the early papers on the cone of positive semidefinite matrices is by Bohnenblust [52]. Horn and Johnson's book [179] contains many useful results on matrix analysis that are relevant to our discussion. Also see, Bhatia [43] and Higham [172]. For introductory material on semidefinite optimization, see Alizadeh [5], Goemans [140], Vandenberghe and Boyd [354] and Todd [344]. Network matrices in the form we mentioned were introduced by Tutte [352]. Results such as Theorem 1.33 are the outcome of works of Fulkerson [134, 135] and Lovász [244]. For Total Dual Integrality, see Edmonds and Giles [112]. Theorem 1.46 is from Cook, Gerards, Schrijver and Tardos [93]. For other related results, see Chvátal [87], Chvátal, Cook and Hartmann [88], Eisenbrand [114], Bockmayr, Eisenbrand, Hartmann and Schulz [51], Eisenbrand and Schulz [115]. Such results have some connections to Hoffman [177], also see Güler, Hoffman and Rothblum [161] and Ho and Tunçel [176].

The Strong Perfect Graph Theorem was recently proved by Chudnovsky, Robertson, Seymour and Thomas [82, 83].

For more information on symmetric cones, see Faraut and Korányi [120]; for an exposure to the combinatorics of polytopes, see Ziegler [377], for an exposure to discrete convex analysis, see Murota [263].

CHAPTER 2

# Duality Theory

Given an optimization problem which is to minimize a function subject to some constraints, how do we evaluate the quality of a feasible solution? One way would be to have a tight lower bound on the optimal objective function value if well-defined. If the optimum value is not well-defined then we would like to have easy to evaluate certificates of this fact. These questions can be answered by a suitable notion of *the dual problem.* In LP theory, the properties of the dual are wonderful. Only some of these wonderful properties remain valid when we move to the settings of SDP and more general convex optimization. As a result, the duality theory for SDP is much richer than that of LP, more interesting, more powerful and in some way perhaps even more elegant.

## 2.1 Dual Cones

Let $K \subseteq \mathbb{R}^d$ be a convex cone. Recall that its dual is

$$K^* := \left\{ s \in \mathbb{R}^d : \langle x, s \rangle \geq 0, \ \forall x \in K \right\}.$$

Note that $K^*$ is always closed (intersection of closed sets) and convex (intersections of convex sets). Using the results of this chapter (for instance using Theorem 2.8 which will be proven from the first principles, independent of the results of this section), we can prove the following three fundamental results.

**Theorem 2.1** *For every convex cone $K$ in $\mathbb{R}^d$, we have*

$$K^{**} = \mathrm{cl}(K).$$

*Moreover, a convex cone $K$ is closed iff $K = K^{**}$.*

In fact, put slightly differently, $K$ is a closed convex cone iff $K = K^{**}$. Now, recall our formulation of primal and dual convex optimization problems with cone constraints involving $K$ and $K^*$ in the previous chapter. LP and SDP special cases are based on symmetric cones ($\mathbb{R}^n_+$ and $\Sigma^n_+$ respectively); so, in these cases $K = K^* = K^{**}$. In general, we have the following property.

**Theorem 2.2** *Let $K \subset \mathbb{R}^d$ be a pointed, closed convex cone with nonempty interior. Then so is $K^*$.*

In many instances, we intersect convex sets (or cones) and consider their duals in our arguments. Another common algebraic operation is the *Minkowski sum* of two sets:

$$K_1 + K_2 := \{u + v : u \in K_1, v \in K_2\}.$$

For pairs of convex cones, these two operations are "dual to each other" in the sense of the next theorem.

**Theorem 2.3** *Let $K_1, K_2 \subseteq \mathbb{R}^d$ be nonempty convex cones. Then*

- $(K_1 + K_2)^* = K_1^* \cap K_2^*$;
- $(\text{cl}\,(K_1) \cap \text{cl}\,(K_2))^* = \text{cl}\,(K_1^* + K_2^*)$.

*If $K_1$ and $K_2$ are closed and* $\text{relint}(K_1) \cap \text{relint}(K_2) \neq \emptyset$ *in the above then*

$$(K_1 \cap K_2)^* = K_1^* + K_2^*.$$

## 2.2 Polars of (Compact) Sets

Let $K \subset \mathbb{R}^d$ be a pointed closed convex cone with nonempty interior. Let $\bar{s} \in \text{int}\,(K^*)$. Define

$$G := \{x \in K : \langle \bar{s}, x \rangle \leq 1\}.$$

Then $G$ is a compact convex set in $\mathbb{R}^d$ which possesses all the information about $K$ in a very nice way. Sometimes it is slightly more convenient to deal with $G$ and sometimes it is the other way around. So, we will use both.

Now, let $G \subseteq \mathbb{R}^d$. Duality for sets is also very nice. We define the *polar of $G$* as

$$G^o := \left\{s \in \mathbb{R}^d : \langle x, s \rangle \leq 1, \forall x \in G\right\}.$$

**Theorem 2.4** *Let $G \subset \mathbb{R}^d$ be a compact convex set such that $0 \in \text{int}(G)$. Then so is $G^o$. Moreover, $G^{oo} = G$.*

We also have the analogous result to Theorem 2.3 in the current setting.

**Theorem 2.5** *Let $G_1, G_2 \subset \mathbb{R}^d$ be compact convex sets such that $0 \in \text{int}(G_1) \cap \text{int}(G_2)$. Then*

- $[\text{conv}\,(G_1 \cup G_2)]^o = G_1^o \cap G_2^o$;
- $(G_1 \cap G_2)^o = \text{conv}\,(G_1^o \cup G_2^o)$.

## 2.3 Conjugates of (Convex) Functions

Another very closely related domain to study duality is that of (convex) functions. For a given function $f : \mathbb{R}^d \to \mathbb{R} \cup \{\infty\}$, we define its Legendre-Fenchel conjugate as

$$f_*(s) := \sup\left\{-\langle s, x \rangle - f(x) : x \in \mathbb{R}^d\right\}.$$

This notation and definition are slightly different than the usual ones in convex analysis. (There $f^*$ is used and $f^*(s) = f_*(-s)$; of course, the definition of dual cone is also slightly different in most references in convex analysis.)

Given $f$ as above, the *epigraph of $f$* is defined as

$$\text{epi}(f) := \left\{\begin{pmatrix} t \\ x \end{pmatrix} \in \mathbb{R} \oplus \mathbb{R}^d : f(x) \leq t\right\}.$$

It is easy to prove that

**Theorem 2.6** *Let $f$ be defined as above. Then, $f$ is convex iff* $\text{epi}(f)$ *is a convex set.*

Therefore, studying convex sets is essentially equivalent to studying convex functions, as the results for one can be translated via the above correspondence to the results about the other. In particular, we can express the theorems of Sections 2.1 and 2.2 in terms of convex functions and their conjugates (the reader not familiar with these correspondences should try this exercise).

## 2.4 A Strong Duality Theorem via the Hahn-Banach Theorem

The Hahn-Banach Theorem is one of the most important tools of functional analysis. The most common form of the Hahn-Banach Theorem allows extensions of bounded linear operators from a subspace of a vector space to the whole vector space. Perhaps more importantly, it proves that every normed vector space has sufficiently many linear functionals (dual elements) defined on it so that it makes the space of linear functionals (the dual space) very useful and interesting. This very general issue of *having enough dual elements for a specific purpose* will arise in many disguises throughout the rest of the chapters. Shortly, we will discuss the Strong Duality Theorem for SDP and then try to extend the theorem to the general case where we do not require any constraint qualification (in that case, not having *enough* dual variables will be the initial hurdle). Towards the end of the monograph, when we discuss Successive Convex Relaxation Methods, Polynomial Optimization Problems and Sums-of-Squares proofs of nonnegativity, such issues will again be paramount.

Perhaps the second most common form of the Hahn-Banach theorem is stated as the *Hahn-Banach Separation Theorem* (see for example Rudin [309]), and this separation theorem is usually referred to as *Mazur's geometric form of the Hahn-Banach Theorem.*

**Theorem 2.7** *(Hahn-Banach Separation Theorem) Let $\mathbb{E}$ be a topological vector space ($\mathbb{E}^*$ is its dual) and $G_1, G_2 \subset \mathbb{E}$. Suppose $G_1$ and $G_2$ are nonempty, convex and disjoint. Then*

(a) *if $G_1$ is open, then there exist a continuous linear functional $\ell \in \mathbb{E}^*$ and $\alpha \in \mathbb{R}$ such that*

$$\sup\{\ell(u) : u \in G_1\} < \alpha \leq \inf\{\ell(v) : v \in G_2\};$$

(b) *if $G_1$ is compact, $G_2$ is closed and $\mathbb{E}$ is locally convex, then there exist a continuous linear functional $\ell \in \mathbb{E}^*$ and $\alpha \in \mathbb{R}$ such that*

$$\sup\{\ell(u) : u \in G_1\} < \alpha < \inf\{\ell(v) : v \in G_2\}.$$

In the above, we stated the theorem with the specification that the underlying scalar field is $\mathbb{R}$. (When the underlying scalar field is $\mathbb{C}$, we simply replace $\ell(u)$ and $\ell(v)$ by their real parts.)

Below, we derive finite-dimensional counterparts of the above theorem from the first principles. These theorems serve a similarly central role in finite-dimensional optimization.

**Theorem 2.8** *(Separation Theorem) Let $G \subset \mathbb{R}^d$ be a nonempty, closed convex set. Suppose $0 \notin G$. Then there exist $a \in \mathbb{R}^d \backslash \{0\}$ and $\alpha \in \mathbb{R}_{++}$ such that*

$$G \subseteq \left\{x \in \mathbb{R}^d : a^T x \geq \alpha\right\}.$$

**Proof** Let $\bar{x} \in G$. Define

$$\bar{G} := \{x \in G : \|x\|_2 \leq \|\bar{x}\|_2\}.$$

Let $a \in \bar{G}$ be the point in $\bar{G}$ with the minimum norm. (Such a point $a \in \mathbb{R}^d$ exists, since $\bar{G} \neq \emptyset$, $\bar{G}$ is compact, $\|x\|_2^2$ is continuous on $\bar{G}$; in fact $a$ is unique since $\|x\|_2^2$ is strictly convex on $\bar{G}$.) Define $\alpha := a^T a > 0$. Note that for every $x \in G$, the line segment $[a, x]$ lies completely in $G$. Thus, for every $x \in G$ and for every $\lambda \in (0, 1]$,

$$\begin{aligned}[\lambda x + (1-\lambda)a] \in G \quad &\Rightarrow \quad \|\lambda x + (1-\lambda)a\|_2^2 \geq \|a\|_2^2 = \alpha \\ &\Rightarrow \quad \|\lambda(x-a) + a\|_2^2 \geq \alpha \\ &\Rightarrow \quad \lambda^2\|x-a\|_2^2 + 2\lambda a^T(x-a) \geq 0 \\ &\Rightarrow \quad a^T(x-a) \geq -\frac{\lambda}{2}\|x-a\|_2^2.\end{aligned}$$

Now, we take limits as $\lambda \to 0^+$ and conclude $a^T(x-a) \geq 0$. Therefore, for every $x \in G$, we have $a^T x \geq \alpha$. □

**Corollary 2.9** *Let $G_1$, $G_2 \subset \mathbb{R}^d$ be disjoint, nonempty closed convex sets. If $G_1$ or $G_2$ is bounded then there exists $a \in \mathbb{R}^d \setminus \{0\}$ such that*

$$\inf\left\{a^T x : x \in G_1\right\} > \sup\left\{a^T x : x \in G_2\right\}.$$

**Proof** Let $G_0 := G_1 - G_2$. $G_0$ is nonempty, convex. Since $G_1$ and $G_2$ are disjoint, $0 \notin G_0$. We will prove that $G_0$ is closed. Then applying the previous theorem will conclude the proof. Let $\left\{g_0^{(k)}\right\}$ be a sequence in $G_0$ such that $g_0^{(k)} \to \overline{g}_0$. Then $g_0^{(k)} = g_1^{(k)} - g_2^{(k)}$ for some $g_1^{(k)} \in G_1$ and $g_2^{(k)} \in G_2$. Without loss of generality, we can assume that $G_2$ is bounded. Hence $G_2$ is compact. By refining to a subsequence if necessary, we can assume $g_2^{(k)} \to \overline{g}_2 \in G_2$. Then the corresponding refined subsequence $g_1^{(k)} \to (\overline{g}_0 - \overline{g}_2) =: \overline{g}_1$. Since $G_1$ is closed, $\overline{g}_1 \in G_1$. Therefore, $\overline{g}_0 = (\overline{g}_1 - \overline{g}_2) \in G_0$. Hence, $G_0$ is closed, as desired. Applying the previous theorem, we conclude that there exist $a \in \mathbb{R}^d \setminus \{0\}$ and $\alpha \in \mathbb{R}_{++}$ such that

$$\begin{aligned}& a^T g_0 \geq \alpha > 0, \ \forall g_0 \in (G_1 - G_2) \\ \Rightarrow \quad & a^T g_1 \geq \alpha + a^T g_2, \ \forall g_1 \in G_1, g_2 \in G_2 \\ \Rightarrow \quad & \inf\left\{a^T x : x \in G_1\right\} > \sup\left\{a^T x : x \in G_2\right\}\end{aligned}$$

as desired. □

**Remark 2.10** We need at least one of $G_1$, $G_2$ bounded. Consider

$$G_1 := \left\{x \in \mathbb{R}^2_{++} : x_2 \geq \frac{1}{x_1}\right\}$$

and

$$G_2 := \left\{x \in \mathbb{R}^2 : x_1 > 0, x_2 \leq -\frac{1}{x_1}\right\}.$$

Then the only viable candidate for $a \in \mathbb{R}^2 \setminus \{0\}$ is $\begin{pmatrix} 0 \\ 1 \end{pmatrix}$ which leads to

$$\inf\left\{a^T x : x \in G_1\right\} = \sup\left\{a^T x : x \in G_2\right\}.$$

**Theorem 2.11** *Let $G \subset \mathbb{R}^d$ be a nonempty convex set. Suppose $0 \notin G$. Then there exists $a \in \mathbb{R}^d \setminus \{0\}$ such that*

$$G \subseteq \left\{x \in \mathbb{R}^d : a^T x \geq 0\right\}.$$

**Proof** For every $x \in G$, we define the unit hemisphere in $\mathbb{R}^d$ intersecting the halfspace $\{x\}^*$:

$$HS(x^*) := \left\{s \in \mathbb{R}^d : x^T s \geq 0, \ \|s\|_2 = 1\right\}.$$

We would like to find $a \in \bigcap_{x \in G} HS(x^*)$. By definition, $HS(x^*)$ is a hemisphere in $\mathbb{R}^d$ and therefore is compact. Hence, if every finite collection of $HS(x^*)$'s (for $x \in G$) has a nonempty intersection, then so do all $HS(x^*)$ (where the intersection runs over all elements of $G$). Let $x^{(1)}, x^{(2)}, \ldots, x^{(k)}$ be an arbitrary collection of points in $G$. Consider $\operatorname{conv}\left(x^{(1)}, x^{(2)}, \ldots, x^{(k)}\right) \subseteq G$. The set $\operatorname{conv}\left(x^{(1)}, x^{(2)}, \ldots, x^{(k)}\right)$ is a nonempty, compact, convex set not containing 0. Therefore, by the separation theorem, there exists $a \in \mathbb{R}^d \setminus \{0\}$, such that

$$\operatorname{conv}\left(x^{(1)}, x^{(2)}, \ldots, x^{(k)}\right) \subseteq \left\{x \in \mathbb{R}^d : a^T x \geq 0\right\}.$$

We set $a := \frac{a}{\|a\|_2}$. Therefore, every finite collection of $HS(x^*)$'s (for $x \in G$) has a nonempty intersection implying, there exists $a \in \mathbb{R}^d \setminus \{0\}$ such that

$$G \subseteq \left\{x \in \mathbb{R}^d : a^T x \geq 0\right\}.$$

□

*Aside:* We used a standard fact from elementary real analysis in the above proof. Let $H_t, t \in T$ be an arbitrary collection of closed sets and let $\hat{H}$ be a compact set. If the intersection of all $H_t$ has no intersection with $\hat{H}$, then the union of the complements of $H_t$ (open sets) cover $\hat{H}$. Since $\hat{H}$ is compact, any open cover of it must have a finite subcover (Heine-Borel characterization of compactness). Upon complementing this subcover, we see that the intersection of these finitely many $H_t$ has no intersection with $\hat{H}$ either. In the above proof, we took $H_t := \{x\}^*$, $T := G$ and $\hat{H}$ as the unit hypersphere in $\mathbb{R}^d$.

**Corollary 2.12** *Let $G_1, G_2 \subset \mathbb{R}^d$ be nonempty convex sets such that $G_1 \cap G_2 = \emptyset$. Then there exists $a \in \mathbb{R}^d \setminus \{0\}$ such that*

$$\inf\left\{a^T x : x \in G_1\right\} \geq \sup\left\{a^T x : x \in G_2\right\}.$$

**Proof** We simply apply the previous theorem to $G := G_1 - G_2$. □

Now, we are ready to begin discussing a *Strong Duality Theorem* for SDP. In the statement of our theorem, a *constraint qualification* is necessary. This is what we define next. Recall,

$$\begin{array}{llrcl}
(P) & \inf & \langle C, X\rangle & & \\
& \text{subject to:} & \mathcal{A}(X) & = & b, \\
& & X & \succeq & 0,
\end{array}$$

$$\begin{array}{llrcrcl}
(D) & \sup & b^T y & & & & \\
& \text{subject to:} & \mathcal{A}^*(y) & + & S & = & C, \\
& & & & S & \succeq & 0.
\end{array}$$

**Definition 2.13** We say that $(P)$ *satisfies the Slater condition*, or $(P)$ *has a Slater point* if there exists $\bar{X} \in \Sigma^n$ such that $\mathcal{A}(\bar{X}) = b$ and $\bar{X} \succ 0$. Similarly, we say that $(D)$ *satisfies the Slater condition*, or $(D)$ *has a Slater point* if there exist $\bar{S} \in \Sigma^n$ and $\bar{y} \in \mathbb{R}^m$ such that $\mathcal{A}^*(\bar{y}) + \bar{S} = C$ and $\bar{S} \succ 0$.

The next theorem has different assumptions and conclusions about $(P)$ and $(D)$. We are assuming that we are interested in the optimal solutions of the primal problem; hence, the attainment of the optimal objective value of $(P)$ is more critical to us. (If it is the attainment of the optimal objective value of $(D)$ that is more important, then we switch the roles of $(P)$ and $(D)$.)

**Theorem 2.14** *(A Strong Duality Theorem) Suppose* $(D)$ *has a Slater point. If the objective value of* $(D)$ *is bounded from above then* $(P)$ *attains its optimum value and the optimum values of* $(P)$ *and* $(D)$ *coincide.*

**Proof** Suppose there exist $\bar{S} \in \Sigma_{++}^n$, $\bar{y} \in \mathbb{R}^m$ such that $\mathcal{A}^*(\bar{y}) + \bar{S} = C$. Let $z^* := \sup\left\{b^T y : \mathcal{A}^*(y) \preceq C\right\}$. We can assume without loss of generality that $b \neq 0$. If $b = 0$, then $\bar{X} := 0$ is an optimal solution of $(P)$, because $(\bar{y}, \bar{S})$ is feasible in $(D)$ with objective value zero, which matches the objective value of $\bar{X}$ which is feasible in $(P)$. Consider

$$G_1 := \left\{S \in \Sigma^n : S = C - \mathcal{A}^*(y) \text{ for some } y \in \mathbb{R}^m, b^T y \geq z^*\right\}.$$

Note that $G_1 \neq \emptyset$. There exists a sequence $\left\{\left(y^{(k)}, S^{(k)}\right)\right\} \subset \mathbb{R}^m \oplus \Sigma^n$ such that $\mathcal{A}^*\left(y^{(k)}\right) + S^{(k)} = C$, $S^{(k)} \succeq 0$, and $b^T y^{(k)} \to z^*$. This last statement implies that there exists $\left(\hat{y}, \hat{S}\right) \in \mathbb{R}^m \oplus \Sigma^n$ such that $\mathcal{A}^*(\hat{y}) + \hat{S} = C$ and $b^T \hat{y} = z^*$ (a linear function over an affine subspace attains its limit; note that $\hat{S} \notin \Sigma_+^n$ is allowed; there are many elementary ways of proving this, an unusual proof is to apply Theorem 1.4 to the LP:

$$\max\left\{b^T y : b^T y \leq z^*, \mathcal{A}^*(y) + S = C, y \in \mathbb{R}^m, S \in \Sigma^n\right\}).$$

Now, let us define $G_2 := \Sigma_{++}^n$. We claim that $G_1 \cap G_2 = \emptyset$. We prove this claim by contradiction. Suppose there exists $\bar{y} \in \mathbb{R}^m$ such that $\mathcal{A}^*(\bar{y}) \prec C$, $b^T \bar{y} \geq z^*$; then $\hat{y} := \bar{y} + \epsilon b$ for some $\epsilon > 0$ satisfies $\mathcal{A}^*(\hat{y}) \prec C$ and $b^T \hat{y} > z^*$ (here we used the fact that $b \neq 0$). Now we arrived at a contradiction, since $z^*$ is the optimal value of $(D)$. We apply the last corollary to $G_1$ and $G_2$. Then, there exists $\tilde{X} \in \Sigma^n \setminus \{0\}$ such that

$$\sup\left\{\langle \tilde{X}, S\rangle : S \in G_1\right\} \leq \inf\left\{\langle \tilde{X}, S\rangle : S \in \Sigma_{++}^n\right\}.$$

Since $G_1 \neq \emptyset$, we conclude that $\inf_{S \in \Sigma_{++}^n} \langle \tilde{X}, S\rangle$ is bounded from below. Since $\Sigma_{++}^n$ is a cone, this yields $\langle \tilde{X}, S\rangle \geq 0$, for every $S \in \Sigma_{++}^n$. So, $\langle \tilde{X}, S\rangle \geq 0$, for every $S \in \operatorname{cl}\left(\Sigma_{++}^n\right) = \Sigma_+^n$. Hence, $\tilde{X} \in \Sigma_+^n$, by Proposition 1.10 part (f). We proved $\inf_{S \in \Sigma_{++}^n} \langle \tilde{X}, S\rangle \geq 0$. Sending $\left\{S^{(k)}\right\} \subset \Sigma_{++}^n$ to the zero matrix, we see that the infimum is equal to zero. Therefore, $\sup_{S \in G_1} \langle \tilde{X}, S\rangle \leq 0$. This is equivalent to

$$\langle \tilde{X}, C\rangle - \langle \tilde{X}, \mathcal{A}^*(y)\rangle \leq 0, \text{ for every } y \in \mathbb{R}^m \text{ such that } b^T y \geq z^*.$$

Equivalently,

$$\mathcal{A}(\tilde{X})^T y \geq \langle C, \tilde{X}\rangle, \text{ for every } y \in \mathbb{R}^m \text{ such that } b^T y \geq z^*.$$

So, $\mathcal{A}(\tilde{X})^T y$ is bounded below on $\left\{y \in \mathbb{R}^m : b^T y \geq z^*\right\}$. For instance, LP duality implies $\mathcal{A}(\tilde{X}) = \alpha b$, for some $\alpha \geq 0$. If $\alpha = 0$ then $\mathcal{A}(\tilde{X}) = 0$ implies $\langle C, \tilde{X}\rangle \leq 0$. Since $\tilde{X} \in \Sigma_+^n$, $\tilde{X} \neq 0$, and there exists $\left(\bar{y}, \bar{S}\right) \in \mathbb{R}^m \oplus \Sigma^n$ such that $\mathcal{A}^*(\bar{y}) + \bar{S} = C$, $\bar{S} \succ 0$, we have

$$0 \geq \langle C, \tilde{X}\rangle = \langle \mathcal{A}^*(\bar{y}) + \bar{S}, \tilde{X}\rangle = \underbrace{\mathcal{A}(\tilde{X})^T \bar{y}}_{=0} + \underbrace{\langle \tilde{X}, \bar{S}\rangle}_{>0} > 0,$$

(we used Proposition 1.11 (f)) a contradiction. So, we can assume $\alpha > 0$. We define $\bar{X} := \frac{1}{\alpha}\tilde{X} \in \Sigma_+^n$. Also, $\mathcal{A}(\bar{X}) = b$. Hence, $\bar{X}$ is feasible in $(P)$. Finally,

$$\begin{aligned} & \mathcal{A}(\tilde{X})^T y \geq \langle C, \tilde{X}\rangle, \quad \text{for all } y \in \mathbb{R}^m \text{ such that } b^T y \geq z^* \\ \Rightarrow \quad & \langle C, \bar{X}\rangle \leq z^*. \end{aligned}$$

Therefore, by the weak duality relation, $\bar{X}$ is an optimal solution of $(P)$. □

**Remark 2.15** Note that

(i) under the assumption that $(D)$ has a Slater point, the above theorem implies that the set of optimal solutions of $(P)$ is compact;
(ii) under the assumptions of the above theorem, the optimum value of $(D)$ may not be attained even though it is guaranteed to exist;
(iii) this theorem and its proof generalize to the conic convex optimization setting.

Noting that the dual of $(D)$ is equivalent to $(P)$, we have the following corollary.

**Corollary 2.16** *If $(P)$ has a feasible solution and $(D)$ has a Slater point then $(P)$ attains its optimal objective value and the optimal objective values of $(P)$ and $(D)$ are the same.*

**Corollary 2.17** *If $(P)$ and $(D)$ both have Slater points, then they both attain their optimal objective values and the optimal objective values of $(P)$ and $(D)$ coincide.*

Consider

$$
\begin{array}{llrcl}
(P) & \inf & \left\langle \begin{pmatrix} 1 & 0 \\ 0 & 0 \end{pmatrix}, X \right\rangle & & \\
& \text{subject to:} & \left\langle \begin{pmatrix} 0 & 1 \\ 1 & 0 \end{pmatrix}, X \right\rangle & = & 2, \\
& & X & \succeq & 0.
\end{array}
$$

Its dual is

$$
\begin{array}{llrcl}
(D) & \sup & 2y & & \\
& \text{subject to:} & \begin{pmatrix} 1 & -y \\ -y & 0 \end{pmatrix} & \succeq & 0.
\end{array}
$$

Then $X(\epsilon) := \begin{pmatrix} \epsilon & 1 \\ 1 & \frac{1}{\epsilon} \end{pmatrix}$ is feasible in $(P)$ for every $\epsilon > 0$ and $\langle C, X(\epsilon)\rangle \to 0$ as $\epsilon \to 0$. However, $\langle C, X\rangle = 0$ implies $X_{11} = 0$ which in turn implies $X_{12} = X_{21} = 0$ (since $X$ is positive semidefinite). This contradicts the equality constraint in the primal $X_{12} + X_{21} = 2$. Therefore, the optimal value of $(P)$ is zero but it is not attained. For $(D)$, we see that $\bar{y} := 0$ is the only feasible solution (hence is also optimal). For this example, $\bar{X} := I + \bar{e}\bar{e}^T$ is a Slater point for $(P)$ and the optimal value of $(D)$ is attained. Nevertheless, $(D)$ does not have a Slater point and even though $(P)$ has a well-defined optimal value, it is not attained.

Problem $(P)$ is equivalent to $\inf\{X_{11} : X_{11}X_{22} \geq 1, X_{11} \geq 0, X_{22} \geq 0\}$. Even though the feasible region of this problem is a closed set, its projection onto $\begin{pmatrix} \mathbb{R} \\ 0 \end{pmatrix}$ is not.

Another example (for $\gamma > 0$):

$$
C := \begin{pmatrix} 0 & 1 & 0 \\ 1 & 0 & 0 \\ 0 & 0 & 0 \end{pmatrix}, A_1 := \begin{pmatrix} 0 & 0 & 0 \\ 0 & 1 & 0 \\ 0 & 0 & 0 \end{pmatrix}, A_2 := \begin{pmatrix} 0 & 0 & 1 \\ 0 & 0 & 0 \\ 1 & 0 & 0 \end{pmatrix},
$$

$$
A_3 := \begin{pmatrix} 0 & -1 & 0 \\ -1 & 0 & 0 \\ 0 & 0 & 2 \end{pmatrix}, b := \begin{pmatrix} 0 \\ 0 \\ 2\gamma \end{pmatrix}.
$$

Note that $X^* := \begin{pmatrix} 0 & 0 & 0 \\ 0 & 0 & 0 \\ 0 & 0 & \gamma \end{pmatrix}$ is an optimal solution in $(P)$ with $\langle C, X^* \rangle = 0$ and $y^* := \begin{pmatrix} 0 \\ 0 \\ -1 \end{pmatrix}$ is an optimal solution in $(D)$ with $b^T y^* = -2\gamma$. Both optimal values are attained but there is a duality gap of $2\gamma$. Of course, neither problem has Slater points.

We can also give examples of SDP problems where neither $(P)$ nor $(D)$ attains the common optimal value:

$$n := 4, m := 3, C := e_1 e_2^T + e_2 e_1^T + e_4 e_4^T,$$
$$A_1 := -e_1 e_1^T, A_2 := -e_2 e_2^T, A_3 := e_3 e_4^T + e_4 e_3^T,$$
$$b := [0, -1, 2]^T.$$

By modifying this example, we can also give a primal-dual pair such that there is a finite positive duality gap and neither problem attains its (finite) optimal value:

$$n := 5, m := 4, C := e_1 e_2^T + e_2 e_1^T + e_4 e_4^T + e_5 e_5^T,$$
$$A_1 := -e_1 e_1^T, A_2 := -e_2 e_2^T, A_3 := e_3 e_4^T + e_4 e_3^T, A_4 := e_1 e_3^T + e_3 e_1^T + e_5 e_5^T,$$
$$b := [0, -1, 2, 1]^T.$$

Another case that is impossible in the linear optimization setting but can happen in SDP is that $(P)$ has a finite optimal value that is unattained and $(D)$ is infeasible:

$$n := 4, m := 4, C := e_1 e_2^T + e_2 e_1^T,$$
$$A_1 := -e_1 e_1^T, A_2 := -e_2 e_2^T, A_3 := e_1 e_3^T + e_3 e_1^T, A_4 := e_1 e_4^T + e_4 e_1^T + e_3 e_3^T,$$
$$b := [0, -1, 1, 0]^T.$$

A useful and intriguing property of the LP problems is the following (Goldman-Tucker Theorem of strict complementarity):

**Theorem 2.18** *Suppose $(LP)$ and $(LD)$ have feasible solutions. Then $(LP)$ and $(LD)$ have optimal solutions $x^*, s^*$ respectively, such that*

$$x^* + s^* > 0.$$

This theorem does not generalize to SDP in its full power. There are examples of SDP problems which do not have the *strict complementarity property.* There are even examples with $(P)$ and $(D)$ both having Slater points as well as unique optimal solutions such that the strict complementarity property fails for $(P)$ and $(D)$.

**Example 2.19** Consider the SDP problem with the following data:

$$C := \begin{pmatrix} -1 & -1 & -1 \\ -1 & 0 & 1 \\ -1 & 1 & 0 \end{pmatrix}, A_1 := \begin{pmatrix} 1 & 0 & 0 \\ 0 & 0 & 0 \\ 0 & 0 & 0 \end{pmatrix}, A_2 := \begin{pmatrix} 0 & 0 & 1 \\ 0 & 1 & 0 \\ 1 & 0 & 0 \end{pmatrix},$$

$$A_3 := \begin{pmatrix} 0 & 1 & 0 \\ 1 & 0 & 0 \\ 0 & 0 & 1 \end{pmatrix}, b := \begin{pmatrix} 1 \\ 0 \\ 0 \end{pmatrix}.$$

This SDP problem has a unique primal optimal solution $X^*$ and a unique dual optimal solution $(y^*, S^*)$, there is no duality gap; however, $X^* + S^* \not\succ 0$.

We say that *strict complementarity fails* for a given SDP, if there does not exist a pair of optimal solutions $(\bar{X}, \bar{S})$ (for $(P)$ and $(D)$ respectively) satisfying $\bar{X} + \bar{S} \succ 0$.

It is worth noting that for the above SDP,

$$\bar{X} := \begin{bmatrix} 1 & -\frac{1}{4} & -\frac{1}{4} \\ -\frac{1}{4} & \frac{1}{2} & 0 \\ -\frac{1}{4} & 0 & \frac{1}{2} \end{bmatrix} \succ 0$$

is a Slater point for $(P)$ and $\bar{y}_1 := -4$, $\bar{y}_2 := -3$, $\bar{y}_3 := -2$ yields

$$\bar{S} := C - \mathcal{A}^*(\bar{y}) = \begin{bmatrix} 3 & 1 & 2 \\ 1 & 3 & 1 \\ 2 & 1 & 2 \end{bmatrix} \succ 0$$

a Slater point for $(D)$. Therefore, in the sense of the Strong Duality Theorem and Corollary 2.17, the example is well-behaved. In the domain of such SDP problems (when both problems $(P)$ and $(D)$ have Slater points), we can ask "*how common are the instances failing strict complementarity, like the one in the last example?*" While working on this question, it is convenient to parameterize the data in the space $\mathcal{GM}_m \oplus \Sigma_+^n \oplus \Sigma_+^n$ representing the data space for $(\mathcal{A}, b, c)$. Here, $\mathcal{GM}_m$ denotes the (Grassmannian) manifold of all linear subspaces of dimension $m$ in $\Sigma^n$ (representing the range of the linear transformation $\mathcal{A}^*$) and we represent $b$ by those $\tilde{X} \in \Sigma^n$ which lie in the affine space

$$\left\{\tilde{X} \in \Sigma^n : \mathcal{A}(\tilde{X}) = b\right\};$$

we represent $C$ by those $\tilde{S}$ which lie in the affine space

$$\left\{\tilde{S} \in \Sigma^n : \tilde{S} = C - \mathcal{A}^*(y), \text{ for some } y \in \mathbb{R}^m\right\}.$$

Alizadeh, Haeberly and Overton [7] showed that the SDPs failing strict complementarity are not common at all. Their notion of the failure of strict complementarity is slightly different than ours; also they assume that the data satisfy the Slater condition for the primal and the dual (as we stated above); however, their results can be used to prove the following version of the theorem as well.

**Theorem 2.20** *For every $m, n \in \mathbb{Z}_{++}$, the set of data points for SDP, in the space $\mathcal{GM}_m \oplus \Sigma_+^n \oplus \Sigma_+^n$, for which strict complementarity fails has measure zero relative to the set of data points $(\mathcal{A}, b, c)$ for which there is no duality gap. In particular, for every $\epsilon > 0$, there exists a countable family of sets in $\mathcal{GM}_m \oplus \Sigma_+^n \oplus \Sigma_+^n$, of total measure less than $\epsilon$, which cover all the data points failing strict complementarity; moreover, the set of data points satisfying strict complementarity has a positive measure.*

This theorem generalizes to all convex optimization problems in conic form.

## 2.5 Linear Consequences, Proving Unboundedness, Strong Infeasibility Certificates

Consider

$$\begin{array}{llrcl}(P) & \inf & \left\langle \begin{pmatrix} 0 & 1 \\ 1 & 0 \end{pmatrix}, X \right\rangle & & \\ & \text{subject to:} & \left\langle \begin{pmatrix} 1 & 0 \\ 0 & 0 \end{pmatrix}, X \right\rangle & = & 1, \\ & & X & \succeq & 0. \end{array}$$

Its dual is

$$\begin{array}{llrcl}(D) & \sup & & y & \\ & \text{subject to:} & \begin{pmatrix} y & 0 \\ 0 & 0 \end{pmatrix} & \preceq & \begin{pmatrix} 0 & 1 \\ 1 & 0 \end{pmatrix}. \end{array}$$

Let $X(\alpha) := \begin{pmatrix} 1 & \alpha \\ \alpha & \alpha^2 \end{pmatrix}$. $X(\alpha)$ is feasible for every $\alpha \in \mathbb{R}$. Therefore $(P)$ is unbounded. For an unbounded LP problem in an analogous form, we have a nice proof of unboundedness (a feasible point and a direction $d$ such that $Ad = 0$, $d \geq 0$, $c^T d < 0$). However, there is no such *linear unboundedness proof* for this example. That is, there does not exist $D \in \Sigma^n_+$ such that $\mathcal{A}(D) = 0$ and $\mathrm{Tr}(CD) < 0$. Note that for this example, $\mathrm{Tr}(A_1 D) = 0$ and $D \succeq 0$ imply $D_{11} = D_{12} = D_{21} = 0$ and $D_{22} \geq 0$. But for every such $D$, we must have $\mathrm{Tr}(CD) = 0$. Therefore, no *LP-like* unboundedness proof can exist for unbounded SDPs in general.

Indeed, such an issue also arises in infeasibility detection. Consider the convex set in $\mathbb{R}^2$ defined by the constraint

$$\begin{pmatrix} 1 & X_{21} \\ X_{21} & X_{22} \end{pmatrix} \succeq 0.$$

This is the region:

$$\left\{ \begin{pmatrix} X_{21} \\ X_{22} \end{pmatrix} \in \mathbb{R}^2 : X_{22} \geq X_{21}^2 \right\}.$$

Let the objective function be $\inf\{2X_{21}\}$. Any ray of the form

$$\begin{pmatrix} 0 \\ 0 \end{pmatrix} + \begin{pmatrix} -\alpha \\ \beta \end{pmatrix}$$

for $\alpha > 0, \beta > 0$, must leave the parabolic region. Then $(D)$ is infeasible, since the dual constraint

$$\begin{pmatrix} -y & 1 \\ 1 & 0 \end{pmatrix} \succeq 0$$

is impossible to satisfy. However, $(D)$ is *almost feasible*. That is, for every $\epsilon > 0$, there exists $C' \in \Sigma^n$ such that $\|C - C'\| < \epsilon$ and $\mathcal{A}^*(y) \preceq C'$ is feasible.

**Theorem 2.21** *Suppose* $\mathcal{A} : \Sigma^n \to \mathbb{R}^m$ *and* $C \in \Sigma^n$ *are given. Then*

(a) *if there exists* $D \in \Sigma^n$ *such that* $D \succeq 0$, $\mathcal{A}(D) = 0$, $\mathrm{Tr}(CD) < 0$ *then there does not exist* $y \in \mathbb{R}^m$ *such that* $\mathcal{A}^*(y) \preceq C$*;*

(b) *if there does not exist* $D \in \Sigma^n$ *such that* $D \succeq 0$, $\mathcal{A}(D) = 0$, $\mathrm{Tr}(CD) < 0$ *then* $(D)$ *is almost feasible.*

**Proof** We can prove part (a) in a similar way to the proof of the weak duality relation. Suppose $D$ is as promised and we have $\bar{y} \in \mathbb{R}^m$ such that $\mathcal{A}^*(\bar{y}) \preceq C$. Then

$$0 \leq \langle C, D\rangle - \langle \mathcal{A}^*(\bar{y}), D\rangle = \langle C, D\rangle - \bar{y}^T \mathcal{A}(D) = \langle C, D\rangle < 0.$$

We arrived at a contradiction. To prove part (b), we set up a primal-dual pair of SDPs and apply the Strong Duality Theorem. Consider

$$\begin{array}{llrcl} (D_1) & \sup & & \eta & \\ & \text{subject to:} & \mathcal{A}^*(y) + \eta I & \preceq & C, \\ & & \eta & \leq & 0. \end{array}$$

Its dual is

$$\begin{array}{llrcl} (P_1) & \inf & \langle C, X\rangle & & \\ & \text{subject to:} & \mathcal{A}(X) & = & 0, \\ & & \mathrm{Tr}(X) & \leq & 1, \\ & & X & \succeq & 0. \end{array}$$

Let $\bar{y} := 0$, $\bar{\eta} := -\|C\|_2 - 1$. This proves that $(D_1)$ has a Slater point. $\bar{X} := 0$ is feasible in $(P_1)$. Therefore the Strong Duality Theorem for SDP applies. $(P_1)$ must attain its optimal value and the objective values are the same. Suppose there does not exist $D \in \Sigma^n$ such that $D \succeq 0$, $\mathcal{A}(D) = 0$, $\mathrm{Tr}(CD) < 0$. Then the optimal objective value of $(P_1)$ is zero. By the Strong Duality Theorem, the optimal objective value of $(D_1)$ is also zero. Either $(D_1)$ attains this value (i.e., $\mathcal{A}^*(y) \preceq C$ has a solution) or it doesn't (but then there must exist a sequence $\left\{y^{(k)}, \eta_k\right\}$ such that

$$\mathcal{A}^*\left(y^{(k)}\right) + \eta_k I \preceq C, \quad \text{and } \eta_k \to 0^-).$$

Therefore, $\mathcal{A}^*(y) \preceq C$ is almost feasible. □

As a consequence, we have

**Theorem 2.22** *There exists $D \in \Sigma^n$ such that $D \succeq 0$, $\mathcal{A}(D) = 0$, $\mathrm{Tr}(CD) < 0$ iff $\mathcal{A}^*(y) \preceq C$ is not almost feasible.*

**Proof** Part (b) of Theorem 2.21 establishes one direction. For the other direction, suppose there exists $D \in \Sigma^n$ such that $D \succeq 0$, $\mathcal{A}(D) = 0$, $\mathrm{Tr}(CD) < 0$. We may choose (in addition) $D$ such that $\mathrm{Tr}(CD) = -1$. Then, for every $C' \in \Sigma^n$ such that $\|C - C'\| < \frac{1}{\|D\|}$, $\mathcal{A}^*(y) \preceq C'$ is infeasible (its infeasibility is certified by the same $D$ above). Therefore, $\mathcal{A}^*(y) \preceq C$ is not almost feasible. □

An interpretation of the above theorem is that it characterized when

$$\mathcal{A}(D) = 0, D \succeq 0 \Rightarrow \langle C, D\rangle \geq 0.$$

Alternatively, it characterized when the linear inequality $\langle C, D\rangle \geq 0$ is a *consequence of the system* $\{\mathcal{A}(D) = 0, D \succeq 0\}$. The next fundamental question along these lines, which has ties to duality theory is:

**When is a quadratic inequality a consequence of other quadratic inequalities?**

Let us start with the simplest version:

**When is a quadratic inequality a consequence of another?**

The answer to this question is the so-called $\mathcal{S}$-Lemma:

**Theorem 2.23** *Let $A, C \in \Sigma^n$. Assume $\lambda_1(A) > 0$. Then, the following are equivalent:*

(I) *for every* $h \in \mathbb{R}^n$, $h^T Ah \geq 0 \implies h^T Ch \geq 0$;
(II) *there exists* $y \geq 0$ *such that* $yA \preceq C$.

We can prove the above theorem, utilizing the duality theory developed so far.

**Proof** Suppose $A, C \in \Sigma^n$ with $\lambda_1(A) > 0$ such that (II) holds for $\bar{y} \in \mathbb{R}_+$. Then $\bar{y}A \preceq C$ and $\bar{y} \geq 0$ imply that for every $h \in \mathbb{R}^n$ with $h^T Ah \geq 0$, we have

$$0 \leq \bar{y}h^T Ah \leq h^T Ch \text{ and (I) follows.}$$

Suppose $A, C \in \Sigma^n$ with $\lambda_1(A) > 0$ such that (I) holds. Then

$$\min \left\{h^T Ch : h^T Ah \geq 0\right\} = 0. \tag{2.1}$$

The above minimum value is also equal to

$$\min \left\{\langle C, X\rangle : \langle A, X\rangle \geq 0, X \succeq 0, \operatorname{rank}(X) \leq 1\right\}.$$

If we can prove that the SDP relaxation of this problem (obtained by removing the rank restriction constraint) also has the optimal objective value equal to zero with dual attainment, then we are done. We have the SDP relaxation

$$(P) \min \left\{\langle C, X\rangle : \langle A, X\rangle \geq 0, X \succeq 0\right\}$$

ands its dual

$$(D) \max \left\{0 : yA \preceq C, y \geq 0\right\}.$$

Since $\lambda_1(A) > 0$, $(P)$ has a Slater point. So, if we prove that the objective function value of $(P)$ is bounded from below, then by the Strong Duality Theorem, $(D)$ has an optimal solution and the optimal objective values of $(P)$ and $(D)$ are the same. In particular, there exists $\bar{y} \geq 0$ such that $\bar{y}A \preceq C$. We will prove that the optimal objective value of $(P)$ is zero, which will finish the proof. For a contradiction, suppose not, i.e., there exists $\bar{X} \succeq 0$ such that $\langle A, \bar{X}\rangle \geq 0$ and $\langle C, \bar{X}\rangle < 0$. Apply Theorem 1.8 to $\bar{X}$, let $r := \operatorname{rank}(\bar{X})$, and rewrite the current problem in the space $\Sigma^r$. So, we have $\hat{X} \in \Sigma^r_{++}$ such that $\langle \bar{A}, \hat{X}\rangle \geq 0$ and $\langle \bar{C}, \hat{X}\rangle < 0$. Note that $r \geq 2$ (if $r = 0$, then $\bar{X} = 0$, a contradiction to $\langle C, \bar{X}\rangle < 0$; if $r = 1$, then $\bar{X} = hh^T$ for some $h \in \mathbb{R}^n$ such that $h^T Ah \geq 0$ and $h^T Ch < 0$, a contradiction to (2.1)). Then, there exists $D \in \Sigma^r \setminus \{0\}$ such that $\langle \bar{A}, D\rangle = 0$ and $\langle \hat{X}, D\rangle = 0$. If $\langle C, D\rangle > 0$, replace $D$ by $(-D)$ to make sure that $\langle C, D\rangle \leq 0$. Now, $\hat{X} + \alpha D$ satisfies

$$\langle \bar{A}, \hat{X} + \alpha D\rangle \geq 0, \text{ and } \langle \bar{C}, \hat{X} + \alpha D\rangle < 0, \text{ for every } \alpha \geq 0.$$

We choose the largest $\alpha > 0$ such that $\hat{X} + \alpha D \succeq 0$. Such $\alpha$ exists, since $D \neq 0$, $\hat{X} \succ 0$ and $\langle \hat{X}, D\rangle = 0$. Moreover, $\left(\hat{X} + \alpha D\right)$ has at most $(r-1)$ positive eigenvalues. So, we repeat this process until $r = 0$ or $r = 1$ at which time we arrive at a contradiction. Therefore, the result follows. □

Indeed, as we mentioned before the last theorem, a much more interesting question is:

**When is the quadratic inequality $h^T Ch \geq 0$ a consequence of the system $h^T A_1 h \geq 0, h^T A_2 h \geq 0, \cdots, h^T A_m h \geq 0$?**

The above question led to very interesting research results. See for instance, Ben-Tal and Nemirovski [34] and Pólik and Terlaky [291].

## 2.6 Slater Condition, Borwein-Wolkowicz approach

Suppose we know that there exists $\bar{X} \in \Sigma^n$ feasible in $(P)$ and that $(P)$ is not unbounded. Is it possible to define a dual that in addition to the usual properties we expect from primal-dual pairs, this pair will have the property that

$$\text{infimum} = \text{supremum?}$$

Of course, in addition we would like to guarantee that at least one of the problems $(P)$ or $(D)$ (especially the one for which we want an optimal solution), attains its optimal objective value. So, we want *strong duality* to hold. Let us introduce some geometric notions that will be useful in this theory.

**Definition 2.24** Let $K \subseteq \mathbb{R}^d$ be a closed convex cone. A convex cone $G \subseteq K$ is a *face of* $K$ if for every $u, v \in K$ such that $(u + v) \in G$, we have $u \in G$, $v \in G$.

A face $G$ of $K$ is *exposed* if there exists $a \in \mathbb{R}^d \setminus \{0\}$ such that

$$G = \{x \in K : \langle a, x \rangle = 0\} \text{ and } K \subseteq \left\{x \in \mathbb{R}^d : \langle a, x \rangle \leq 0\right\},$$

i.e., $G$ is the intersection of $K$ with one of its supporting hyperplanes.

A face $G$ of $K$ is a *proper face of* $K$ if

$$\{0\} \subset G \subset K.$$

Notice that our earlier definition of a face of a polyhedron is precisely the above definition of exposed faces. Indeed, if we were to define a face of a polyhedral cone using the above definition for a general convex cone, we could easily prove that every proper face of every polyhedral cone is exposed.

**Theorem 2.25** (a) *Every nonempty face $G$ of $\Sigma^n_+$ is characterized by a unique subspace $L \subset \mathbb{R}^n$ such that*

$$G = \left\{X \in \Sigma^n_+ : \operatorname{Null}(X) \supseteq L\right\},$$

$$\operatorname{relint}(G) = \left\{X \in \Sigma^n_+ : \operatorname{Null}(X) = L\right\}.$$

(b) *Every proper face $G$ of $\Sigma^n_+$ is exposed.*

(c) *The cone $\Sigma^n_+$ is* projectionally exposed. *That is, every nonempty face $G$ of $\Sigma^n_+$ can be expressed as*

$$G = (I - Q)\Sigma^n_+(I - Q),$$

*where $Q \in \Sigma^n$ is the projection onto the unique subspace $L$ defining $G$.*

Note that every proper face of $\Sigma^n_+$ is isomorphic to $\Sigma^k_+$ for some $k < n$. For instance, once we "find" the minimal face $\bar{G}$ containing the feasible region, we can apply a linear isomorphism (actually an automorphism of $\Sigma^n_+$ works) so that the image of $\bar{G}$ is

$$\left\{\begin{pmatrix} \hat{X} & 0 \\ 0 & 0 \end{pmatrix} \in \Sigma^n : \hat{X} \in \Sigma^k_+\right\}.$$

Now, we can rewrite $(P)$ as

$$\begin{array}{lllll} (\tilde{P}) & \inf & \langle C, X \rangle & & \\ & \text{subject to:} & \mathcal{A}(X) & = & b, \\ & & X & \in & \bar{G}. \end{array}$$

Clearly the feasible region and the set of optimal solutions of $(P)$ and $(\tilde{P})$ are the same. But since the algebraic description is different, the (algebraic) dual is also different.

A *poor* way of seeing this is to find $\tilde{\mathcal{A}} : \Sigma^n \to \mathbb{R}^p$ such that the primal problem is equivalent to

$$\begin{array}{llrcl} (\tilde{P}) & \inf & \langle C, X\rangle & & \\ & \text{subject to:} & \mathcal{A}(X) & = & b, \\ & & \tilde{\mathcal{A}}(X) & = & 0, \\ & & X & \in & \Sigma^n_+. \end{array}$$

Then the dual is

$$\begin{array}{llrcl} (\tilde{D}) & \sup & b^T y & & \\ & \text{subject to:} & \mathcal{A}^*(y) + \tilde{\mathcal{A}}^*(\tilde{y}) & \preceq & C, \end{array}$$

whose feasible region (when projected onto the $y$ variables) is potentially larger than that of $(D)$. Therefore, adding redundant equations to the primal can only make the dual feasible region larger. While this makes no difference in the special case of LP, in nonlinear optimization (including SDP) there is a significant difference.

Let us consider another example:

$$C := \begin{pmatrix} 1 & 0 & 0 \\ 0 & 0 & 0 \\ 0 & 0 & 0 \end{pmatrix}, A_1 := \begin{pmatrix} 0 & 0 & 0 \\ 0 & 1 & 0 \\ 0 & 0 & 0 \end{pmatrix}, A_2 := \begin{pmatrix} 1 & 0 & 0 \\ 0 & 0 & 1 \\ 0 & 1 & 0 \end{pmatrix}, b := \begin{pmatrix} 0 \\ 1 \end{pmatrix}.$$

Then for every feasible solution of $(D)$, we have $y_2 = 0$ and $y_1 \leq 0$. So, the set of feasible solutions of $(D)$ (in the $S$-space) is

$$\left\{ \begin{pmatrix} 1 & 0 & 0 \\ 0 & S_{22} & 0 \\ 0 & 0 & 0 \end{pmatrix} : S_{22} \geq 0 \right\}.$$

Moreover, every feasible solution of $(D)$ is optimal in $(D)$. For the primal, it is also true that the set of feasible solutions and the set of optimal solutions coincide, which is (in the $X$-space):

$$\left\{ \begin{pmatrix} 1 & 0 & X_{31} \\ 0 & 0 & 0 \\ X_{31} & 0 & X_{33} \end{pmatrix} : X_{33} \geq X_{31}^2 \right\}.$$

Even though both $(P)$ and $(D)$ have finite optimum objective values, both of these values are attained, there is a duality gap of 1.

Now, suppose we add to $(P)$ the following redundant constraint:

$$\langle A_3, X\rangle = b_3, \text{ where } A_3 := \begin{pmatrix} 0 & 0 & 0 \\ 0 & 0 & 1 \\ 0 & 1 & 0 \end{pmatrix}, \text{ and } b_3 := 0.$$

The primal SDP problem $(P)$ still has the same set of feasible solutions, the same set of optimal solutions and the same optimal value of 1. However, the dual constraints become:

$$\begin{pmatrix} (1-y_2) & 0 & 0 \\ 0 & -y_1 & -(y_2+y_3) \\ 0 & -(y_2+y_3) & 0 \end{pmatrix} \succeq 0.$$

The latter constraint is equivalent to:

$$y_1 \leq 0, \quad y_2 \leq 1, \quad y_2 = -y_3.$$

Since we are maximizing $y_2$, the new dual has $y_2 = 1$ and $y_3 = -1$ at every optimal solution. The set of optimal solutions of the new dual is (in the $S$-space):

$$\left\{ \begin{pmatrix} 0 & 0 & 0 \\ 0 & S_{22} & 0 \\ 0 & 0 & 0 \end{pmatrix} : S_{22} \geq 0 \right\}.$$

Clearly, there is no duality gap between the primal and the new dual.

Borwein and Wolkowicz [59] showed how to arrive from $(P)$ to $(\tilde{P})$ in finitely many (at most $n$) steps, in each step strictly reducing the dimension of the current cone. An insightful, geometric result in this theory is the following.

**Lemma 2.26** *Let $G$ be a proper face of $\Sigma_+^n$. Then $\left(\Sigma_+^n + G^\perp\right)$ is closed. However, $\left(\Sigma_+^n + \operatorname{span}(G)\right)$ is not closed.*

In the above result,

$$G^\perp := \{S \in \Sigma^n : \langle X, S \rangle = 0, \forall X \in G\},$$

and $\operatorname{span}(G)$ is the smallest linear subspace in $\Sigma^n$ containing $G$.

To arrive at the minimal face $\bar{G}$, Borwein and Wolkowicz repeatedly utilize the following key lemma (which is a theorem of the alternative):

**Lemma 2.27** *Let $\mathcal{A} : \Sigma^n \to \mathbb{R}^m$ be a linear transformation and let $b \in \mathbb{R}^m$ be given. Then, exactly one of the following two systems has a solution:*

(I) $\mathcal{A}(X) = b$, $X \in \Sigma_{++}^n$;

(II) $\mathcal{A}^*(y) \in \Sigma_{++}^n \setminus \{0\}$, $b^T y = 0$.

It follows from the above lemma that either $(P)$ has a Slater point or we can find a nonzero, symmetric positive semidefinite matrix $\bar{S}$ which is in the range of $\mathcal{A}^*$ such that $\langle \bar{S}, X \rangle = 0$ for every feasible solution of $(P)$. Suppose $(P)$ does not have a Slater point. Then by Lemma 2.27, there exists $\bar{u} \in \mathbb{R}^m$ such that

$$\mathcal{A}^*(\bar{u}) \in \Sigma_{++}^n \setminus \{0\}, \text{ and } b^T \bar{u} = 0.$$

Let $\bar{S} := \mathcal{A}^*(\bar{u})$. Now, for every feasible $X$ in $(P)$, we have

$$\langle \bar{S}, X \rangle = \langle \mathcal{A}^*(\bar{u}), X \rangle = \bar{u}^T \mathcal{A}(X) = b^T \bar{u} = 0.$$

Moreover, since $\bar{S} \succeq 0$ and $\bar{S} \neq 0$, the set

$$\{X \in \Sigma_+^n : \langle \bar{S}, X \rangle = 0\}$$

is a proper face of $\Sigma_+^n$. We can replace $(P)$ by an equivalent primal problem which explicitly includes the linear equation $\langle \bar{S}, X \rangle = 0$. If the new primal problem (now over $\Sigma_+^k$, for some $k < n$) does not have a Slater point then we apply Lemma 2.27 to the new problem. Since every proper face of $\Sigma_+^n$ is isomorphic to $\Sigma_+^k$ for $k < n$ and in each step we strictly reduce $k$, we are guaranteed to terminate with the minimal face in at most $n$ steps.

Let us reconsider

$$(\tilde{P}) \qquad \begin{array}{rrcl} \inf & \langle C, X \rangle & & \\ \text{subject to:} & \mathcal{A}(X) & = & b, \\ & X & \in & \bar{G}. \end{array}$$

Since $\bar{G}$ is the minimal face of $\Sigma_+^n$ containing the feasible region, the Slater Condition holds for $(\tilde{P})$. So, the Strong Duality Theorem can be applied.

As SDP became re-popularized in 1990's, Ramana asked (and answered) the question of how to arrive at $(\tilde{P})$ in a way that is efficient from a computational complexity viewpoint. Given the problem

$$\begin{array}{llll}(D) & \sup & b^T y & \\ & \text{subject to:} & \mathcal{A}^*(y) & \preceq C,\end{array}$$

Ramana proposed the following dual (named Extended Lagrangian-Slater Dual):

$$\begin{array}{lllll}(ELSD) & \inf & \langle C, U+W\rangle & & \\ & \text{subject to:} & \mathcal{A}(U+W) & = & b, \\ & & W & \in & \mathcal{W}_n, \\ & & U & \succeq & 0.\end{array}$$

The set $\mathcal{W}_n$ admits an SDP representation as follows:

$$\begin{aligned}\mathcal{C}_k \; := \; & \left\{(U_1, W_1; U_2, W_2; \ldots; U_k, W_k) \in \Sigma^n \oplus \mathbb{R}^{n\times n} \oplus \cdots \oplus \Sigma^n \oplus \mathbb{R}^{n\times n} : W_0 := 0,\right. \\ & \mathcal{A}(U_i + W_{i-1} + W_{i-1}^T) = 0, \langle C, U_i + W_{i-1} + W_{i-1}^T\rangle = 0, \\ & \left. U_i \succeq W_i W_i^T, \forall i \in \{1, 2, \ldots, k\}\right\},\end{aligned}$$

$$\mathcal{W}_k := \left\{W_k + W_k^T : (U_1, W_1; U_2, W_2; \ldots; U_k, W_k) \in \mathcal{C}_k, \text{ for some } U_1, W_1, \ldots, U_k\right\}.$$

Noticing the following Schur complement structure, it is easy to see that $\mathcal{C}_k$ is a convex set:

$$U_i \succeq W_i W_i^T \iff \begin{pmatrix} I & W_i^T \\ W_i & U_i \end{pmatrix} \succeq 0.$$

Now, suppose that $h \in \text{Null}(U_i)$. Then $U_i \succeq W_i W_i^T$ implies $h^T W_i W_i^T h \leq 0$, whence $h \in \text{Null}(W_i^T)$. We proved $\text{Null}(W_i^T) \supseteq \text{Null}(U_i)$. Note the connection to Theorem 2.25. The definitions and the above properties imply that $\mathcal{W}_k$ is a subspace.

**Theorem 2.28** *If $(D)$ has a finite optimal value then $(ELSD)$ is feasible and the optimal objective values of $(D)$ and $(ELSD)$ coincide, $(ELSD)$ attains its optimal value.*

Recall Theorem 2.22. An alternative way of stating that theorem is

*Let $\mathcal{A} : \Sigma^n \to \mathbb{R}^m$, $C \in \Sigma^n$ be given. Then exactly one of the following systems has a solution:*

(I) *there exists $D \in \Sigma^n$ such that $D \succeq 0$, $\mathcal{A}(D) = 0$, $\text{Tr}(CD) < 0$;*
(II) *$\mathcal{A}^*(y) \preceq C$ is almost feasible.*

So, Theorem 2.22 generalized a theorem of the alternative to the SDP setting by keeping the infeasibility certificate "the same" and modifying (weakening) the requirements on the "solution" of the system $\mathcal{A}^*(y) \preceq C$. The next theorem is a complement of the Theorem 2.22 in the sense that it keeps the requirements on the "solution" $y$ the same as it was in the polyhedral setting but modifies (weakens) the requirements on the infeasibility certificate $D$.

**Theorem 2.29** *Let $\mathcal{A} : \Sigma^n \to \mathbb{R}^m$, $C \in \Sigma^n$ be given. Then exactly one of the following systems has a solution:*

(I) $\mathcal{A}^*(y) \preceq C$,
(II) $\mathcal{A}(U+W) = 0$, $W \in \mathcal{W}_n$, $U \succeq 0$, *and* $\langle C, U+W\rangle = -1$.

Now, we turn to the consequences of such results for the computational complexity of Semidefinite Optimization. When system (I) has a feasible solution, we can certify such fact by exhibiting a $\bar{y} \in \mathbb{R}^m$ such that $\mathcal{A}^*(\bar{y}) \preceq C$. If (I) does not have any solutions then by Theorem 2.29, we can certify such a fact by exhibiting a $\bar{U} \in \Sigma_+^n$ and a $\bar{W} \in \Sigma^n$ such that $\mathcal{A}(\bar{U}+\bar{W}) = 0$, $\bar{W} \in \mathcal{W}_n$, and $\langle C, \bar{U}+\bar{W}\rangle = -1$.

This implies the next two theorems:

**Theorem 2.30** *In the* real number computational model, *the problem of deciding SDP feasibility is in $\mathcal{NP} \cap$ co-$\mathcal{NP}$.*

**Theorem 2.31** *In the Turing machine model, if SDP feasibility is in $\mathcal{NP}$ then it is in $\mathcal{NP} \cap$ co-$\mathcal{NP}$.*

It turns out that the approach of Ramana is closely related to the work of Borwein and Wolkowicz. (The duals obtained are essentially the same, including those obtained for smaller $k$ in the intermediate steps.)

The next example proves that in general, there are at least two steps required in the above ELSD. Consider

$$n := m := 3, \; C := 0,$$

$$A_1 := \begin{pmatrix} 0 & 0 & 0 \\ 0 & 1 & 0 \\ 0 & 0 & 0 \end{pmatrix}, \; A_2 := \begin{pmatrix} 1 & 0 & 0 \\ 0 & 0 & 1 \\ 0 & 1 & 0 \end{pmatrix}, \; A_3 := \begin{pmatrix} 0 & 1 & 0 \\ 1 & 0 & 0 \\ 0 & 0 & 0 \end{pmatrix}, \; b := \begin{pmatrix} 0 \\ 0 \\ 1 \end{pmatrix}.$$

Then $(D)$ has optimal objective value zero and it is attained. However, $(P)$ is infeasible. In ELSD with $k = 1$,
$\text{Tr}(A_1(U + W + W^T)) = 0$ implies $U_{22} = 0$. $\text{Tr}(A_3(U + W + W^T)) = 1$ implies $U_{12} = U_{21} = 1/2$. Since $U \succeq 0$, we have a contradiction. Therefore, ELSD is infeasible and at least two steps of the above procedure is required.

We can further generalize this example to generate a family of instances with $n = m$ which requires at least $n - 1 = m - 1$ steps (layers in ELSD): For every $n \geq 3$, let $m := n$, $C := 0$, $b := e_2$, $A_1 := e_1 e_1^T$, $A_2 := e_1 e_2^T + e_2 e_1^T$,

$$A_{2+i} := e_{i+1} e_{i+1}^T + e_1 e_{2+i}^T + e_{2+i} e_1^T, \forall i \in \{1, 2, \ldots, n-2\}.$$

The minimum number of layers required in ELSD (or equivalently the number of steps required by Borwein-Wolkowicz construction) is interesting elsewhere too. Sturm [338] calls this number *degree of singularity* and proves error bounds which are better for the smaller degrees of singularity.

## 2.7 When does the Slater Condition Hold in SDP Relaxations?

Suppose $c \in \mathbb{R}^n$ is given and we are interested in solving (and/or obtaining good lower bounds for) the following linear optimization problem over a **nonconvex** set of feasible solutions $F \subset \mathbb{R}^n$:

$$\inf\left\{c^T x : \; x \in F\right\}. \tag{2.2}$$

First, we will focus on the above problem (2.2). However, given $c \in \mathbb{R}^n$, $C \in \Sigma^n$, we can also handle the nonconvex quadratic objective functions in our same framework below:

$$\inf\left\{c^T x + x^T C x : \; x \in F\right\}. \tag{2.3}$$

**2.7.1 Homogeneous Equality Form.** For this subsection, we assume that the set of feasible solutions $F$ and the linear transformation $\mathcal{A}$ have the property that

$$F = \left\{ x \in \mathbb{R}^n : \mathcal{A}\begin{pmatrix} 1 & x^T \\ x & xx^T \end{pmatrix} = 0 \right\},$$

where $\mathcal{A} : \Sigma^{n+1} \to \mathbb{R}^m$ is a linear transformation. We call this form of expressing the set of feasible solutions $F$, *the Homogeneous Equality Form.* Note that the name we gave to this form is based on the superficial fact that the right hand side is zero. Indeed, in the matrix variable we forced one of the entries to be 1 which allows us to express a wide range of problems in this form.

A very wide range of optimization problems (including some very difficult ones in finite dimensional spaces) can be put into this form. For instance, given $Q^{(i)} \in \Sigma^n$, $q^{(i)} \in \mathbb{R}^n$, $\gamma_i \in \mathbb{R}$, for $i \in \{1, 2, \ldots, m\}$,

$$\left\langle \begin{pmatrix} \gamma_i & \left(q^{(i)}\right)^T \\ q^{(i)} & Q^{(i)} \end{pmatrix}, \begin{pmatrix} 1 & x^T \\ x & xx^T \end{pmatrix} \right\rangle = 0, \quad i \in \{1, 2, \ldots, m\}$$

models a system of quadratic equations. We can also put every system of finitely many quadratic *inequalities* into the Homogeneous Equality Form:

$$\begin{aligned} \left\langle \begin{pmatrix} \gamma & q^T & 0 \\ q & Q & 0 \\ 0 & 0^T & 1 \end{pmatrix}, \begin{pmatrix} 1 & x^T & \tilde{s} \\ x & xx^T & \tilde{s}x \\ \tilde{s} & \tilde{s}x^T & \tilde{s}^2 \end{pmatrix} \right\rangle &= 0 \\ \iff \gamma + 2q^T x + x^T Q x + \tilde{s}^2 &= 0 \\ \iff \gamma + 2q^T x + x^T Q x &\leq 0. \end{aligned}$$

Since any finite system of polynomial inequalities and equations can be expressed as a system of quadratic inequalities (with the use of new variables), we have

**Proposition 2.32** *Every finite system of polynomial inequalities and equations can be put into Homogeneous Equality Form.*

Indeed, since the set of quadratic equations $x_j^2 - x_j = 0$, for every $j \in \{1, 2, \ldots, n\}$ models the constraint $x \in \{0, 1\}^n$, any combinatorial optimization problem as well as any mixed integer 0,1 programming problem can be put into Homogeneous Equality Form.

We define a relaxation in the lifted space:

$$\hat{\mathcal{P}} := \left\{ \begin{pmatrix} 1 & x^T \\ x & X \end{pmatrix} \in \Sigma^{n+1} : \mathcal{A}\begin{pmatrix} 1 & x^T \\ x & X \end{pmatrix} = 0, \begin{pmatrix} 1 & x^T \\ x & X \end{pmatrix} \succeq 0 \right\}.$$

That is, $F$ is a projection onto the $x$-space of the intersection of $\hat{\mathcal{P}}$ with rank-1 matrices. Let's focus on another set in this lifted space:

$$\mathcal{F} := \operatorname{conv}\left\{ \begin{pmatrix} 1 & x^T \\ x & xx^T \end{pmatrix} \in \Sigma^{n+1} : x \in F \right\}.$$

Ideally, we would like to have a description of $\mathcal{F}$, since its projection onto the $x$-space yields

$$\left\{ x \in \mathbb{R}^n : \begin{pmatrix} 1 & x^T \\ x & X \end{pmatrix} \in \mathcal{F} \right\} = \operatorname{conv}(F). \tag{2.4}$$

To see this let $\bar{x} \in \operatorname{conv}(F)$. Then, by Carathédory's theorem, there exists $v^{(1)}, v^{(2)}, \ldots, v^{(n+1)} \in F$ and $\lambda \in \mathbb{R}_+^{n+1}$ such that

$$\bar{x} = \sum_{i=1}^{n+1} \lambda_i v^{(i)} \text{ and } \bar{e}^T \lambda = 1.$$

By the definition of $\mathcal{F}$, $\begin{pmatrix} 1 & \bar{x}^T \\ \bar{x} & \sum_{i=1}^{n+1} \lambda_i v^{(i)} \left(v^{(i)}\right)^T \end{pmatrix} \in \mathcal{F}$. Hence, $\bar{x}$ is in the set described in the right hand side of the above claimed equation (2.4). For the converse, let $\bar{x}$ be an element of the set in the right hand side of the above claimed equation (2.4). Then, by the definition of $\mathcal{F}$ and Carathédory's theorem, there exist $k = O\left(n^2\right)$, $v^{(1)}, v^{(2)}, \ldots, v^{(k)} \in F$, and $\lambda \in \mathbb{R}_+^k$ such that

$$\bar{x} = \sum_{i=1}^{k} \lambda_i v^{(i)} \text{ and } \bar{e}^T \lambda = 1.$$

Whence, $\bar{x} \in F$ as desired.

Clearly, we have $\hat{\mathcal{P}} \supseteq \mathcal{F}$. Therefore, $\hat{\mathcal{P}}$ is a relaxation of $\mathcal{F}$. Finally, note that

$$\inf\left\{c^T x : \ x \in F\right\} = \inf\left\{\left\langle \begin{pmatrix} 0 & \frac{1}{2}c^T \\ \frac{1}{2}c & 0 \end{pmatrix}, \begin{pmatrix} 1 & x^T \\ x & X \end{pmatrix} \right\rangle : \begin{pmatrix} 1 & x^T \\ x & X \end{pmatrix} \in \mathcal{F}\right\}.$$

Hence, $\hat{\mathcal{P}}$ together with the above-mentioned objective function, gives a relaxation of the original nonconvex optimization problem (2.2). For the nonconvex optimization problems with a quadratic objective function as in form (2.3), we note

$$\begin{aligned} &\inf\left\{c^T x + x^T C x : \ x \in F\right\} \\ = \ &\inf\left\{\left\langle \begin{pmatrix} 0 & \frac{1}{2}c^T \\ \frac{1}{2}c & C \end{pmatrix}, \begin{pmatrix} 1 & x^T \\ x & X \end{pmatrix} \right\rangle : \begin{pmatrix} 1 & x^T \\ x & X \end{pmatrix} \in \mathcal{F}\right\}. \end{aligned}$$

Next, we show that our SDP relaxation is guaranteed to behave well under the duality theory.

**Theorem 2.33** *Suppose $F$ and $\hat{\mathcal{P}}$ are given as in the Homogeneous Equality Form above and that* $\operatorname{conv}(F)$ *is full dimensional. Then, the Slater condition holds for $\hat{\mathcal{P}}$.*

**Proof** Suppose the feasible solution set $F$ above is such that $\operatorname{conv}(F)$ is full dimensional. Then there must exist an affinely independent set of vectors $\{v^{(1)}, v^{(2)}, \ldots, v^{(n+1)}\} \subseteq F$. We then let

$$V_\lambda := \sum_{i=1}^{n+1} \lambda_i \begin{pmatrix} 1 \\ v^{(i)} \end{pmatrix} \begin{pmatrix} 1, & \left(v^{(i)}\right)^T \end{pmatrix},$$

for $\lambda > 0$ such that $\bar{e}^T \lambda = 1$. Then $V_\lambda \in \mathcal{F} \subseteq \hat{\mathcal{P}}$. Moreover, by the definition of affine independence,

$$\left\{ \begin{pmatrix} 1 \\ v^{(1)} \end{pmatrix}, \begin{pmatrix} 1 \\ v^{(2)} \end{pmatrix}, \ldots, \begin{pmatrix} 1 \\ v^{(n+1)} \end{pmatrix} \right\} \text{ is linearly independent.}$$

Thus, by Proposition 1.11 part (c), we have $V_\lambda \in \left(\hat{\mathcal{P}} \cap \Sigma_{++}^n\right)$. Therefore, $V_\lambda$ is a Slater point for $\hat{\mathcal{P}}$. $\square$

In many applications, in natural domains for formulations of $F$, $\operatorname{conv}(F)$ may not be full-dimensional. Next, we address the more general situation by removing the assumption on the full dimensionality of $\operatorname{conv}(F)$.

2.7.1.1 *What if* $\operatorname{conv}(F)$ *is not full dimensional?* Suppose $\dim(\operatorname{conv}(F)) = d < n$. Then our last theorem no longer applies, also the proof technique of the theorem always generates singular positive semidefinite matrices $V_\lambda$. To remedy the situation, we do the obvious thing and focus on the smallest affine space containing $\operatorname{conv}(F)$. We find $L \in \mathbb{R}^{d\times n}$, $\ell \in \mathbb{R}^n$ such that $\operatorname{rank}(L) = d$ and

$$x \in F \Rightarrow x = \ell + L^T y \quad \text{for some } y \in \mathbb{R}^d.$$

Is it reasonable to assume that we know the $\dim(\operatorname{conv}(F))$ and $L$ as well as $\ell$? The answer from a theoretical worst-case viewpoint would be "no!" Of course, we can embed many hard optimization problems as feasibility problems and even just deciding whether $F = \emptyset$ could be hard enough (equivalently, deciding whether $\dim(F) \geq 0$ —note that $\dim(\emptyset) = -1$).

However, the answer from a practical viewpoint is "usually yes!" For example, in combinatorial optimization, one of the first things we do when we start studying a polytope is to determine its dimension. Next, we usually study its facets and for this purpose, if $\operatorname{conv}(F)$ is not full-dimensional, then we compute $L$ and $\ell$. Therefore, in many applications, $\dim(F)$, $L$ and $\ell$ are readily available in the combinatorial optimization literature. In many other applications, these are easy to derive.

Let's define a linear transformation $\mathcal{L} : \Sigma^{n+1} \to \Sigma^{d+1}$,

$$\mathcal{L}(Z) := \begin{pmatrix} 1 & \ell^T \\ 0 & L \end{pmatrix} Z \begin{pmatrix} 1 & 0^T \\ \ell & L^T \end{pmatrix}.$$

This map $\mathcal{L}$ and its adjoint:

$$\mathcal{L}^*(W) = \begin{pmatrix} 1 & 0^T \\ \ell & L^T \end{pmatrix} W \begin{pmatrix} 1 & \ell^T \\ 0 & L \end{pmatrix}$$

are all we need to define new variables, to reduce dimension and most importantly to ensure the existence of Slater points for the transformed problem. Now, we can define $\bar{\mathcal{A}} : \Sigma^{d+1} \to \mathbb{R}^m$ which will take the role of the linear map $\mathcal{A}$ in the smaller dimensional space:

$$\bar{\mathcal{A}}(W) := \mathcal{A}\left(\mathcal{L}^*(W)\right).$$

With the above definitions, we can express the set of feasible solutions in an alternative form:

$$F = \left\{\ell + L^T y : \; y \in \mathbb{R}^d, \; \bar{\mathcal{A}}\begin{pmatrix} 1 & y^T \\ y & yy^T \end{pmatrix} = 0\right\}.$$

Based on the above construction, we can define a counterpart of $F$ in the $d$-dimensional space:

$$F_{\mathcal{L}} := \left\{y \in \mathbb{R}^d : \bar{\mathcal{A}}\begin{pmatrix} 1 & y^T \\ y & yy^T \end{pmatrix} = 0\right\}.$$

Then in the lifted space, ideally we would like to get our hands on:

$$\mathcal{F}_{\mathcal{L}} := \operatorname{conv}\left\{\begin{pmatrix} 1 & y^T \\ y & yy^T \end{pmatrix} \in \Sigma^{d+1} : \; y \in F_{\mathcal{L}}\right\}.$$

The latter leads to the relaxation

$$\hat{\mathcal{P}}_{\mathcal{L}} := \left\{\begin{pmatrix} 1 & y^T \\ y & Y \end{pmatrix} \in \Sigma^{d+1} : \; \bar{\mathcal{A}}\begin{pmatrix} 1 & y^T \\ y & Y \end{pmatrix} = 0, \; \begin{pmatrix} 1 & y^T \\ y & Y \end{pmatrix} \succeq 0\right\}.$$

**Theorem 2.34** *The Slater condition holds for $\hat{\mathcal{P}}_{\mathcal{L}}$.*

**Proof** By construction, $\dim(\mathrm{conv}(F_{\mathcal{L}})) = d$, i.e., $\mathrm{conv}(F_{\mathcal{L}})$ is full dimensional. Therefore, by Theorem 2.33, Slater condition holds for $\hat{\mathcal{P}}_{\mathcal{L}}$. □

Now that we obtained an SDP relaxation of the feasible solution set in the lifted space, we need to construct a linear objective function in this space which will lead to an SDP relaxation. First, we observe that $\inf\{c^T x + x^T Cx : x \in F\}$ is equal to

$$\inf\left\{\left\langle\begin{pmatrix} 0 & \frac{1}{2}c^T \\ \frac{1}{2}c & C \end{pmatrix}, \begin{pmatrix} 1 & x^T \\ x & X \end{pmatrix}\right\rangle : \begin{pmatrix} 1 & x^T \\ x & X \end{pmatrix} \in \mathcal{F}\right\}.$$

By our construction of $\bar{\mathcal{A}}$, $L$, and $\ell$ the latter is also equal to

$$\inf\left\{c^T\left(\ell + L^T y\right) + \left(\ell + L^T y\right)^T C\left(\ell + L^T y\right) : y \in \mathbb{R}^d,\ \bar{\mathcal{A}}\begin{pmatrix} 1 & y^T \\ y & yy^T \end{pmatrix} = 0\right\}.$$

Using the definition of $F_{\mathcal{L}}$ we can rewrite the last infimum as

$$\inf\left\{\left\langle\mathcal{L}\begin{pmatrix} 0 & \frac{1}{2}c^T \\ \frac{1}{2}c & C \end{pmatrix}, \begin{pmatrix} 1 & y^T \\ y & yy^T \end{pmatrix}\right\rangle : y \in F_{\mathcal{L}}\right\}.$$

Now, recall the definition of $\mathcal{F}_{\mathcal{L}}$; this is the object we would ideally like to have our hands on, in the lifted space. Therefore, the last infimum is also equal to

$$\inf\left\{\left\langle\mathcal{L}\begin{pmatrix} 0 & \frac{1}{2}c^T \\ \frac{1}{2}c & C \end{pmatrix}, \begin{pmatrix} 1 & y^T \\ y & Y \end{pmatrix}\right\rangle : \begin{pmatrix} 1 & y^T \\ y & Y \end{pmatrix} \in \mathcal{F}_{\mathcal{L}}\right\}.$$

So, we have the following SDP relaxation:

$$\begin{array}{lrcl} \inf & \left\langle\mathcal{L}\begin{pmatrix} 0 & \frac{1}{2}c^T \\ \frac{1}{2}c & C \end{pmatrix}, \begin{pmatrix} 1 & y^T \\ y & Y \end{pmatrix}\right\rangle & & \\ \text{subject to:} & \bar{\mathcal{A}}\begin{pmatrix} 1 & y^T \\ y & Y \end{pmatrix} & = & 0, \\ & \begin{pmatrix} 1 & y^T \\ y & Y \end{pmatrix} & \succeq & 0. \end{array}$$

Moreover, it is guaranteed that the above problem has a Slater point.

Now, let us talk about $L$ and $\ell$. Note that in general, there may be infinitely many $L$ and $\ell$ describing the (same) affine hull of $F$. So, we should use this flexibility to our advantage and choose those $L$ and $\ell$ which help us in our applications. Usually, in many large-scale applications, the most important property to preserve is sparsity. I.e., if $\mathcal{A}$ has a very sparse representation and, the $L$ and $\ell$ that we pick lead to a completely dense representation for $\bar{\mathcal{A}}$, we may be unhappy with the approach.

To make the discussion more concrete, let's move from the abstract representation of the linear map $\mathcal{A}$ to an explicit representation with respect to a chosen basis. Let $A_i \in \Sigma^n$ be such that $[\mathcal{A}(X)]_i = \langle A_i, X\rangle$ for $i \in \{1, 2, \ldots, m\}$. Then

$$\mathcal{L}(A_i) = \begin{pmatrix} 1 & \ell^T \\ 0 & L \end{pmatrix} A_i \begin{pmatrix} 1 & 0^T \\ \ell & L^T \end{pmatrix}, \quad i \in \{1, 2, \ldots, m\}.$$

So, now, assume that each $A_i$ is very sparse. Then among all possible choices of $L$ and $\ell$, we want to pick the one that makes $\mathcal{L}(A_i)$ as sparse as possible for every $i$. For an application of this kind of idea, see [189].

**2.7.2 Nonhomogeneous Equality Form.** Let us consider a form more closely related to the form of popular SDP relaxations of the maximum cut and related combinatorial optimization problems. For this section, we assume that we have $F$ given in *Nonhomogeneous Equality Form*; namely,

$$F = \left\{x \in \mathbb{R}^n : \mathcal{A}(xx^T) = b\right\},$$

where $\mathcal{A} : \Sigma^n \to \mathbb{R}^m$ is a given linear transformation as before and $b \in \mathbb{R}^m$ is also given. There is an obvious choice for an SDP relaxation:

$$\hat{\mathcal{P}} := \{X \in \Sigma^n : \ \mathcal{A}(X) = b, \ X \succeq 0\}.$$

**Theorem 2.35** *Suppose there exists a linearly independent set of vectors $\{v^{(1)}, v^{(2)}, \ldots, v^{(n)}\} \subseteq F$, and $F$ is given in the Nonhomogeneous Equality Form. Then $\hat{\mathcal{P}}$ has Slater points.*

**Proof** Suppose $\{v^{(1)}, v^{(2)}, \ldots, v^{(n)}\} \subseteq F$ is linearly independent. For every $\lambda \in \mathbb{R}^n_{++}$ such that $\bar{e}^T\lambda = 1$, define

$$V_\lambda := \sum_{i=1}^n \lambda_i v^{(i)} \left(v^{(i)}\right)^T.$$

Then

$$\mathcal{A}(V_\lambda) = \sum_{i=1}^n \lambda_i \mathcal{A}\left(v^{(i)} \left(v^{(i)}\right)^T\right) = \sum_{i=1}^n \lambda_i b = b.$$

Moreover, by Proposition 1.11 part (c), $V_\lambda \succ 0$. Therefore, $\hat{\mathcal{P}}$ has Slater points. □

What kind of objective functions are easily handled in this case? Here are some examples: If we have a combinatorial optimization problem which asks for a maximum (or minimum) cardinality or maximum (or minimum) weight set of certain type and the feasible solutions are encoded by 0,1 characteristic vectors, then the objective function $c^Tx \geq 0$ for every $x \in F$. Then, we can equivalently minimize (or maximize) $(c^Tx)^2$ over $F$, yielding the objective function $\langle cc^T, X\rangle$ for the SDP relaxation. Handling the pure quadratic objective functions is also straightforward. In this case, in problem (2.3), we have $c := 0$. The objective function $x^TCx$ in (2.3) would be modeled by the objective function $\langle C, X\rangle$ for the SDP relaxation. The typical example is a way of obtaining the most popular SDP relaxation for Max Cut which can be obtained by using $\operatorname{diag}(xx^T) = \bar{e}$ for the general form $\mathcal{A}(xx^T) = b$ above.

This Nonhomogeneous Equality Form is also very general. We already noticed $\operatorname{diag}(xx^T) = \bar{e}$ could be a part of the constraints $\mathcal{A}(xx^T) = b$. Clearly, we can formulate systems of quadratic equations in which every monomial is degree two except the constant.

We can also formulate systems of quadratic inequalities in which every monomial is degree two except the constant by using additional "slack" variables:

$$\begin{aligned}
\left\langle \begin{pmatrix} Q & 0 \\ 0^T & 1 \end{pmatrix}, \begin{pmatrix} xx^T & \tilde{s}x \\ \tilde{s}x^T & \tilde{s}^2 \end{pmatrix} \right\rangle &= \beta \\
\iff x^TQx + \tilde{s}^2 &= \beta \\
\iff x^TQx &\leq \beta.
\end{aligned}$$

**2.7.3 Homogeneous Convex Inequality Form.** To generalize our setting to nonlinear constraints, we consider once more the representation of the set of feasible solutions in the lifted space:

$$\mathcal{F} := \text{conv}\left\{ \begin{pmatrix} 1 & x^T \\ x & xx^T \end{pmatrix} \in \Sigma^{n+1} : x \in F \right\}.$$

Now, we want a map $\mathcal{A} : \Sigma^{n+1} \to \mathbb{R}^m$, not necessarily linear, such that

$$F = \left\{ x \in \mathbb{R}^n : \mathcal{A}\begin{pmatrix} 1 & x^T \\ x & xx^T \end{pmatrix} \leq 0 \right\}.$$

We also want $\mathcal{A}$ to have some form of convexity, namely

$$\mathcal{A}(\lambda U + (1-\lambda)V) \ \leq \ \lambda \mathcal{A}(U) + (1-\lambda)\mathcal{A}(V),$$

for every $U, V \in \Sigma_+^{n+1}$ and for every $\lambda \in [0,1]$. Finally, we want $\mathcal{A}$ to be good enough of a formulation of $F$ so that it can identify the interior of convex hull of $F$:

$$\text{int}\,(\text{conv}(F)) = \text{conv}\left\{ x \in \mathbb{R}^n : \ \mathcal{A}\begin{pmatrix} 1 & x^T \\ x & xx^T \end{pmatrix} < 0 \right\} \neq \emptyset.$$

Let's call such a form the *Homogeneous Convex Inequality Form.* Our construction above is only for $F$ whose convex hull is full dimensional. If this assumption fails (i.e., $\dim(\text{conv}(F)) = d < n$), then we first apply our technique from our earlier discussion, find the smallest affine space containing $F$, identify the best choices for $L$ and $\ell$ and then continue as above in the transformed space.

Now, it is easy to propose a convex relaxation of the feasible solution set in the lifted space:

$$\hat{\mathcal{P}} := \left\{ \begin{pmatrix} 1 & x^T \\ x & X \end{pmatrix} \in \Sigma^{n+1} : \ \mathcal{A}\begin{pmatrix} 1 & x^T \\ x & X \end{pmatrix} \leq 0, \ \begin{pmatrix} 1 & x^T \\ x & X \end{pmatrix} \succeq 0 \right\}.$$

The corresponding convex relaxation problem is

$$\begin{array}{lrcl} \inf & \left\langle \begin{pmatrix} 0 & \frac{1}{2}c^T \\ \frac{1}{2}c & C \end{pmatrix}, \begin{pmatrix} 1 & x^T \\ x & X^T \end{pmatrix} \right\rangle & & \\ \text{subject to:} & \mathcal{A}\begin{pmatrix} 1 & x^T \\ x & X \end{pmatrix} & \leq & 0, \\ & \begin{pmatrix} 1 & x^T \\ x & X \end{pmatrix} & \succeq & 0. \end{array}$$

The above problem is not necessarily an SDP since $\mathcal{A}$ is not necessarily a linear map. However, this convex relaxation and/or its dual may still be useful in some applications.

In this general setting, we can construct Slater points for the above convex optimization problem as follows: Since we may assume that $\text{conv}(F)$ is full dimensional, there exists an affinely independent set $\left\{v^{(1)}, v^{(2)}, \ldots, v^{(n+1)}\right\} \subseteq F$. We define $V_\lambda$ as in the proof of Theorem 2.33. Then, it is easy to check that $\mathcal{A}(V_\lambda) \leq 0$ and by construction, $V_\lambda \succ 0$.

Also, for every $\lambda \in \mathbb{R}^n_{++}$ such that $\bar{e}^T\lambda = 1$, we have

$$v_\lambda := \sum_{i=1}^{n+1} \lambda_i v^{(i)} \in \text{int}\,(\text{conv}(F)).$$

Moreover, by definition,

$$\mathcal{A}\left(\begin{pmatrix} 1 \\ v_\lambda \end{pmatrix}\begin{pmatrix} 1, & v_\lambda^T \end{pmatrix}\right) < 0.$$

Now, we can define a feasible point for the convex relaxation which will be easily shown to be a Slater point:

$$\bar{V_\lambda} := \frac{1}{2}V_\lambda + \frac{1}{2}\begin{pmatrix} 1 \\ v_\lambda \end{pmatrix}\begin{pmatrix} 1, & v_\lambda^T \end{pmatrix}.$$

Clearly, $\bar{V_\lambda} \succ 0$ and $\mathcal{A}\left(\bar{V_\lambda}\right) < 0$. Therefore, $\bar{V_\lambda}$ is a Slater point for the above given convex relaxation. We showed:

**Theorem 2.36** *Suppose $F$, $\mathcal{A}$ are such that $F$ can be expressed in the Homogeneous Convex Inequality Form based on $\mathcal{A}$ and conv$(F)$ is full dimensional. Then, the Slater condition holds for $\hat{\mathcal{P}}$.*

**2.7.4 Adding Inequalities to a Convex Relaxation.** In many applications, we may want to utilize SDP relaxations in a branch-and-bound or branch-and-cut scheme. In some other applications, we may want to optimize over the $x$-space projection of the SDP relaxation intersected with a polyhedral relaxation (which itself may have exponentially many facets but admit an efficient separation oracle). Such situations usually require adding valid linear inequalities for $F$ in the space of the SDP relaxations. In some settings, we may have quadratic inequalities.

Suppose $Q \in \Sigma^n$, $q \in \mathbb{R}^n$, $\gamma \in \mathbb{R}$ are given so that the inequality

$$x^TQx + 2q^Tx + \gamma \leq 0$$

is a valid inequality for $F$ (perhaps violated by some points in the current convex relaxation that we have). Then the constraint

$$\left\langle \begin{pmatrix} \gamma & q^T \\ q & Q \end{pmatrix}, \begin{pmatrix} 1 & x^T \\ x & X \end{pmatrix} \right\rangle \leq 0$$

can be added to the SDP relaxation or the current convex relaxation (since it is valid for $\mathcal{F}$). The new relaxation, after the addition of the above valid inequality, still has $V_\lambda$ (in the previous subsection, $\bar{V_\lambda}$) as a Slater point.

## 2.8 Bibliographical Notes

Duality theory is extremely important and useful in optimization. Most books on optimization contain significant amount of duality theory or its applications. See for instance, Rockafellar [305], Rockafellar and Wets [306], Barvinok [24], Ben-Tal and Nemirovski [34], Borwein and Lewis [55], Nesterov and Nemirovski [275], the Handbook of SDP [366]. Also see Borwein and Wolkowicz [57], Duffin [109], Kretschmer [215], Ramana [299], Ramana, Tunçel, Wolkowicz [300], Shapiro [323], Tunçel [347], and Wolkowicz [365]. For applications of our approach in Section 2.7, see Benson and Ye [33], Wolkowicz and Zhao [367] and Zhao, Karisch, Rendl and Wolkowicz [376].

Many of the examples we used are well-known and can be found in the Handbook of SDP [366], Alizadeh, Haeberly and Overton [7], Freund [128], Luo, Sturm and Zhang [248], Ramana [299] and [349].

For various properties of $\Sigma^n_+$ from a linear algebraic viewpoint, see Barker [21], Barker and Carlson [22], Hill and Waters [173], Lim [229], Loewy and Schneider [236], Saunders and Schneider [312] and the references therein.

As we pointed out in Lemma 2.26, there is a very fundamental connection between strong duality and the closure of the Minkowski sum of a linear subspace and a convex cone. Such connections go back at least to Guignard [156]. Also see some more recent related work in [58, 59, 56, 40, 41, 335, 126, 290, 54].

Convexity and duality are covered in the above-mentioned references as well as in many others in many different contexts; see, Berman [38], Berman and Plemons [39], Marshall and Olkin [249], Ben-Israel, Ben-Tal and Zlobec [37], Brøndsted [67], Anderson and Nash [9], Hiriart-Urruty and Lemaréchal [174, 175], Webster [363], Schneider [315], Nesterov [272].

Separation theorems (either separating two disjoint convex sets, or a point from a convex set) play a central role in duality theory as well as in convex analysis and optimization. The origins of such theorems go back at least to Minkowski (the late 1800's and the early 1900's). There are many versions of these theorems in many different settings. See for instance, Klee [199, 198, 197, 196], Gale and Klee [136], Klee, Maluta and Zanco [200].

The work on $S$-Lemma goes back at least to Yakubovich [370]; for more recent work, see the survey [291] and Jeyakumar, Lee and Li [182].

Theorem 2.33 (Theorem 3.1 in [347]) is related to Lemma 3 of Mitchell [256]. The strict complementarity result we quoted for SDP (Theorem 2.20) is due to Alizadeh, Haeberly and Overton [7]. Their paper also contains results pertaining to degeneracy/nondegeneracy as well as the uniqueness of optimal solutions in SDPs. Shortly after that, these results were generalized to all convex optimization problems in conic form [288]. For some related results in the setting of *nonlinear* SDP problems, see Shapiro [323]. For connections between the positive duality gaps for the pair $(P)$, $(D)$ and the failure of strict complementarity in the homogenizations of $(P)$ and $(D)$ see [349].

CHAPTER 3

# Ellipsoid Method

$E \subset \mathbb{R}^d$ is an *ellipsoid* if there exist $c \in \mathbb{R}^d$ (determining the center) and $A \in \Sigma_{++}^d$ (determining the size and the shape) such that

$$E := E(A, c) := \left\{x \in \mathbb{R}^d : (x - c)^T A^{-1}(x - c) \leq 1\right\}.$$

We can alternatively express the ellipsoid as the image of the unit ball in $\mathbb{R}^d$ (denoted by $B_d(0,1)$) under an affine mapping as follows:

$$E(A, c) = A^{1/2} B_d(0, 1) + c.$$

The volume of the ellipsoid is proportional to the square-root of the determinant of the positive definite matrix determining its shape:

$$\mathrm{vol}(E(A, c)) = \sqrt{\det(A)}\,\mathrm{vol}(B_d(0, 1)).$$

The volume of the $d$-dimensional unit ball is

$$\mathrm{vol}(B_d(0, 1)) = \frac{\pi^{d/2}}{\Gamma\left(\frac{d}{2} + 1\right)},$$

where $\Gamma(x) := \int_0^\infty \mathrm{e}^{-t} t^{x-1} dt$, for $x > 0$.

We first study the ellipsoid method as an algorithm which computes a point in an implicitly described convex set. In this basic setting, it is easy to see that the ellipsoid method is a beautiful and theoretically very powerful generalization of the *bisection method* from $\mathbb{R}$ to $\mathbb{R}^d$, for an arbitrary $d$.

Let us think about a simple problem first. Suppose we are given possibly a very large interval $E_0$ on the real line. Our task is to find a point inside an unknown interval $P$. We do know that $P \subseteq E_0$. Moreover, we are provided with an *oracle* to help us.

The oracle accepts as input any $\bar{x} \in \mathbb{R}$ and correctly outputs one of the following:

- $\bar{x} \in P$;
- $\bar{x} > x, \ \forall x \in P$;
- $\bar{x} < x, \ \forall x \in P$.

We can apply bisection to complete this task. We compute the mid-point of the interval and feed it to the oracle. If we get lucky (the mid-point is in $P$) then we are done. Otherwise, based on the oracle's response, we can throw away half of the current interval $E_0$ (it certainly does not contain any point from $P$) and continue by computing the mid-point of the remaining piece of $E_0$ which contains $P$. We repeat the above procedure until we are satisfied (either we have a $\bar{x} \in P$ or we have a proof that the size of the interval $P$ is extremely small since the size of the current interval containing it is very small).

In the above context, bisection is optimal. To see the optimality of bisection in the above context, we appeal to *Information Complexity*. We need to show that for every algorithm, operating under the circumstances stated above, there exists an instance of the problem on which the algorithm requires $\Omega(\ln(1/\epsilon))$ iterations. Instead of trying to fix the instance ahead of time, we will give the answers (as the oracle). As the algorithm under review progresses, all we have to be sure of is that there is at least one problem instance which is consistent with all the answers we gave during each run of the algorithm. Thus, if an algorithm asks about some $\bar{x} \in \mathbb{R}$ that is not the midpoint of the current interval (say the current interval is $[a, b]$), then we will answer

- $\bar{x} < x, \ \forall x \in P$, if $\bar{x} < \frac{a+b}{2}$;
- $\bar{x} > x, \ \forall x \in P$, if $\bar{x} > \frac{a+b}{2}$

so as to leave more than half of the current interval in all situations. We showed that in the worst case, no algorithm can be faster than bisection in the above context.

Now, we are ready to present the basic ingredients of the ellipsoid method which generalizes bisection to arbitrary dimensional sets.

## 3.1 Ingredients: Separation Oracles, Inscribed and Circumscribed Ellipsoids

To appreciate the full theoretical power of the *Ellipsoid Method*, we will move away from the *explicit* description of the convex optimization problem at hand. Instead, we will assume that we are given access to a *separation oracle*. Let $G \subset \mathbb{R}^d$ be the convex set we are interested in optimizing over or less ambitiously just finding a point inside the set $G$. We define $\delta$-relaxation of $G$ as follows:

$$\operatorname{relax}(G, \delta) := \left\{u \in \mathbb{R}^d : \|u - x\|_2 \leq \delta, \text{ for some } x \in G\right\}.$$

Note that it follows from the convexity of $G$ and the above definition that $\operatorname{relax}(G, \delta)$ is convex for every $\delta \in \mathbb{R}$.

A *weak separation oracle for* $G$ takes as input $\bar{x} \in \mathbb{Q}^d$, $\delta \in \mathbb{Q}_{++}$. It either outputs "$\bar{x} \in \operatorname{relax}(G, \delta)$" or $a \in \mathbb{Q}^d$ such that $\|a\|_\infty = 1$ and

$$\langle a, \bar{x}\rangle \geq \langle a, x\rangle - \delta, \quad \forall x \in \operatorname{relax}(G, \delta).$$

**Theorem 3.1** *For every compact convex set in $\mathbb{R}^d$ with nonempty interior, there exists a unique minimal volume ellipsoid containing that set. Moreover, shrinking that ellipsoid (around its center) by a factor of at most $d$ gives an ellipsoid contained in the convex set.*

The unique ellipsoid described in the above theorem is usually called the *Löwner-John ellipsoid*. The factor $d$ in the above theorem is the best possible. (The $d$-dimensional simplex proves the claim for every $d$.) Also, for every compact convex set in $\mathbb{R}^d$ with nonempty interior, there exists a unique maximal volume ellipsoid contained in that set.

**Remark 3.2** Let $G \subset \mathbb{R}^d$ be a compact, convex set with nonempty interior. Let $\bar{A} \in \Sigma^d_{++}$, $c \in \mathbb{R}^d$ such that

$$\bar{E} := \left\{x \in \mathbb{R}^d : (x - c)^T \bar{A}(x - c) \leq 1\right\}$$

contains $G$. Such $\bar{A}$ and $c$ exist since $G$ is bounded (e.g., let $M := \max\{\|x\|_2 : x \in G\} + 1$ and define $c := 0$, $\bar{A} := \frac{1}{M^2} I$, then $\bar{E} \supseteq G$). So, we want to solve the optimization problem:

$$\begin{array}{llrl}
(P_{\bar{A}}) & \min & -\ln\left(\det(\bar{A})\right) & \\
& \text{subject to:} & (x-c)^T\bar{A}(x-c) \leq 1, & \forall x \in G, \\
& & (\bar{A} \in \Sigma^d_{++}, \quad c \in \mathbb{R}^d). &
\end{array}$$

Even though the problem has feasible solutions, its objective function is convex (in fact it is strictly convex—see the first page of the next chapter—in $\bar{A}$), its optimal value is bounded from below (the volume of $\bar{E}$ has to be at least as large as the largest ball—whose volume is clearly positive—contained in $G$), the LHS of the constraints contain cubic polynomials (not a convex optimization problem).

We lift the problem to $\mathbb{R}^{d+1}$. As we did before, let's denote the new coordinate by 0 and embed $G$ (and the corresponding $\mathbb{R}^d$) as the hyperplane

$$\{x \in \mathbb{R}^{d+1} : x_0 = 1\}$$

in $\mathbb{R}^{d+1}$. Consider the optimization problem:

$$\begin{array}{llrl}
(P_A) & \min & -\ln\left(\det(A)\right) & \\
& \text{subject to:} & (1, x^T)\begin{pmatrix} \alpha & a^T \\ a & A \end{pmatrix}\begin{pmatrix} 1 \\ x \end{pmatrix} \leq 1, & \forall x \in G, \\
& & \begin{pmatrix} \alpha & a^T \\ a & A \end{pmatrix} \succeq 0, & \\
& & (A \in \Sigma^d_{++}, \quad a \in \mathbb{R}^d, \alpha \in \mathbb{R}). &
\end{array}$$

$(P_A)$ is a convex optimization problem, it has feasible solutions, its objective function value is bounded from below. In fact $(P_{\bar{A}})$ and $(P_A)$ are very closely related. Let $(\bar{A}, c)$ be a feasible solution of $(P_{\bar{A}})$. Let

$$A := \bar{A}, \quad a := -\bar{A}c, \quad \alpha := c^T\bar{A}c.$$

Then it is easy to see that $(A, a, \alpha)$ is a feasible solution of $(P_A)$ with the same objective value.

Conversely, let $(A, a, \alpha)$ be an optimal solution of $(P_A)$. Then we claim that $\alpha = a^TA^{-1}a$ (otherwise, $\alpha > a^TA^{-1}a$ and we can reduce $\alpha$ and update $A \rightarrow (A + \epsilon aa^T)$ for $\epsilon$ positive but small enough, and still stay feasible, contradicting the optimality of $(A, a, \alpha)$). Now, setting

$$\bar{A} := A, \quad c := -A^{-1}a,$$

we see that we have a feasible solution of $(P_{\bar{A}})$ with the same objective value.

Using the strict convexity of $-\ln(\det(\cdot))$ on $\Sigma^d_{++}$ uniqueness of the optimal solution of $(P_A)$ (and hence of $(P_{\bar{A}})$) can be established.

Let

$$E := \left\{x \in \mathbb{R}^d : (x-c)^TA^{-1}(x-c) \leq 1\right\},$$

and

$$\tilde{E} := \left\{x \in E : a^Tx \leq a^Tc\right\},$$

for some $a \in \mathbb{R}^d \setminus \{0\}$. We will assume $d \geq 2$. We would like to construct the smallest volume ellipsoid $E_+$ containing the half-ellipsoid $\tilde{E}$. Let $c_+ \in \mathbb{R}^d$ and $A_+ \in \Sigma^d_{++}$ denote the center and the positive definite matrix determining $E_+$. Then

$$c_+ = c - \frac{1}{(d+1)\sqrt{a^TAa}}Aa,$$

$$A_+ = \frac{d^2}{d^2-1}\left[A - \frac{2}{(d+1)a^TAa}Aaa^TA\right].$$

First, let's observe that $A_+ \succ 0$ if $A \succ 0$. Let $h \in \mathbb{R}^d \setminus \{0\}$. Then

$$\begin{aligned} h^TA_+h &= \frac{d^2}{d^2-1}h^TAh - \frac{2d^2}{(d-1)(d+1)^2a^TAa}(a^TAh)^2 \\ &\geq \frac{d^2}{d^2-1}\left\|A^{1/2}h\right\|_2^2 - \frac{2d^2\left\|A^{1/2}a\right\|_2^2\left\|A^{1/2}h\right\|_2^2}{(d-1)(d+1)^2\left\|A^{1/2}a\right\|_2^2} \\ &= \left(\frac{d}{d+1}\right)^2\left\|A^{1/2}h\right\|_2^2 \end{aligned}$$

which is clearly positive. Therefore, $A_+$ is positive definite if $A$ is. We can explicitly compute the volume of $E_+$ in terms of the volume of $E$, since $A_+$ is a rank-1 update of $A$. When we take the image of $\mathbb{R}^d$ under the mapping $A^{-1/2}$, our ellipsoid $E$ becomes the unit ball. Under this mapping our update formula becomes

$$I_+ = \frac{d^2}{d^2-1}\left[I - \frac{2}{(d+1)\bar{a}^T\bar{a}}\bar{a}\bar{a}^T\right],$$

where $\bar{a} := A^{1/2}a$. The eigenvalues of $\left[I - \frac{2}{(d+1)\bar{a}^T\bar{a}}\bar{a}\bar{a}^T\right]$ are $\frac{d-1}{d+1}$ and 1 (with multiplicity $(d-1)$). Therefore,

$$\det(I_+) = \left(\frac{d}{d+1}\right)^{d+1}\left(\frac{d}{d-1}\right)^{d-1} = \left(\frac{d^2}{d^2-1}\right)^{d-1}\frac{d^2}{(d+1)^2}.$$

Hence the volume of $E_+$ can be related to the volume of $E$ as follows:

$$\operatorname{vol}(E_+) = \left(\frac{d}{\sqrt{d^2-1}}\right)^{d-1}\frac{d}{d+1}\operatorname{vol}(E).$$

The following well-known fact is called the *Sherman-Morrison-Woodbury formula.*

**Lemma 3.3** *Let $A \in \Sigma^d$ be nonsingular. Also let $u, v \in \mathbb{R}^d$ be given. Then $(A + uv^T)$ is nonsingular iff $(1 + u^TA^{-1}v) \neq 0$. If the latter condition holds then*

$$\left(A + uv^T\right)^{-1} = A^{-1} - \frac{1}{1+u^TA^{-1}v}A^{-1}uv^TA^{-1}.$$

**Proof** Direct computation. □

**Theorem 3.4** *We have*

$$\tilde{E} \subseteq E_+ \text{ and } \ln\left(\frac{\operatorname{vol}(E_+)}{\operatorname{vol}(E)}\right) \leq -\frac{1}{2d}.$$

**Proof** In the previous lemma, let $u := \frac{\sqrt{2}}{\sqrt{(d+1)a^TAa}}Aa$ and $v := -u$. Indeed,

$$1 + u^TA^{-1}v = \frac{d-1}{d+1} \neq 0.$$

Therefore,

$$A_+^{-1} = \frac{d^2-1}{d^2}\left[A^{-1} + \frac{2}{(d-1)a^TAa}aa^T\right].$$

By a direct computation, we show that

$$\left(x-c+\frac{1}{(d+1)\sqrt{a^TAa}}Aa\right)^T A_+^{-1}\left(x-c+\frac{1}{(d+1)\sqrt{a^TAa}}Aa\right)\leq 1,$$

for every $x\in\mathbb{R}^d$ such that $(x-c)^TA^{-1}(x-c)\leq 1$ and $a^Tx\leq a^Tc$. (This computation is easier to do in the space where $E$ is the unit ball; the formula for $A_+$ is also easier to derive in that space.) For the second part of the statement of the theorem, it suffices to prove that

$$(d+1)\ln\left(\frac{d}{d+1}\right)+(d-1)\ln\left(\frac{d}{d-1}\right)\leq -\frac{1}{d}.$$

We expand the logarithm terms to infinite series and compute

$$\begin{aligned}
&(d+1)\ln\left(\frac{d}{d+1}\right)+(d-1)\ln\left(\frac{d}{d-1}\right)\\
&= -(d+1)\ln\left(1+\frac{1}{d}\right)-(d-1)\ln\left(1-\frac{1}{d}\right)\\
&= (d+1)\sum_{k=1}^{\infty}(-1)^k\frac{1}{kd^k}+(d-1)\sum_{k=1}^{\infty}\frac{1}{kd^k}\\
&= d\sum_{k=1}^{\infty}\frac{1}{kd^{2k}}-\sum_{k=1}^{\infty}\frac{2}{(2k-1)d^{2k-1}}\\
&= -\sum_{k=1}^{\infty}\frac{1}{k(2k-1)d^{(2k-1)}}\\
&< -\frac{1}{d},
\end{aligned}$$

as desired. □

## 3.2 Complexity Analysis

Suppose $\mathrm{vol}(E_0)=R$ and we want $E_k$ such that $\mathrm{vol}(E_k)\leq\epsilon$. Then by the last theorem,

$$\ln\left(\frac{\mathrm{vol}(E_k)}{\mathrm{vol}(E_0)}\right)\leq -\frac{k}{2d},$$

and $O\left(d\ln\left(\frac{R}{\epsilon}\right)\right)$ iterations suffice. Similarly, if $R$ is the radius of the initial ball (whose image under the mapping $x\mapsto A_0^{-1/2}x$ contains the set) and we want the stopping criterion to be that the radius of the current ball (whose image under the mapping $x\mapsto A_k^{-1/2}x$ contains $G$) is at most $\epsilon$, then using

$$\left(\frac{\mathrm{vol}(E_k)}{\mathrm{vol}(E_0)}\right)=d\ln\left(\frac{R}{\epsilon}\right),$$

we find that $O\left(d^2\ln\left(\frac{R}{\epsilon}\right)\right)$ iterations suffice.

**Theorem 3.5** *Let $G\subset\mathbb{R}^d$ be a convex set such that $G\subseteq B_d(0,R)$. Suppose we are also given $\epsilon\in\mathbb{Q}_{++}$. If we have access to a weak separation oracle for $G$, then in polynomial time (polynomial in $(d,\ln(R/\epsilon))$), we can compute $\bar{x}\in\mathrm{relax}(G,\epsilon)$ or prove that $\mathrm{vol}(G)\leq\epsilon$.*

These results can be extended to optimizing a convex function over $G$. We need one more ingredient to deal with the objective function. Let $f : \mathbb{R}^d \to \mathbb{R}$ be a convex function.

**Definition 3.6** A *subgradient oracle for* $f$ takes as input $\bar{x} \in \mathbb{R}^d$ and returns in polynomial time (polynomial in $d$ and for a proper definition, size$(\bar{x})$) $f(\bar{x})$ and $h \in \mathbb{R}^d$ such that

$$f(x) \geq f(\bar{x}) + h^T(x - \bar{x}), \quad \forall x \in \mathbb{R}^d.$$

Suppose we are interested in solving the convex optimization problem

$$\inf \{f(x) : x \in G\},$$

where $f : \mathbb{R}^d \to \mathbb{R}$ is a convex function.

**Theorem 3.7** *Let $G \subset \mathbb{R}^d$ be a convex set and $r, R \in \mathbb{Q}_{++}$ be given such that*

$$B_d(\tilde{x}, r) \subseteq G \subseteq B_d(0, R), \quad \textit{for some } \tilde{x} \in \mathbb{R}^d$$

*(here, $\tilde{x}$ is* **not** *given). Let $\epsilon \in \mathbb{Q}_{++}$ be also given. Suppose that a subgradient oracle for $f$ and a weak separation oracle for $G$ are available. Then after*

$$O\left(d^2\left[\ln\left(\frac{R}{r}\right) + \ln\left(\frac{\mu_0}{\epsilon}\right)\right]\right)$$

*iterations, the ellipsoid method returns a feasible solution $\bar{x} \in G$ such that*

$$f(\bar{x}) \leq \inf \{f(x) : x \in G\} + \epsilon.$$

*In the above, $\mu_0 := \epsilon + \sup_{x \in E_0}\{f(x)\} - \inf_{x \in E_0}\{f(x)\}$.*

## 3.3 Applications to SDP Problems

Suppose $(P)$ and $(D)$ have Slater points $\bar{X}$ and $(\bar{y}, \bar{S})$ respectively. Then we can compute $r$ and $R$ to apply the last theorem. We can replace our primal problem $(P)$ by

$$\begin{array}{llrcl} (\tilde{P}) & \inf & \langle C, X\rangle & & \\ & \text{subject to:} & \mathcal{A}(X) & = & b, \\ & & \langle \bar{S}, X\rangle & \leq & 2\langle \bar{X}, \bar{S}\rangle, \\ & & X & \succeq & 0. \end{array}$$

We first state a well-known fact.

**Lemma 3.8** *Let $X, S \in \Sigma^n$. Then*

$$\sum_{i=1}^{n} \lambda_i(X)\lambda_{n-i+1}(S) \leq \langle X, S\rangle \leq \sum_{i=1}^{n} \lambda_i(X)\lambda_i(S).$$

Using the above lemma and elementary arguments, we can easily establish the following theorem.

**Theorem 3.9** (a) *The problems $(P)$ and $(\tilde{P})$ attain their optimal values.*

(b) *The optimal solution sets of $(P)$ and $(\tilde{P})$ are the same.*

(c) *Let $G$ denote the feasible region of $(\tilde{P})$. Then $G$ is convex and bounded. Moreover,*

$$B_G(\bar{X}, \lambda_n(\bar{X})) \subseteq G \subseteq B_G\left(0, \frac{2\langle \bar{X}, \bar{S}\rangle}{\lambda_n(\bar{S})}\right),$$

*where $B_G$ denotes the unit ball in the affine space spanned by $G$.*

(d) *We have*

$$\sup\{\langle C, X\rangle : X \in G\} - \inf\{\langle C, X\rangle : X \in G\} \leq \frac{4n\|C\|_2\langle \bar{X}, \bar{S}\rangle}{\lambda_n(\bar{S})}.$$

Also, we note here that the separation problem for $\Sigma_+^d$ can be solved in polynomial time via the computation of $LDL^T$ decomposition (with proper pivoting) of the given symmetric matrix.

## 3.4 Applications to Combinatorial Optimization Problems

Since 1979, one of the most important applications of the ellipsoid method has been in the classification of optimization problems in terms of whether they admit a polynomial time algorithm. Since the ellipsoid method does not require an explicit description of the convex set $G$ or the objective function $f$, its theoretical power has been far reaching.

For instance, in combinatorial optimization, if we suspect that our problem can be solved in polynomial time, then we might try to come up with a compact LP formulation. Failing that, we might come up with an exponential size formulation. Then, if we can construct a weak separation oracle for the feasible region, we would have a polynomial time algorithm for optimizing a linear function (or in general, a convex function which admits a subgradient oracle) over this feasible region in polynomial time (even though the formulation is of exponential size).

As a simple example, we consider the *Subtour Elimination Polytope* (SEP) of the *Traveling Salesman Problem* (TSP).

Let $G = (V, E)$ be a given undirected graph (we will describe the *symmetric TSP* here) and $c \in \mathbb{R}^E$ the costs on the edges of $G$. The TSP is to find a *Hamiltonian cycle* in $G$ of minimum total cost. Given $U \subseteq V$, $\delta(U)$ denotes the set of edges in $E$ with one end point in $U$ and the other outside of $U$. Then an integer programming formulation for TSP is as follows:

$$\begin{array}{llrcll}
(TSP) & \min & c^T x & & & \\
 & \text{subject to:} & x(\delta(v)) & = & 2, & \forall v \in V \\
 & & x(\delta(U)) & \geq & 2, & \forall U \subset V,\ 2 \leq |U| \leq |V| - 2, \\
 & & 0 \leq x \leq \bar{e}, & & x \text{ is integral.} &
\end{array}$$

A well-known LP relaxation is obtained from the above formulation by removing the integrality constraints.

$$\begin{array}{llrcll}
(SEP) & \min & c^T x & & & \\
 & \text{subject to:} & x(\delta(v)) & = & 2, & \forall v \in V \\
 & & x(\delta(U)) & \geq & 2, & \forall U \subset V,\ 2 \leq |U| \leq |V| - 2, \\
 & & 0 \leq x \leq \bar{e}. & & &
\end{array}$$

We also denote by *SEP* (the *Subtour Elimination Polytope*), the feasible region of the above relaxation. Now, we outline how a well-known separation oracle can be designed for this polytope *SEP*.

Let $\bar{x} \in \mathbb{R}^E$ be given. Indeed, it is easy to check directly (and in polynomial time) that $\bar{x}$ satisfies $x \in [0,1]^E$ and $x(\delta(v)) = 2$, for every $v \in V$. So we can assume that these constraints are satisfied by $\bar{x}$. Next, we construct a directed graph $G' := (V', E')$. We let $V' := V$ and $E' := \{(u,v),(v,u) : \{u,v\} \in E\}$. We assign the *capacity* $\bar{x}_{uv}$ to both $(u,v)$ and $(v,u)$. Note that the capacity of the *cut* $U$ in $G'$ is simply $\bar{x}(U)$. Therefore, there exists $U \subset V$ such that $\bar{x}(U) < 2$ iff some cut in $G'$ has capacity strictly less than 2. By the maximum flow-minimum cut theorem, this is true iff there exists a pair of nodes $s, t \in V'$ such that the value of the maximum flow from $s$ to $t$ in $G'$ is less than 2. Hence, we can solve the separation problem for $SEP$, by simply solving $\binom{n}{2}$ maximum flow problems on $G'$. This implies that the convex set $SEP$ admits a weak separation oracle.

In fact, there are more efficient algorithms for the above purpose. For instance, an $O(|V||E| + |V|^2 \ln(|V|))$ algorithm exists for computing the global minimum cut in $G'$. Also, there are very efficient randomized algorithms for this purpose.

There are also explicit formulations of polynomial size polyhedra whose projection equals $SEP$. This gives another way to see that the linear (or convex) optimization problem over $SEP$ can be solved in polynomial time. However, it is not clear that this can always be done. That is, the existence of a weak separation oracle may not imply the existence of explicit compact representations by polynomial size polyhedra. Regardless of the answer to such questions, it is clear that finding a weak separation oracle is no harder than finding a combinatorial algorithm (running in polynomial time) which solves the whole optimization problem or finding an explicit compact representation by a polynomial size polyhedron.

An important point for our example is that the weights $(\bar{x}_e)$ on the edges are nonnegative. If this assumption is violated, then the underlying optimum cut problem is called the *maximum cut problem* which is $\mathcal{NP}$-hard in general. We will address this problem in Chapter 5.

A general framework of polynomially solvable combinatorial optimization problems is the *submodular function minimization*. As early as 1981, it was known (via the theory of ellipsoid method) that these problems could be solved in polynomial time. However, after much effort, purely combinatorial algorithms were first developed in 1999. See Iwata, Fleischer and Fujishige [181] as well as Schrijver [318].

## 3.5 Equivalence of Separation and Optimization

An important consequence of the ellipsoid method and its properties is that it proves

$$\text{Separation over } G \Rightarrow \text{ Optimization over } G,$$

in our context. The converse of the above implication is also true and it follows from duality. As it should be clear from our study so far, separating and supporting hyperplanes for a convex set have fundamental connections to the *dual solutions* for a suitably chosen dual problem.

Suppose we can solve the linear optimization problem over $G$ in polynomial time. Assume $G \subseteq \mathbb{R}^d$ is closed, convex with nonempty interior and that $0 \in G$. Then we can solve the separation problem over $G^*$ in polynomial time. Given $u \in \mathbb{R}^d$, we solve

$$\sup\left\{\langle u, x\rangle : x \in G\right\}.$$

Let $\bar{x} \in \mathbb{R}^d$ be an approximately optimal solution. If $\langle u, \bar{x}\rangle > (1+\epsilon)$ then we have a separating hyperplane. Otherwise, $u \in \operatorname{relax}(G^*, \epsilon)$. Therefore, we have a weak separation oracle for $G^*$. Hence, using the ellipsoid method, we can solve the linear optimization problem over $G^*$ in polynomial time. Now, we apply our argument to $G^*$ to conclude that we have a weak separation oracle for $G^{**}$. But under our assumptions, $G^{**} = G$. Therefore, we proved

$$\text{Separation over } G \iff \text{ Optimization over } G,$$

under our assumptions.

## 3.6 Bibliographical Notes

Until 1979, very few mathematicians in the West knew about the ellipsoid method. Khachiyan [193], in 1979, proved that the Ellipsoid Method can be adapted to solve linear programming problems in polynomial time (hence settling a long outstanding problem). This announcement caused an unprecedented reaction from the media.

The Ellipsoid Method was originally proposed by Yudin and Nemirovski [374] and Shor [332] in 1970's. This original method was designed to deal with essentially *any* convex optimization problem that can be posed in a finite dimensional space by the potential usage of oracles (an important point which should be emphasized is that the functions involved in defining the feasible solution set, the objective function need **not** be differentiable). For a nice exposure to Ellipsoid Method for linear programming problems, see the survey paper by Bland, Goldfarb and Todd [47].

Shortly after Khachiyan's result, it was established that this method is a very powerful tool in determining the computational complexity status (hence the degree of difficulty) of various combinatorial optimization problems. A good reference is the book by Grötschel, Lovász and Schrijver [153].

There are also more recent complexity analyses of the ellipsoid method which use other condition measures to bound the number of iterations required in the worst case. See Freund and Vera [130] as well as Freund [127].

There are many elegant and generally simple algorithms for convex optimization which only use subgradient information. Such algorithms are called *First-Order Algorithms*. Ellipsoid Method may be considered to loosely belong to the same general family of algorithms for convex optimization. Some algorithms from this big family have impressive theoretical properties (like the Ellipsoid Method); see, for instance, [268, 269, 270, 271, 266] and the references therein. Moreover, many of them can be very useful in practice for very large-scale problems (for instance those problems that are out-of-reach of the Second-Order Algorithms and) which require only moderate accuracy of the solution.

In the late 1980's, Ellipsoid Method was applied to some problems in System and Control Theory (see the book by Boyd et al. [63]). These problems were small but relatively difficult convex optimization problems. Since late 1980's and early 1990's interior-point methods consistently took over the solution process. These are the methods we discuss and analyze next.

CHAPTER 4

# Primal-Dual Interior-Point Methods

Suppose $X^{(0)} \succ 0$ and $y^{(0)} \in \mathbb{R}^m$, $S^{(0)} \succ 0$ are given such that $\mathcal{A}(X^{(0)}) = b$ and $\mathcal{A}^*(y^{(0)})+S^{(0)} = C$. Assume that $\mathcal{A}$ is surjective. A corollary of the duality theorem for SDP (Corollary 2.17) implies that both $(P)$ and $(D)$ attain their optimal values and the optimal values of $(P)$ and $(D)$ coincide. We will describe an algorithm that generates a sequence $\left\{(X^{(k)}, y^{(k)}, S^{(k)})\right\}$ such that $X^{(k)} \succ 0$, $S^{(k)} \succ 0$, $\mathcal{A}(X^{(k)}) = b$ and $\mathcal{A}^*(y^{(k)}) + S^{(k)} = C$ and $\langle X^{(k)}, S^{(k)}\rangle$ approaching 0.

The modern theory of interior-point methods, reformulates the *difficult constraints* by utilizing a *strictly convex function* (below, we will use the well-known fact that a twice continuously differentiable function is strictly convex if its Hessian is positive definite) with very specific properties. In our case, the difficult constraint is "$X \succeq 0$."

We define $f : \Sigma^n_{++} \to \mathbb{R}$,

$$f(X) := -\ln(\det(X)).$$

$f$ is a strictly convex function on $\Sigma^n_{++}$ and for any sequence $\{X^{(k)}\} \subset \Sigma^n_{++}$, converging to a point on the boundary of $\Sigma^n_{+}$, we have

$$f(X^{(k)}) \to +\infty.$$

**Proposition 4.1** *For the above definition of $f$, for every $X \in \Sigma^n_{++}$ and $H \in \Sigma^n$, we have*

$$\begin{aligned}
\langle f^{'}(X), H\rangle &= -\mathrm{Tr}(X^{-1}H),\\
\langle f^{''}(X)H, H\rangle &= \mathrm{Tr}(X^{-1}HX^{-1}H),\\
f^{'''}(X)[H,H,H] &= -2\mathrm{Tr}\left[(X^{-1/2}HX^{-1/2})^3\right].
\end{aligned}$$

**Proof** Let $X \in \Sigma^n_{++}$ and $H \in \Sigma^n$. Then, we compute $\langle f^{'}(X), H\rangle$ by evaluating

$$\lim_{\alpha\to 0} \frac{-\ln\det(X+\alpha H) + \ln\det(X)}{\alpha}.$$

Note that

$$\begin{aligned}
-\ln\det(X+\alpha H)+\ln\det(X) &= -\ln\det\left[X^{1/2}\left(I+\alpha X^{-1/2}HX^{-1/2}\right)X^{1/2}\right]\\
&\quad +\ln\det(X)\\
&= -\ln\det\left(I+\alpha X^{-1/2}HX^{-1/2}\right) - \ln\det(X)\\
&\quad +\ln\det(X)\\
&= -\ln\det\left(I+\alpha\bar{H}\right),
\end{aligned}$$

where $\bar{H} := X^{-1/2}HX^{-1/2}$. Now, let us analyze $\det\left(I+\alpha\bar{H}\right)$. Expanding the determinant, expressing it as a univariate polynomial in $\alpha$ and collecting the constant and the linear terms in $\alpha$ exactly, we obtain for $\alpha \to 0$,

$$\det\left(I+\alpha\bar{H}\right) = 1 + \operatorname{Tr}\left(\bar{H}\right)\alpha + o(\alpha).$$

Hence,

$$\begin{aligned}\langle f^{'}(X), H\rangle &= \lim_{\alpha\to 0}\frac{-\ln\left[1+\operatorname{Tr}\left(\bar{H}\right)\alpha + o(\alpha)\right]}{\alpha}\\ &= \lim_{\alpha\to 0}\frac{-\operatorname{Tr}\left(\bar{H}\right)\alpha + o(\alpha)}{\alpha}\\ &= -\operatorname{Tr}\left(X^{-1/2}HX^{-1/2}\right).\end{aligned}$$

To compute $\langle f^{''}(X)H, H\rangle$, we find the derivative of $f^{'}(X)$ by evaluating $(X+\alpha H)^{-1}$ $-X^{-1}$ as $\alpha \to 0$ as a univariate polynomial in $\alpha$ and collect the constant and linear terms exactly:

$$\begin{aligned}(X+\alpha H)^{-1} - X^{-1} &= \left[X\left(I+\alpha X^{-1}H\right)\right]^{-1} - X^{-1}\\ &= \left(I+\alpha X^{-1}H\right)^{-1}X^{-1} - X^{-1}\\ &= \left(I-\alpha X^{-1}H + \alpha^2\cdots\right)X^{-1} - X^{-1}\\ &= -\alpha X^{-1}HX^{-1} + o(\alpha).\end{aligned}$$

Hence,

$$\langle f^{''}(X)H, H\rangle = \langle X^{-1}HX^{-1}, H\rangle = \operatorname{Tr}\left(X^{-1}HX^{-1}H\right) = \operatorname{Tr}\left[\left(X^{-1/2}HX^{-1/2}\right)^2\right].$$

Finally, for the third derivative of $f$, let us compute

$$\langle f^{''}(X+\alpha H)H, H\rangle - \langle f^{''}(X)H, H\rangle \text{ as } \alpha \to 0:$$

$$\begin{aligned}&\operatorname{Tr}\left[H(X+\alpha H)^{-1}H(X+\alpha H)^{-1}\right] - \operatorname{Tr}\left(HX^{-1}HX^{-1}\right)\\ =\ &\operatorname{Tr}\left[H\left(I+\alpha X^{-1}H\right)^{-1}X^{-1}H\left(I+\alpha X^{-1}H\right)^{-1}X^{-1}\right] - \operatorname{Tr}\left(HX^{-1}HX^{-1}\right)\\ =\ &\operatorname{Tr}\left[H\left(I-\alpha X^{-1}H+o(\alpha)\right)X^{-1}H\left(I-\alpha X^{-1}H+o(\alpha)\right)X^{-1}\right]\\ &-\operatorname{Tr}\left(HX^{-1}HX^{-1}\right)\\ =\ &\operatorname{Tr}\left(HX^{-1}HX^{-1} - 2\alpha HX^{-1}HX^{-1}HX^{-1}\right) - \operatorname{Tr}\left(HX^{-1}HX^{-1}\right) + o(\alpha)\\ =\ &-2\alpha\operatorname{Tr}\left(HX^{-1}HX^{-1}HX^{-1}\right) + o(\alpha).\end{aligned}$$

Therefore,

$$f^{'''}(X)[H,H,H] = -2\operatorname{Tr}\left(HX^{-1}HX^{-1}HX^{-1}\right) = -2\operatorname{Tr}\left[\left(X^{-1/2}HX^{-1/2}\right)^3\right].$$

□

Since $\langle f^{''}(X)H, H\rangle = \operatorname{Tr}\left[(X^{-1/2}HX^{-1/2})^2\right] \geq 0$ (equality holds iff $H = 0$), we immediately verify that $f$ is strictly convex.

Note that

$$\langle f^{''}(X)H, H\rangle = \operatorname{Tr}\left[\left(X^{-1/2}HX^{-1/2}\right)^2\right] = \text{s2vec}\,(H)^T\left(X^{-1}\overset{s}{\otimes}X^{-1}\right)\text{s2vec}\,(H).$$

## 4.1 Central Path

For each $\mu > 0$ we define

$$(P_\mu) \qquad \inf \ \tfrac{1}{\mu}\langle C, X\rangle + f(X)$$
$$\text{subject to:} \qquad \mathcal{A}(X) = b.$$

In the above $X \succ 0$ is implied, since $\Sigma^n_{++}$ is the domain of $f$. We established that $f$ is strictly convex and tends to infinity for every sequence converging to a boundary point of $\Sigma^n_{++}$. For every fixed $\mu > 0$ these properties are inherited by the objective function of $(P_\mu)$. Therefore, for every $\mu > 0$, $(P_\mu)$ has a unique solution in $\Sigma^n_{++}$. We can characterize the unique solution via necessary and sufficient conditions for optimality applied to the convex optimization problem $(P_\mu)$:

$$\mathcal{A}(X) = b, \qquad X \succ 0,$$
$$-\mathcal{A}^*(y) - X^{-1} + \frac{1}{\mu}C = 0.$$

Let's do the substitutions $y := \mu y$ and $S := \mu X^{-1}$. We obtain the system

$$\begin{aligned} \mathcal{A}(X) &= b, \ X \succ 0, \\ \mathcal{A}^*(y) + S &= C, \\ S &= \mu X^{-1}. \end{aligned}$$

For each $\mu > 0$, the unique solution of the above system $(X(\mu), y(\mu), S(\mu))$ defines *the primal-dual central path*:

$$\left\{(X(\mu), y(\mu), S(\mu)) \in \Sigma^n_{++} \oplus \mathbb{R}^m \oplus \Sigma^n_{++} : \mu > 0\right\}.$$

Indeed, given $y(\mu)$, $S(\mu)$ is redundant and vice versa (assuming $\mathcal{A}$ is surjective). *The primal central path* is

$$\left\{X(\mu) \in \Sigma^n_{++} : \mu > 0\right\},$$

and *the dual central path* is

$$\{y(\mu) \in \mathbb{R}^m : \mu > 0\} \text{ or } \left\{S(\mu) \in \Sigma^n_{++} : \mu > 0\right\},$$

possibly both, combined in a direct sum.

Very many algorithms which utilize the properties of the central path have been proposed. These algorithms take an approximate solution of the above system (for some $\bar{\mu} > 0$) and using a Newton-type search direction, try to find another feasible primal-dual pair maintaining positive definiteness (hence the name *interior-point method*) such that the new primal-dual pair approximately solves the central path equations for $\hat{\mu} < \bar{\mu}$. Note that

$$\langle X(\mu), S(\mu)\rangle = \langle X(\mu), \mu\,(X(\mu))^{-1}\rangle = n\mu.$$

Therefore, decreasing $\mu$ is equivalent to decreasing $\left(\langle C, X\rangle - b^T y\right)$, the duality gap.

For an arbitrary pair (not necessarily on the central path) of primal-dual feasible solutions $X$, $S$, we define "their $\mu$" as

$$\mu(X, S) := \frac{\langle X, S\rangle}{n}.$$

Whenever it is clear from the context, we will simply write $\mu$ for $\mu(X, S)$.

We define a measure of *centrality*

$$\psi(X, S) := n \ln\left(\frac{\langle X, S\rangle}{n}\right) - \ln(\det(X)) - \ln(\det(S)).$$

**Theorem 4.2** *For every* $(X, S) \in \Sigma^n_{++} \oplus \Sigma^n_{++}$,

$$\psi(X, S) \geq 0.$$

*Moreover, the equality holds above iff* $S = \mu X^{-1}$ *with* $\mu := \frac{\langle X,S\rangle}{n} > 0$.

**Proof** For every $(X, S) \in \Sigma^n_{++} \oplus \Sigma^n_{++}$, we have

$$\langle X, S\rangle - \ln(\det(X)) + \sup_{X\in\Sigma^n_{++}} \{-\langle S, X\rangle + \ln(\det(X))\} \geq 0.$$

The equality holds above iff $X = S^{-1}$ (we used the strict convexity of $f(X)$ and the first order necessary and sufficient conditions for a minimizer). Hence we have for every $(X, S) \in \Sigma^n_{++} \oplus \Sigma^n_{++}$,

$$\langle X, S\rangle - \ln(\det(X)) - \ln(\det(S)) - n \geq 0,$$

with equality iff $S = X^{-1}$. Since $\Sigma^n_{++}$ is a cone, we can replace $X$ by $\alpha_1 X$ for $\alpha_1 > 0$ and $S$ by $\alpha_2 S$ for any $\alpha_2 > 0$. Thus, we obtain

$$\alpha_1\alpha_2\langle X, S\rangle - n\ln(\alpha_1\alpha_2) - \ln(\det(X)) - \ln(\det(S)) - n \geq 0,$$

for every $(X, S) \in \Sigma^n_{++} \oplus \Sigma^n_{++}$, and every $\alpha_1 > 0$, $\alpha_2 > 0$. Moreover, the equality holds above iff $S = \frac{1}{\alpha_1\alpha_2}X^{-1}$. Let's substitute $t := \alpha_1\alpha_2$. Then the left hand side of the inequality is a strictly convex function of $t$; in particular, the left hand side is minimized with respect to $t$ uniquely at $t = \frac{n}{\langle X,S\rangle}$. Therefore, we obtain

$$n\ln\left(\frac{\langle X, S\rangle}{n}\right) - \ln(\det(X)) - \ln(\det(S)) \geq 0,$$

for every $(X, S) \in \Sigma^n_{++} \oplus \Sigma^n_{++}$. Moreover, the equality holds above iff $S = \mu X^{-1}$, for $\mu := \frac{\langle X,S\rangle}{n}$. □

**Remark 4.3** In the above proof, we implicitly used another notion of *dual*. Namely the *conjugate* of a function. Let $g : \mathbb{R}^d \to \mathbb{R} \cup \{+\infty\}$. Recall that we defined the *Legendre-Fenchel conjugate of* $g$ as

$$g_*(s) := \sup_{x\in\text{dom}(g)} \{-\langle s, x\rangle - g(x)\},$$

where

$$\text{dom}(g) := \left\{x \in \mathbb{R}^d : g(x) < +\infty\right\}$$

is the *domain of* $g$. We showed in the above proof that $f$ differs from its Legendre-Fenchel conjugate only by a constant. So in that sense, just like the cone $\Sigma^n_+$, the function $f$ is also self-dual.

**Remark 4.4** This last theorem generalizes the arithmetic-geometric mean inequality. Conversely, the basic arithmetic-geometric mean inequality gives a big chunk of the proof when applied to the eigenvalues of $X^{1/2}SX^{1/2}$. Let $\lambda_1, \lambda_2, \ldots, \lambda_n$

denote the eigenvalues of $X^{1/2}SX^{1/2}$. Then

$$\begin{aligned}\psi(X,S) &= n\ln\left(\frac{\sum_{j=1}^n \lambda_j}{n}\right) - \ln\left(\prod_{j=1}^n \lambda_j\right)\\ &= n\ln\left(\frac{\sum_{j=1}^n \lambda_j/n}{\left(\prod_{j=1}^n \lambda_j\right)^{1/n}}\right)\\ &= n\ln\left(\frac{\text{arithmetic mean}(\lambda_1,\lambda_2,\ldots,\lambda_n)}{\text{geometric mean}(\lambda_1,\lambda_2,\ldots,\lambda_n)}\right)\\ &\geq n\ln(1).\end{aligned}$$

The last inequality holds as equality iff $\lambda_1 = \lambda_2 = \cdots = \lambda_n = \mu$.

## 4.2 Primal-Dual Potential Function

Given $(\hat{X},\hat{S})$ and $(\bar{X},\bar{S})$ a pair of primal-dual feasible and interior pairs, we would like to have a simple way of comparing them. We have two criteria:

- smaller the duality gap $\langle X, S\rangle$ is the better,
- smaller the distance to the central path (that is, the value of $\psi(X,S)$) is the better.

The next function, called *the primal-dual potential function*, serves such a purpose and allows us to design and perform the complexity analysis directly on it.

$$\phi_q(X,S) := q\ln(\langle X,S\rangle) + \psi(X,S),$$

where $q > 0$ (we will take $q := \sqrt{n}$ in our analysis).

**Theorem 4.5** *Suppose we have $\left(X^{(0)}, S^{(0)}\right)$ feasible for $(P)$ and $(D)$ such that*

$$\psi(X^{(0)}, S^{(0)}) \leq \sqrt{n}\ln\left(\frac{1}{\epsilon}\right), \text{ for some } \epsilon \in (0,1).$$

*If we generate $(X^{(k)}, S^{(k)})$ feasible in $(P)$ and $(D)$ such that*

$$\phi_{\sqrt{n}}(X^{(k)}, S^{(k)}) \leq \phi_{\sqrt{n}}(X^{(k-1)}, S^{(k-1)}) - \delta, \text{ for every } k \geq 1,$$

*for some $\delta > 0$ an absolute constant, then for some $\bar{k} = O(\sqrt{n}\ln(1/\epsilon))$, we have*

$$\langle X^{(k)}, S^{(k)}\rangle \leq \epsilon\langle X^{(0)}, S^{(0)}\rangle, \text{ for every } k \geq \bar{k}.$$

**Proof** Suppose $(X^{(k)}, S^{(k)})$ satisfies the hypothesis of the theorem, for every $k \geq 1$. Then on the one hand, we have

$$\phi_{\sqrt{n}}(X^{(k)}, S^{(k)}) - \phi_{\sqrt{n}}(X^{(0)}, S^{(0)}) \leq -k\delta;$$

on the other hand we have

$$\begin{aligned}\phi_{\sqrt{n}}(X^{(k)}, S^{(k)}) - \phi_{\sqrt{n}}(X^{(0)}, S^{(0)}) &= \sqrt{n}\ln\left(\frac{\langle X^{(k)}, S^{(k)}\rangle}{\langle X^{(0)}, S^{(0)}\rangle}\right) + \underbrace{\psi(X^{(k)}, S^{(k)})}_{\geq 0}\\ &\quad - \underbrace{\psi(X^{(0)}, S^{(0)})}_{\leq \sqrt{n}\ln(1/\epsilon)}\\ &\geq \sqrt{n}\ln\left(\frac{\langle X^{(k)}, S^{(k)}\rangle}{\langle X^{(0)}, S^{(0)}\rangle}\right) - \sqrt{n}\ln(1/\epsilon).\end{aligned}$$

Therefore,

$$-k\delta \geq \sqrt{n}\ln\left(\frac{\langle X^{(k)}, S^{(k)}\rangle}{\langle X^{(0)}, S^{(0)}\rangle}\right) - \sqrt{n}\ln(1/\epsilon).$$

As a result, for every $k \geq \frac{2\sqrt{n}}{\delta}\ln(1/\epsilon)$, we have $\langle X^{(k)}, S^{(k)}\rangle \leq \epsilon\langle X^{(0)}, S^{(0)}\rangle$ as desired. □

## 4.3 Algorithm and Computational Complexity Analysis

Based on the last theorem, our problem of producing an approximately optimal solution pair is reduced to decreasing the potential function value by a constant, in every iteration. We want to update $(X^{(k)}, S^{(k)})$ to $(X^{(k+1)}, S^{(k+1)})$ such that $\mathcal{A}(X) = b$, $\mathcal{A}^*(y) + S = C$, $X \succ 0$, $S \succ 0$ are all maintained; moreover, $\langle X, S\rangle$ is decreased and $\psi(X, S)$ is not increased a lot (in comparison to the duality gap). We can express the update from $(X^{(k)}, S^{(k)})$ to $(X^{(k+1)}, S^{(k+1)})$ by a pair of *search directions* $D_X$, $D_S \in \Sigma^n$ respectively and a step size $\alpha \in \mathbb{R}_{++}$. We drop the iterate numbers for this part of our study and define

$$X(\alpha) := X + \alpha D_X, \text{ and } S(\alpha) := S + \alpha D_S.$$

To maintain the feasibility of the iterates, the search directions must satisfy

$$\mathcal{A}(D_X) = 0 \text{ and } \mathcal{A}^*(d_y) + D_S = 0,$$

for some $d_y \in \mathbb{R}^m$; i.e., $D_X$ must be in the null space of $\mathcal{A}(\cdot)$ and $D_S$ must be in the range of $\mathcal{A}^*(\cdot)$.

The dual problem $(D)$ can be easily put into the form of the primal problem $(P)$. Then an algorithm that only accepts input in the form of $(P)$ can be fed the input for $(D)$. We will derive an algorithm that is symmetric between the primal and the dual (*primal-dual symmetry*) in the sense that if the algorithm generates the primal-dual sequence $\left(X^{(k)}, S^{(k)}\right)$ for a specific input $(P)$, it will generate the sequence $\left(S^{(k)}, X^{(k)}\right)$ for the input dual to the same $(P)$ (in the form of our primal problem).

Notice that if we take the current iterate $\left(X^{(k)}, S^{(k)}\right)$ and apply an automorphism $T \in \mathrm{Aut}(\Sigma^n_+)$ to transform the pair to $\left(T(X^{(k)}), T^{-*}(S^{(k)})\right)$ we could redefine the data $\mathcal{A}, C$ but we would not change the cone constraints. We would also like to have our algorithm invariant under such mappings. In general, we want the algorithm to be invariant under the symmetries of the cone constraints. We call this property *scale-invariance*.

We will describe an approach which attains these two goals of primal-dual symmetry and scale-invariance. For every $(X, S) \in \Sigma^n_{++} \oplus \Sigma^n_{++}$ we would like to have a self-adjoint, positive definite linear transformation $T : \Sigma^n \to \Sigma^n$ such that

- $T \in \mathrm{Aut}(\Sigma^n_+)$,
- $T(S) = T^{-1}(X) =: V$,
- $T(X^{-1}) = T^{-1}(S^{-1}) = V^{-1}$.

If we can find such transformation $T$, then we can map our primal-space with the mapping $T^{-1}$ and the dual-space with $T$. This modification does not change anything significantly, except that our current primal-dual iterate is mapped onto $(V, V)$. Let's elaborate:

Since $T$ (and therefore $T^{-1}$; because, $T$ is self-adjoint) is an automorphism of the cone of positive semidefinite matrices, the most important part of the problem (for the current interior-point method approach) is unchanged. We define

$$\begin{aligned}\bar{\mathcal{A}}(\cdot) &:= \mathcal{A}(T(\cdot))\\ \bar{C} &:= T(C)\\ \bar{D}_X &:= T^{-1}(D_X),\\ \bar{D}_S &:= T(D_S).\end{aligned}$$

Now, $(P)$ and $(D)$ become

$$\begin{array}{llrcl}(\bar{P}) & \inf & \langle \bar{C}, X\rangle & & \\ & \text{subject to:} & \bar{\mathcal{A}}(X) & = & b,\\ & & X & \in & T^{-1}(\Sigma^n_+) = \Sigma^n_+,\end{array}$$

$$\begin{array}{llrcl}(\bar{D}) & \sup & b^T y & & \\ & \text{subject to:} & \bar{\mathcal{A}}^*(y) + S & = & \bar{C},\\ & & S & \in & T(\Sigma^n_+) = \Sigma^n_+.\end{array}$$

In these *scaled spaces*, the search directions are still orthogonal: $\bar{D}_X$ must lie in the null space of $\bar{\mathcal{A}}(\cdot)$ and $\bar{D}_S$ must lie in the range of $\bar{\mathcal{A}}^*(\cdot)$.

Let's analyze the duality gap. We have

$$\begin{aligned}\langle X(\alpha), S(\alpha)\rangle &= \langle X, S\rangle + \alpha\left[\langle X, D_S\rangle + \langle S, D_X\rangle\right] + \alpha^2\langle D_X, D_S\rangle\\ &= \langle X, S\rangle + \alpha\left[\langle T^{-1}(X), T(D_S)\rangle + \langle T(S), T^{-1}(D_X)\rangle\right]\\ &= \langle X, S\rangle + \alpha\left[\langle V, \bar{D}_S\rangle + \langle V, \bar{D}_X\rangle\right]\\ &= \langle X, S\rangle + \alpha\langle V, (\bar{D}_S + \bar{D}_X)\rangle.\end{aligned}$$

Therefore, if we take as $\bar{D}_X$ and $\bar{D}_S$ the orthogonal projection of $-V$ onto the null space of $\bar{\mathcal{A}}(\cdot)$ and range of $\bar{\mathcal{A}}^*(\cdot)$ respectively, then we will have the best search direction to reduce the duality gap in this setting.

Now, let's turn to the *centrality part* of the potential function. What kind of search direction would improve the barrier function values in this setting? We utilize the following technical lemma which summarizes many of the nice properties of the barrier function $f$.

**Lemma 4.6** *Let $X \in \Sigma^n_{++}$. Suppose $D \in \Sigma^n$ satisfies*

$$\|D\|_X := \langle D, X^{-1}DX^{-1}\rangle^{1/2} \le 1.$$

*Then*

$$f(X) + \langle f'(X), D\rangle \le f(X + D) \le f(X) + \langle f'(X), D\rangle + \frac{\|D\|_X^2}{2(1 - \|D\|_X)^2}.$$

**Proof** The left hand side inequality simply follows from the convexity of $f$. The right hand side inequality can be established by utilizing Taylor's Theorem, together with the properties that

$$\begin{aligned}\langle f''(X)H, H\rangle &= \operatorname{Tr}(X^{-1}HX^{-1}H),\\ f'''(X)[H, H, H] &= -2\operatorname{Tr}\left[(X^{-1/2}HX^{-1/2})^3\right].\end{aligned}$$

□

**Remark 4.7** The condition $\|D_X\|_X \leq 1$ ($< 1$) implies $(X + D) \succeq 0$ ($\succ 0$). This is clear from the statement of the lemma. But it can also be directly observed as follows:

$$1 \geq \|D_X\|_X = \|X^{-1/2} D X^{-1/2}\|_F.$$

Therefore, $\|X^{-1/2} D X^{-1/2}\|_2 \leq 1$. But this is equivalent to $I \pm X^{-1/2} D X^{-1/2} \succeq 0$. If we apply the automorphism $X^{1/2} \cdot X^{1/2}$ of $\Sigma_+^n$ to both sides, we obtain equivalently $X \pm D \succeq 0$.

Let's focus on the first order upper estimate on $f(X(\alpha)) + f(S(\alpha))$ given by the above lemma. We obtain

$$\begin{aligned}
\left[\langle f'(X), D_X\rangle + \langle f'(S), D_S\rangle\right] &= \left[\langle -X^{-1}, D_X\rangle + \langle -S^{-1}, D_S\rangle\right] \\
&= -\left[\langle T(X^{-1}), T^{-1}(D_X)\rangle + \langle T^{-1}(S^{-1}), T(D_S)\rangle\right] \\
&= -\left[\langle V^{-1}, \bar{D}_X\rangle + \langle V^{-1}, \bar{D}_S\rangle\right] \\
&= -\langle V^{-1}, \bar{D}_X + \bar{D}_S\rangle.
\end{aligned}$$

Thus, as in our analysis of the duality gap, if we take as $\bar{D}_X$ and $\bar{D}_S$ the orthogonal projection of $V^{-1}$ onto the null space of $\bar{\mathcal{A}}(\cdot)$ and range of $\bar{\mathcal{A}}^*(\cdot)$ respectively, then we will have the best search direction to reduce the first order term in the upper bound on the value of the barrier terms $f(X) + f(S)$, in this setting.

Therefore, to reduce the value of the potential function, it seems desirable to choose a matrix which is a nonnegative linear combination of $-V$ and $V^{-1}$ and then define $\bar{D}_X$ and $\bar{D}_S$ as the orthogonal projections of this matrix onto the null space of $\bar{\mathcal{A}}(\cdot)$ and the range of $\bar{\mathcal{A}}^*(\cdot)$ respectively. This is precisely what we do next. Let

$$\tilde{U} := -\frac{(n + \sqrt{n})}{\langle X, S\rangle} V + V^{-1}.$$

Note that $\|\tilde{U}\|_F = 0$ iff $V^{-1} = \frac{(n+\sqrt{n})}{\langle X,S\rangle} V$. But the latter leads to a contradiction (that $n = n + \sqrt{n}$) upon taking the inner product with $V$ of both sides. Therefore, $\|\tilde{U}\|_F > 0$ and we define

$$U := \frac{\tilde{U}}{\|\tilde{U}\|_F}.$$

In fact, $\|\tilde{U}\|_F$ is connected to a measure of centrality. Let

$$\tilde{\mu} := \frac{\langle X^{-1}, S^{-1}\rangle}{n} = \frac{\langle V^{-1}, V^{-1}\rangle}{n}.$$

Recall that

$$\mu = \frac{\langle X, S\rangle}{n} = \frac{\langle V, V\rangle}{n}.$$

Then

$$\begin{aligned}
\|\tilde{U}\|_F^2 &= \frac{(n + \sqrt{n})^2}{(n\mu)^2} n\mu - 2\frac{(n + \sqrt{n})}{\mu} + n\tilde{\mu} \\
&= \frac{1}{\mu}\left[n(\mu\tilde{\mu} - 1) + 1\right].
\end{aligned}$$

We have

**Theorem 4.8** *Let $X, S \in \Sigma^n_{++}$. Then*

$$\mu\tilde{\mu} \geq 1.$$

*Equality holds above iff $S = \mu X^{-1}$.*

**Proof** Note that

$$\begin{aligned} \frac{\langle T(S-\mu X^{-1}), T(S-\mu X^{-1})\rangle}{n\mu} &= \frac{1}{n\mu}\langle V-\mu V^{-1}, V-\mu V^{-1}\rangle \\ &= \frac{1}{n\mu}\left(n\mu - 2n\mu + n\mu^2\tilde{\mu}\right) \\ &= \mu\tilde{\mu} - 1. \end{aligned}$$

Clearly the left hand side is always nonnegative. Moreover, it is equal to zero iff $T(S-\mu X^{-1}) = 0$. Since $T$ is positive definite, the latter is true iff $S = \mu X^{-1}$. □

**Remark 4.9** This theorem generalizes the *arithmetic-harmonic mean inequality*. As before, let $\lambda_1, \lambda_2, \ldots, \lambda_n$ denote the eigenvalues of $X^{1/2}SX^{1/2}$. Then

$$\mu\tilde{\mu} = \left(\frac{1}{n}\sum_{j=1}^{n}\lambda_j\right)\left(\frac{1}{n}\sum_{j=1}^{n}\frac{1}{\lambda_j}\right) = \frac{\text{arithmetic mean}(\lambda_1, \lambda_2, \ldots, \lambda_n)}{\text{harmonic mean}(\lambda_1, \lambda_2, \ldots, \lambda_n)}.$$

Therefore, $\mu\tilde{\mu} \geq 1$ and equality holds iff $\lambda_1 = \lambda_2 = \cdots = \lambda_n = \mu$. Also note that $(\mu\tilde{\mu} - 1)$ is $\frac{1}{n\mu}$ times the squared norm of the error in the equation $S = \mu X^{-1}$, where the norm is with respect to the local metric induced by $T^2$. Indeed, since $T = T^*$, the first derivation in the above proof says

$$\frac{1}{n\mu}\langle (S-\mu X^{-1}), T^2(S-\mu X^{-1})\rangle = \mu\tilde{\mu} - 1.$$

**Corollary 4.10** *For every $X, S \in \Sigma^n_{++}$, we have*

$$\|\tilde{U}\|_F^2 \geq \frac{1}{\mu}(> 0).$$

*The equality holds above iff $S = \mu X^{-1}$.*

Then $\bar{D}_X$, $d_y$, $\bar{D}_S$ make up the unique solution of the system:

$$\begin{aligned} \bar{\mathcal{A}}(\bar{D}_X) &= 0 \\ \bar{\mathcal{A}}^*(d_y) + \bar{D}_S &= 0 \\ \bar{D}_X + \bar{D}_S &= U. \end{aligned}$$

By definition,

$$\|\bar{D}_X\|_F^2 + \|\bar{D}_S\|_F^2 = \|U\|_F^2 = 1.$$

Therefore, we immediately conclude that

$$\|\bar{D}_X\|_F \leq 1 \text{ and } \|\bar{D}_S\|_F \leq 1.$$

$T$ can be chosen as $W \cdot W$, where

$$W^2 := S^{-1/2}\left(S^{1/2}XS^{1/2}\right)^{1/2}S^{-1/2}.$$

Then we trivially have

$$T^{-1}(X) = W^{-1}XW^{-1} = WSW = T(S)$$

as desired. Now, we analyze $\|D_X\|_X$ and $\|D_S\|_S$. We have

$$\begin{aligned}
\|D_X\|_X^2 &= \langle D_X, X^{-1}D_XX^{-1}\rangle \\
&= \langle W\bar{D}_XW, X^{-1}W\bar{D}_XWX^{-1}\rangle \\
&= \langle \bar{D}_X, (WX^{-1}W)\bar{D}_X(WX^{-1}W)\rangle \\
&\leq \left\|V^{-1}\cdot V^{-1}\right\|_2 \|\bar{D}_X\|_F^2 \\
&\leq \left\|V^{-1}\cdot V^{-1}\right\|_2 \\
&\leq \frac{1}{(\lambda_n(V))^2}. \qquad (4.1)
\end{aligned}$$

Note that in the above derivation, we encountered the linear transformation

$$T\left[f''(X)\,(T(\cdot))\right]$$

which happens to coincide (in this case) with the linear transformation

$$\left(T(f'(X))\cdot T(f'(X))\right).$$

Similarly,

$$\begin{aligned}
\|D_S\|_S^2 &= \langle D_S, S^{-1}D_SS^{-1}\rangle \\
&= \langle W^{-1}\bar{D}_SW^{-1}, S^{-1}W^{-1}\bar{D}_SW^{-1}S^{-1}\rangle \\
&= \langle \bar{D}_S, (W^{-1}S^{-1}W^{-1})\bar{D}_S(W^{-1}S^{-1}W^{-1})\rangle \\
&\leq \left\|V^{-1}\cdot V^{-1}\right\|_2 \|\bar{D}_S\|_F^2 \\
&\leq \left\|V^{-1}\cdot V^{-1}\right\|_2 \\
&\leq \frac{1}{(\lambda_n(V))^2}. \qquad (4.2)
\end{aligned}$$

Note that in the above analysis, we can directly compute the norm of the operator $V^{-1}\cdot V^{-1}$. Alternatively, we can express the operator using the symmetric Kronecker Product $V^{-1}\overset{s}{\otimes}V^{-1}$ or the usual Kronecker Product $V^{-1}\otimes V^{-1}$ and use the corresponding results on the eigenvalues of the Kronecker products.

To finish the analysis, in addition to the above results, we use the next lemma.

**Lemma 4.11** *Let $\tilde{U}$ and $V$ be defined as above (from a pair of feasible, interior points $(X,S)$). Then*

$$\|\tilde{U}\|_F \geq \frac{\sqrt{3}}{2\lambda_n(V)}.$$

**Proof** Recall, $\tilde{U} = \left(V^{-1} - \frac{1}{\mu}V\right) - \frac{1}{\sqrt{n\mu}}V$. Since $\langle V, V^{-1} - \frac{1}{\mu}V\rangle = 0$, we have

$$\begin{aligned}
\|\tilde{U}\|_F^2 &= \left\|V^{-1} - \frac{1}{\mu}V\right\|_F^2 + \frac{1}{n\mu^2}\|V\|_F^2 \\
&\geq \left(\frac{1}{\lambda_n(V)} - \frac{\lambda_n(V)}{\mu}\right)^2 + \frac{1}{\mu} \\
&= \frac{1}{\lambda_n^2(V)}\left[\left(\frac{\lambda_n^2(V)}{\mu} - \frac{1}{2}\right)^2 + \frac{3}{4}\right].
\end{aligned}$$

Therefore, we have the desired inequality. □

We have,

$$\begin{aligned}
\phi_{\sqrt{n}}(X(\alpha),S(\alpha)) - \phi_{\sqrt{n}}(X,S) &\leq (n+\sqrt{n})\ln\left(\frac{\langle X(\alpha),S(\alpha)\rangle}{\langle X,S\rangle}\right) - \alpha\langle U,V^{-1}\rangle \\
&\quad +\alpha^2\left(\frac{\|D_X\|_X^2}{2(1-\alpha\|D_X\|_X)^2}+\frac{\|D_S\|_S^2}{2(1-\alpha\|D_S\|_S)^2}\right) \\
&\leq \alpha\frac{(n+\sqrt{n})}{\langle X,S\rangle}\langle U,V\rangle - \alpha\langle U,V^{-1}\rangle \\
&\quad +2\alpha^2\left(\frac{(1/\lambda_n(V))^2}{2(1-\alpha(1/\lambda_n(V)))^2}\right) \\
&= -\alpha\|\tilde{U}\|_F + \alpha^2\left(\frac{(1/\lambda_n(V))^2}{(1-\alpha(1/\lambda_n(V)))^2}\right) \\
&\leq -\frac{\sqrt{3}}{2\lambda_n(V)}\alpha + \alpha^2\left(\frac{(1/\lambda_n(V))^2}{(1-\alpha(1/\lambda_n(V)))^2}\right). \quad (4.3)
\end{aligned}$$

For the first inequality above, we used Lemma 4.6. For the second inequality, we used

$$\langle X(\alpha),S(\alpha)\rangle = \langle X,S\rangle + \alpha\langle U,V\rangle,$$

Lemma 4.6 and the relations (4.1), (4.2). For the third and last inequality, we used Lemma 4.11. Now, we will choose a value for $\alpha$. Since we are looking for a minimizer, any value given by any choice of $\alpha$ will be an upper bound on the reduction of the potential value. We choose to set $\alpha$ to $\frac{\lambda_n(V)}{8}$. We obtain

$$\begin{aligned}
\phi_{\sqrt{n}}(X(\alpha),S(\alpha)) - \phi_{\sqrt{n}}(X,S) &\leq -\frac{\sqrt{3}}{16}+\frac{1}{64}\frac{1}{(7/8)^2} && (4.4)\\
&= -\frac{\sqrt{3}}{16}+\frac{1}{49} && (4.5)\\
&< -\frac{1}{12}. && (4.6)
\end{aligned}$$

**Remark 4.12** Combining Lemma 4.11 and Corollary 4.10 we proved

$$\|\tilde{U}\|_F \geq \max\left\{\frac{\sqrt{3}}{2\lambda_n(V)}, \frac{1}{\sqrt{\mu}}\right\}.$$

The above analysis relied only on the first lower bound (which is independent of $\mu$). However, we know that eventually $\mu \to 0^+$. So, as we get closer to the optimal solution set, we can possibly have much better performance if the second term $\frac{1}{\sqrt{\mu}}$ dominates in the above maximum. Indeed, under some mild conditions variants of this algorithm and many other interior-point algorithms have been proven to achieve asymptotic superlinear and quadratic convergence rates for LP as well as for SDP (even though the results require more significant changes in the algorithms and nontrivial assumptions on the data, in the case of SDP).

**ALGORITHM**

Given $(X^{(0)}, S^{(0)})$ feasible in $(P)$ and $(D)$ such that $X^{(0)} \succ 0$, $S^{(0)} \succ 0$. Also given is $\epsilon \in (0,1)$ such that $\psi(X^{(0)}, S^{(0)}) \leq \sqrt{n}\ln(1/\epsilon)$.

$k := 0$.

While $\langle X^{(k)}, S^{(k)}\rangle > \epsilon\langle X^{(0)}, S^{(0)}\rangle$,

$W^2 := \left(S^{(k)}\right)^{-1/2}\left[\left(S^{(k)}\right)^{1/2}X^{(k)}\left(S^{(k)}\right)^{1/2}\right]^{1/2}\left(S^{(k)}\right)^{-1/2}$
$\bar{\mathcal{A}}(\cdot) := \mathcal{A}(W\cdot W)\ \left[\bar{A}_i := WA_iW\right]$
$V := WS^{(k)}W$
$\tilde{U} := -\frac{n+\sqrt{n}}{\langle X^{(k)},S^{(k)}\rangle}V + V^{-1}$
$U := \frac{\tilde{U}}{\|\tilde{U}\|_F}$
Solve the system

$$\begin{aligned}\bar{\mathcal{A}}(\bar{D}_X) &= 0\\ \bar{\mathcal{A}}^*(d_y)+\bar{D}_S &= 0\\ \bar{D}_X+\bar{D}_S &= U.\end{aligned}$$

Compute
$\bar{\alpha} := \operatorname{argmin}\left\{\phi_{\sqrt{n}}(X^{(k)}+\alpha W\bar{D}_XW, S^{(k)}+\alpha W^{-1}\bar{D}_SW^{-1}) : \alpha > 0\right\}.$
Let $X^{(k+1)} := X^{(k)}+\bar{\alpha}W\bar{D}_XW$, $S^{(k+1)} := S^{(k)}+\bar{\alpha}W^{-1}\bar{D}_SW^{-1}$.
$k := k+1.$

end{While}

The above is what is typically called a *potential-reduction algorithm*. Even though we wrote explicit formulas for $W$, $V$, $V^{-1}$ etc., in an implementation one need not (and usually *should not!*) explicitly form these.

We have all the ingredients to prove the following theorem.

**Theorem 4.13** *The above algorithm terminates in at most*

$$24\sqrt{n}\ln(1/\epsilon)$$

*iterations with feasible* $X^{(k)}, S^{(k)}$ *such that*

$$\langle X^{(k)},S^{(k)}\rangle \le \epsilon\langle X^{(0)},S^{(0)}\rangle.$$

**Proof** First note that if $X^{(k)}\succ 0$, $S^{(k)}\succ 0$ then $X^{(k+1)}\succ 0$ and $S^{(k+1)}\succ 0$ unless the algorithm stops with an *exact* optimal solution pair $X^{(k+1)}$, $S^{(k+1)}$. (By the definition of the potential function and $\bar{\alpha}$, we have $\phi_{\sqrt{n}}\left(X^{(k)}+\bar{\alpha}W\bar{D}_XW, S^{(k)}+\bar{\alpha}W^{-1}\bar{D}_SW^{-1}\right) = +\infty$ if $X^{(k+1)}$ or $S^{(k+1)}$ has a zero eigenvalue and $\langle X^{(k+1)},S^{(k+1)}\rangle > 0$.) Next, we establish by induction on $k$ that all the iterates $\left(X^{(k)},S^{(k)}\right)$ are feasible in their respective problems. The claim is clearly true for $k=0$; we will prove it for the $(k+1)$st iterate. Since we are in the $(k+1)$st iteration (the algorithm did not stop at the end of the $k$th iteration), by the above, $X^{(k)}\succ 0$, $S^{(k)}\succ 0$. So, all the steps of the algorithm are well-defined for the $(k+1)$st iteration. Since $\mathcal{A}\left(X^{(k)}\right)=b$, $\mathcal{A}^*\left(y^{(k)}\right)+S^{(k)}=C$, for some $y^{(k)}\in\mathbb{R}^m$ by the induction hypothesis and

$$\bar{D}_X,\bar{D}_S\in\Sigma^n, \mathcal{A}\left(W\bar{D}_XW\right)=0, \mathcal{A}^*\left(d_y\right)+W^{-1}\bar{D}_SW^{-1}=0,\ \text{for some } d_y\in\mathbb{R}^m,$$

we have that $X^{(k+1)},S^{(k+1)}\in\Sigma^n$, $\mathcal{A}\left(X^{(k+1)}\right)=b$, and that $\mathcal{A}^*\left(y^{(k+1)}\right)+S^{(k+1)}=C$ for some $y^{(k+1)}\in\mathbb{R}^m$. The closure of the domain of $\phi$ is $\Sigma^n_+\oplus\Sigma^n_+$ and by the definition of $\bar{\alpha}$ we have $X^{(k+1)}\succeq 0$, $S^{(k+1)}\succeq 0$. Therefore, $\left(X^{(k+1)},S^{(k+1)}\right)$ is a feasible pair as desired. (Note again that $X^{(k+1)}\succ 0$ and $S^{(k+1)}\succ 0$ unless $\bar{\alpha}=\sqrt{n}\langle X^{(k)},S^{(k)}\rangle\|\tilde{U}\|_F$ in which case $\left(X^{(k+1)},S^{(k+1)}\right)$ is an optimal pair.) Now, we use the fact that by (4.3) and (4.4)-(4.6), there exists a step size along the chosen search direction which reduces the potential function value by at least $\frac{1}{12}$ in every iteration. Since the algorithm minimizes the potential function value along this direction, the reduction achieved by the algorithm is at least $\delta := \frac{1}{12}$ in every iteration. We apply Theorem 4.5 with this value of $\delta$ and conclude from the proof of Theorem 4.5 that $\frac{2\sqrt{n}}{\delta} \ln\left(\frac{1}{\epsilon}\right)$ iterations suffice as claimed. □

Even though the algorithm requires the iterates to lie in the interior of the underlying cone constraints, we can relax the initial feasibility assumption by using auxiliary optimization problems.

Clearly, $\mathcal{A}(I) + (b - \mathcal{A}(I)) = b$. So, the auxiliary SDP problem

$$\begin{array}{llrcl} (P_{\text{aux}}) & \inf & \xi & & \\ & \text{subject to:} & \mathcal{A}(X) + \xi(b - \mathcal{A}(I)) & = & b, \\ & & \langle I, X \rangle & \leq & M, \\ & & \xi & \geq & 0, \\ & & X & \succeq & 0, \end{array}$$

(where we added the artificial variable $\xi$ and an artificial constraint $\text{Tr}(X) \leq M$, for some large $M$) has the Slater point $\bar{X} := I$, $\bar{\xi} := 1$. Also, because of the artificial constraint, we have that the feasible region of $(P_{\text{aux}})$ is bounded.

We can apply the ellipsoid method or an interior-point method to $(P_{\text{aux}})$. If we find a feasible solution with $\xi = 0$, then we have a feasible solution of the original problem $(P)$. To guarantee the generation of a Slater point (if one exists) with the ellipsoid method, we would need to use a different formulation (perturbing the positive semidefiniteness constraint by a "small but controlled amount"). If we use an interior-point method and if $(P)$ has a Slater point then we would expect to compute a Slater point for $(P)$ from an optimal solution of $(P_{\text{aux}})$ with $\xi = 0$. However, if the optimum value of $(P_{\text{aux}})$ is positive, then all we can say about $(P)$ is that "$(P)$ does not have any feasible solution satisfying $\text{Tr}(X) \leq M$." While this is just fine for linear programming problems with integer or rational data, for SDP it does not quite work that way.

Let's elaborate... Recall the notation of the first chapter. Given data $(A, b, c)$ with rational entries, we can assume without loss of generality that all entries are integer. Recall the size of the data $(A, b, c)$, denoted by $L$.

**Proposition 4.14** (a) *Let $u \in \mathbb{Z}^n$. Then*

$$\|u\|_1 \leq 2^{\text{size}(u) - \dim(u)}.$$

(b) *Let $A \in \mathbb{Z}^{m \times n}$. Define*

$$\text{maxdet}(A) := \max\{|\det(B)| : B \textit{ is a square submatrix of } A\}.$$

*Then*

$$\text{maxdet(A)} \leq 2^{\text{size}(A) - \dim(A)}.$$

(c) *Let $n \geq m \geq 2$. If $x$ is an extreme point of $\{x \in \mathbb{R}^n : Ax = b, x \geq 0\}$, then*

$$\|x\|_\infty \leq 2^{L - mn}, \quad \|x\|_1 \leq 2^{L-n}.$$

*Moreover, if $\bar{x}$ and $\hat{x}$ are extreme points of $\{x \in \mathbb{R}^n : Ax = b, x \geq 0\}$, then either*

$$c^T\bar{x} = c^T\hat{x},$$

*or*

$$|c^T\bar{x} - c^T\hat{x}| > 2^{-2L}.$$

**Proof** (a) Let $u \in \mathbb{Z}^n$. Then,

$$\begin{aligned}\|u\|_1 &= \sum_{j=1}^{n} |u_j| \\ &\leq \sum_{j:|u_j|\geq 1} (|u_j|+1) \\ &\leq \prod_{j:|u_j|\geq 1} (|u_j|+1) \\ &= \prod_{j=1}^{n} (|u_j|+1) \\ &\leq 2^{\text{size}(u)-\dim(u)}.\end{aligned}$$

(b) By induction on $n$. If $n = 1$, then

$$\text{maxdet}(A) = \|A\|_\infty \leq \|A\|_1 \leq 2^{\text{size}(A)-\dim(A)}.$$

Now, suppose the inequality holds for all matrices with at most $(n-1)$ columns. Let $A \in \mathbb{Z}^{m\times n}$. Write

$$A =: [A_0, A_1],$$

where $A_0$ is the first column of $A$. Also let $A_B$ be any square submatrix of $A$. If the column $A_0$ is not in $A_B$ then we have by the induction hypothesis,

$$|\det(A_B)| \leq 2^{\text{size}(A_1)-\dim(A_1)} \leq 2^{\text{size}(A)-\dim(A)}.$$

Otherwise, $A_0$ is in $A_B$. We expand the $\det(A_B)$ along the column $A_0$, denote by $\bar{A}$ a square submatrix of $A$:

$$\begin{aligned}|\det(A_B)| &\leq \sum_i |A_{i0} \det(\bar{A})| \\ &\leq \|A_0\|_1 \text{maxdet}(A_1) \\ &\leq 2^{\text{size}(A_0)-\dim(A_0)+\text{size}(A_1)-\dim(A_1)} \\ &= 2^{\text{size}(A)-\dim(A)}.\end{aligned}$$

(c) Without loss of generality, we may assume $\text{rank}(A) = m$. Let $x$ be a basic feasible solution of $\{x \in \mathbb{R}^n : Ax = b, x \geq 0\}$. Then there exists a basis matrix $A_B$ among the columns of $A$ such that

$$x_N = 0, x_B = A_B^{-1} b \geq 0.$$

Write

$$x_j = \frac{p_j}{q},$$

where $q = \det(A_B)$ and $p_j$ is a subdeterminant of $[A_B, b]$. From part (b) we have $|p_j| \leq 2^{L-mn}$ and

$$1 \leq |q| \leq 2^{L-mn}.$$

Thus,

$$\|x\|_\infty \leq 2^{L-mn}, \quad \|x\|_1 \leq n\|x\|_\infty \leq 2^{L-mn+n} \leq 2^{L-n}.$$

Moreover, if $\bar{x}$ and $\hat{x}$ are extreme points of $\{x \in \mathbb{R}^n : Ax = b, x \geq 0\}$ such that

$$c^T\bar{x} \neq c^T\hat{x}.$$

So, $c^T\bar{x} = \frac{p_1}{q_1}$ and $c^T\hat{x} = \frac{p_2}{q_2}$ for some integers $p_1, p_2, q_1, q_2$ such that $|q_1 q_2| < 2^{2L}$ and $|p_1 q_2 - p_2 q_1| \geq 1$; whence,

$$|c^T\bar{x} - c^T\hat{x}| = \left|\frac{p_1 q_2 - p_2 q_1}{q_1 q_2}\right| > 2^{-2L}.$$

□

We will denote by $v(LP)$ the optimal value of $(LP)$.

**Corollary 4.15** *If we have a feasible $x$ such that $c^T x - v(LP) \leq 2^{-2L}$, then we can find an exact optimal solution in $O(n^3)$ arithmetic operations.*

**Proof** By Proposition 4.14 (c), any extreme point $\hat{x}$ with $c^T\hat{x} \leq c^T x$ must be optimal. An algorithm to find such $\hat{x}$ follows.

Repeat the following:

- $B := \{j : x_j > 0\}$, $N := \{1, 2, \ldots, n\} \backslash B$.
- Find a linear dependence among the columns $\{A_j : j \in B\}$.
- If none, then STOP, $x$ is an extreme point solution and therefore is optimal.
- Otherwise, we have $d \in \mathbb{R}^n$ such that $Ad = 0$, $d_N = 0$, $d_B \neq 0$.
- Choose the sign of $d$ such that $c^T d < 0$ (or $c^T d = 0$ and $d_B$ has a negative component).
- Let $\alpha := \max\{\alpha > 0 : x + \alpha d \geq 0\}$.
- Set $x := x + \alpha d$.

□

Note that in the above algorithm, the number of arithmetic operations is $O(n^2.$(the number of positive components of original $x$)).

The algorithm in the proof of the last corollary can be easily generalized to much more abstract settings, including SDP. Consider the following algorithm that takes $\bar{X} \in \Sigma^n_+$ such that $\mathcal{A}(\bar{X}) = b$ and $C \in \Sigma^n$ as the input. Let $\hat{X} := \bar{X}$. Repeat the following:

- Let $G \subseteq \Sigma^n_+$ be the smallest face of $\Sigma^n_+$ containing $\hat{X}$. By Theorem 2.25, there exists a subspace $L \subseteq \mathbb{R}^n$ such that
$$G = \left\{X \in \Sigma^n_+ : \text{Null}(X) \supseteq L\right\}.$$
Find such $L$ (e.g., compute a basis for $L$ using $\hat{X}$).
- Find $D \in \Sigma^n \setminus \{0\}$: $\text{Null}(D) \supseteq L$ and $\mathcal{A}(D) = 0$.
- If none exists, then STOP, $\hat{X}$ is an extreme point solution.
- Choose the sign of $D$ such that $\langle C, D\rangle < 0$ (or, $\langle C, D\rangle = 0$ and $D$ has a negative eigenvalue).
- Let $\alpha := \sup\{\alpha > 0 : \hat{X} + \alpha D \succeq 0\}$. (If $\alpha = +\infty$ then the SDP is unbounded; the proof is provided by the feasible ray $\left\{\hat{X} + \alpha D : \alpha \geq 0\right\}$.)
- Set $\hat{X} := \hat{X} + \alpha D$.

Note that the subspace $L$ can be found using $\hat{X}$, and the "direction" $D$ can be found using $L$ and $\mathcal{A}$, based on simple linear algebraic operations (such as Gaussian Elimination). In particular, during the first iteration of the above algorithm, $D$ will be a rational function of $\bar{X}$ and $\mathcal{A}$. In contrast, the line search determining the value of $\alpha$ is of different nature. For instance, $\alpha$ may be irrational. So it is not clear how to implement the above algorithm *exactly* in the Turing Machine model; hence, the real number machine model is more suitable here.

The above algorithm guarantees that if the problem is not unbounded, it will always return an extreme point $\hat{X}$ such that

$$\langle C, \hat{X} \rangle \le \langle C, \bar{X} \rangle.$$

However, no matter how close $\langle C, \bar{X} \rangle$ is to the optimal value of $(P)$, in general, there can be infinitely many extreme points of the feasible region of $(P)$ such that the objective value of every one of these extreme points is strictly better than $\langle C, \bar{X} \rangle$ and strictly worse than the optimal value.

Some other difficulties which arise in SDP are as follows. Suppose the data of $(P)$ are given as $C, A_1, A_2, \ldots, A_m \in \Sigma^n \cap \mathbb{Z}^{n \times n}$ and $b \in \mathbb{Z}^m$. Then we can define $L := \text{size}(\mathcal{A}, b, C)$. The following are possible:

- $(P)$ has optimal solutions, but all feasible solutions of $(P)$ have entries like $2^{2^{\Omega(L)}}$.
- $(P)$ has a unique feasible solution which happens to be irrational (or unique optimal solution which is also irrational).
- The interior of the feasible region of $(P)$ is nonempty, but the largest ball contained in it has radius $2^{-2^{\Omega(L)}}$.

We can construct families of instances for each of the above three categories by exploiting the "quadratic effect" of positive semidefiniteness constraints. For example, consider the SDP whose feasible region is given by the constraints:

$$y_1 \ge 2, \begin{pmatrix} 1 & y_1 \\ y_1 & y_2 \end{pmatrix} \succeq 0, \quad \begin{pmatrix} 1 & y_2 \\ y_2 & y_3 \end{pmatrix} \succeq 0, \quad \cdots, \quad \begin{pmatrix} 1 & y_{n-1} \\ y_{n-1} & y_n \end{pmatrix} \succeq 0.$$

In every feasible solution $y$ of such an SDP, we have $y_n \ge 2^{2^{n-1}} = 2^{2^{\Omega(n)}}$. Modifying this example, it is possible to construct similarly ill-behaved SDPs.

**Open Problem 2:** Given $A_1, A_2, \ldots, A_m \in \Sigma^n \cap \mathbb{Z}^{n \times n}$ and $b \in \mathbb{Z}^m$, the *SDP-Feasibility Problem* is to decide whether there exists $X \in \Sigma^n_+$ such that $\langle A_i, X \rangle = b_i$, for every $i \in \{1, 2, \ldots, m\}$. *Is SDP-feasibility in $\mathcal{P}$?*

It is conceivable that the answer to this question is "YES." It is even conceivable that methods such as the Ellipsoid Method or the Interior-Point Method can be employed for this purpose. If so, then there is a representation problem. Note that an algorithm that explicitly represents the matrix variable $X$ as itself, is hopeless, since storing such $X$ can take exponential space or it may be impossible in a Turing Machine (if the unique solution $X$ is irrational).

There are elegant ways of dealing with the absence of a pair of feasible, interior solutions to start the primal-dual interior-point algorithms.

## 4.4 Auxiliary Self-Dual Problems

One approach is to make up an auxiliary optimization problem which is self-dual and has obvious starting points such that the solution of such an optimization problem reveals (in the worst case only partial) information about whether the original problem has an optimal solution, it is infeasible or unbounded.

Consider the feasibility problem

$$\begin{aligned} \mathcal{A}(X) - \xi b &= \tilde{b}, \\ \mathcal{A}^*(y) + S - \xi C &= \tilde{C}, \\ \langle C, X\rangle - b^T y + \eta &= \tilde{\gamma} \\ X \succeq 0, S \succeq 0, \xi \geq 0, \eta \geq 0. \end{aligned}$$

In the above formulation, we are allowed to choose $\tilde{b}$, $\tilde{C}$ and $\tilde{\gamma}$. We choose them based on the choices for $X^{(0)}$, $(y^{(0)}, S^{(0)})$. Let's take *any* $X^{(0)} \succ 0$, *any* $y^{(0)} \in \mathbb{R}^m$, *any* $S^{(0)} \succ 0$, *any* $\xi_0 > 0$, and *any* $\eta_0 > 0$. Then we let

$$\tilde{b} := \mathcal{A}(X^{(0)}) - \xi_0 b, \quad \tilde{C} := \mathcal{A}^*(y^{(0)}) + S^{(0)} - \xi_0 C, \quad \tilde{\gamma} := \langle C, X^{(0)}\rangle - b^T y^{(0)} + \eta_0.$$

So we can start the primal-dual interior-point algorithm from the initial point $(X^{(0)}, y^{(0)}, S^{(0)})$. The goal of this algorithm is to compute a solution of the following system

$$\begin{aligned} \mathcal{A}(X) - \xi b &= 0, \\ \mathcal{A}^*(y) + S - \xi C &= 0, \\ \langle C, X\rangle - b^T y + \eta &= 0, \\ \langle X, S\rangle + \xi\eta &= 0, \\ X \succeq 0, S \succeq 0, \xi \geq 0, \eta \geq 0. \end{aligned}$$

If we find a solution satisfying $\xi > 0$, then we easily obtain optimal solutions to the original problems $(P)$ and $(D)$. If we instead find a solution with $\eta > 0$, then we can use such a solution to prove that either $(P)$ or $(D)$ is infeasible. Using sufficiently accurate, approximate solutions of the above system, we can reach to similar (but weaker) conclusions. If the solution we find for the above system has $\xi = \eta = 0$ then it may be difficult to reach a conclusion.

## 4.5 Infeasible-Start Algorithms

Another approach is to work in the framework of the algorithm that we discussed (or some other primal-dual interior-point algorithm) but modify the search directions so that the search directions also try eliminate the error in the linear equations defining the primal and dual feasible regions. Instead of having our search directions $D_X$ and $D_S$ lying in the null-space of $\mathcal{A}(\cdot)$ and range of $\mathcal{A}^*(\cdot)$, we ask that they satisfy the following system of equations:

$$\mathcal{A}(D_X) = b - \mathcal{A}(X^{(k)}) \text{ and } \mathcal{A}^*(d_y) + D_S = C - \mathcal{A}^*(y^{(k)}) - S^{(k)}.$$

The analysis becomes more complicated, however, this is one of the popular ways to solve SDP problems in practice. The algorithms need to carefully monitor the progress in attaining feasibility, reducing $\langle X, S\rangle$ as well as the proximity to the "central surface." (Since we allow infeasible iterates, we will be concerned with the distance to the "central surface" rather than the central path.)

For instance, the algorithm should not allow the fast reduction of $\langle X, S\rangle$ unless the iterates are getting to be near feasible at least as fast.

## 4.6 Other Interior-Point Algorithms, General Remarks

The search directions that we discussed are known as the NT direction (for Nesterov-Todd [276]). Some related algorithms were also proposed by Sturm and S. Zhang [339]. These algorithms have been generalized to convex optimization problems over arbitrary convex cones [346].

Other primal-dual algorithms that are useful and popular rely on the search directions proposed by Helmberg, Rendl, Vanderbei, Wolkowicz [164], Kojima, Shindoh, Hara [208], Monteiro [257] (HKM direction, see [164, 208, 257]) and Alizadeh-Haeberly-Overton (AHO direction [6]). All these directions can be defined and treated in a unified way (due to Y. Zhang [375], some other related work is due to Monteiro and Y. Zhang [261]):

Let $P \in \mathbb{R}^{n\times n}$ be nonsingular. Define $H_P : \mathbb{R}^{n\times n} \to \Sigma^n$ as follows

$$H_P(M) := PMP^{-1} + P^{-T}M^TP^T.$$

This $H_P(\cdot)$ is a *symmetrized similarity transformation.* To compute the search direction, we solve the system

$$\begin{aligned} \mathcal{A}(D_X) &= b - \mathcal{A}(X^{(k)}), \\ \mathcal{A}^*(d_y) + D_S &= C - \mathcal{A}^*(y^{(k)}) - S^{(k)}, \\ H_P\left(X^{(k)}D_S + D_XS^{(k)}\right) &= 2\gamma\mu I - H_P\left(X^{(k)}S^{(k)}\right), \end{aligned}$$

where $\gamma \in [0,1]$ a parameter fixed by the user/algorithm and $\mu = \langle X^{(k)}, S^{(k)}\rangle/n$ as before. Choosing $P := I$ gives the AHO direction, $P := (S^{(k)})^{1/2}$ yields the HKM direction, choosing any $P \in \mathbb{R}^{n\times n}$ such that

$$P^TP = (X^{(k)})^{-1/2}\left((X^{(k)})^{1/2}S^{(k)}(X^{(k)})^{1/2}\right)^{1/2}(X^{(k)})^{-1/2}$$

(for instance, $P := \left((X^{(k)})^{1/2}S^{(k)}(X^{(k)})^{1/2}\right)^{1/4}(X^{(k)})^{-1/2})$ gives the NT direction.

For a unified complexity analysis of algorithms based on such search directions, see Kojima, Shida and Shindoh [207].

## 4.7 Further Bibliographical Notes

There are many books on interior-point methods. See Nesterov-Nemirovski [275], Nesterov [272], Renegar [302], Roos-Terlaky-Vial [307], Vanderbei [356], Wright [369] and Ye [373]. The algorithm and the analysis we presented are based on Nesterov-Todd [276, 277] and Tunçel [348, 346]. For other potential reduction based complexity analyses, see Todd [345], Vandenberghe and Boyd [353]. For a discussion of complexity issues in continuous optimization, see Vavasis [359].

The algorithm given in the proof of Corollary 4.15 is called a *rounding algorithm* (referring to rounding the current solution to an extreme point solution), or a *purification algorithm* (referring to purifying the current solution to an extreme point solution), or a *cross-over algorithm* (referring to crossing over from an interior-point method to a boundary based method —such as the simplex method for LP). For a generalization to SDP, see Pataki [286].

For some small classes of SDP problems, we can write tractable formulas to explicitly compute an optimal solution. See, Vanderbei and Yang [357], Wolkowicz [364], Muramatsu [262].

When we discussed the complexity results for ellipsoid method we assumed that the radii $R$ and $r$ of the balls (the former for a ball containing some optimal solutions and the latter for a ball contained in the set of feasible solutions intersected with the first ball) were given. Also, in our discussion of interior-point algorithms we introduced a large constant $M$; but, we left unsettled the question of how large of a value for $M$ would be satisfactory. It is possible to analyze the algorithms in the case of unknown $R, r$, and to express the computational complexity of the algorithms in terms of $R$ and $r$. Freund and Vera [129] prove that if a convex set $G \subseteq \mathbb{R}^d$ is given by a separation oracle, the problem of computing $\bar{x} \in G$ can be solved in $O\left(d\left(\ln(d) + d\ln\left(\frac{1}{r} + \frac{R}{r}\right) + \ln(R+1)\right)\right)$ iterations where each iteration makes one call to the separation oracle (and all the constants hidden by $O(\cdot)$ are at most 2). For a bit complexity analysis of the SDP feasibility problem, see Porkolab and Khachiyan [294].

CHAPTER 5

# Approximation Algorithms Based on SDP

There are very many interesting $\mathcal{NP}$-hard optimization problems which admit $\rho$-approximation algorithms where $\rho$ is an absolute constant $> 1$; but, unless $\mathcal{P} = \mathcal{NP}$ they do not admit anything better. I.e., there exists $\underline{\rho} > 1$ such that the existence of a polynomial-time $\underline{\rho}$-approximation algorithm for the optimization problem implies $\mathcal{P} = \mathcal{NP}$.

We need to at least clarify a bit what we mean by $\rho$-approximation. Suppose that we are given an optimization problem such that its objective value is nonnegative for every feasible solution and every instance of the problem. Moreover, assume that the optimal objective value denoted by $OPT$ is finite for every instance of the optimization problem. Then a $\rho$-approximation algorithm is a polynomial-time algorithm in the sense of the Turing machine model such that the for every instance of the problem, the algorithm delivers a feasible solution $\bar{x}$ such that

$$\max\left\{\frac{OPT}{Obj.Val.(\bar{x})}, \frac{Obj.Val.(\bar{x})}{OPT}\right\} \leq \rho.$$

There are two closely related classes of optimization problems defined in the literature. The earlier one is called *MAXSNP-hard*. An example is the Max Cut problem which we will study soon. The class *MAXSNP* is established through a polynomial-time reduction mechanism. In contrast to the mechanism used in the definition of $\mathcal{NP}$-hard problems, here we also require the reduction to preserve the objective function values within a constant factor. Even though the class MAXSNP was first proposed in late 1980s by Papadimitriou and Yannakakis [281], it is nicely connected to a definition of $\mathcal{NP}$ due to Fagin [119] (this goes back to 1973).

More recent literature uses the notion of $\mathcal{APX}$*-hard* (which uses a notion of $AP$-reducibility). That is, an optimization problem, whose decision version is in $\mathcal{NP}$, is in $\mathcal{APX}$ if the optimization problem admits a polynomial-time $\rho$-approximation algorithm for some absolute constant $\rho$. Then the $\mathcal{APX}$-hard problems are those optimization problems to which every $\mathcal{APX}$ problem is $AP$-reducible. $AP$-reducibility is defined in Crescenzi, Kann, Silvestri and Trevisan [98].

The first two optimization problems we study in this chapter are MAX CUT and MAX 2SAT. Both of these problems admit polynomial-time approximation algorithms with $\rho \leq 1.15$ and they are both $\mathcal{APX}$-hard and $MAXSNP$-hard. So, for each of these problems, there exists a $\underline{\rho} > 1$ such that existence of any polynomial-time approximation algorithm for either problem, beating the corresponding approximation ratio $\underline{\rho}$ would imply $\mathcal{P} = \mathcal{NP}$.

When the non-approximability results were first being obtained in the early 1990s, it seemed that these $\underline{\rho}$ values would be too far from what is actually achievable by polynomial-time approximation algorithms. However, soon after that, the constants in the nonapproximability results kept getting better and approximation algorithms also kept getting better. Now, there are some interesting problems for which there is no gap between the best approximation ratio attainable by a polynomial-time approximation algorithm and the corresponding nonapproximability threshold $\underline{\rho}$. For some others the gap is very small. Majority of the problems have relatively large gaps. (Many open problems await!)

Semidefinite Optimization played a key role in obtaining good approximation algorithms for some problems and has the potential of having more impact in this area.

## 5.1 MAX CUT, Goemans-Williamson Analysis

Let $G := (V, E)$ be an undirected graph. Let $w \in \mathbb{R}_+^E$ be given, $n := |V|$. Every $U \subseteq V$ in $G$ determines a *cut*: $(U, V \setminus U)$ in $G$ (we can just refer to this cut as "cut $U$"). We use the notation

$$\delta(U) := \{\{i,j\} \in E : i \in U, j \in V \setminus U\}$$

to denote *the set of edges crossing the cut* $U$. The *maximum cut problem* is to find $U \subseteq V$ such that $\left(\sum_{\{i,j\}\in\delta(U)} w_{ij}\right)$ is maximum. This problem is $\mathcal{NP}$-hard and *MAXSNP-hard* as we mentioned earlier.

There is a very simple heuristic algorithm which is a polynomial-time 2-approximation algorithm for MAX CUT:
Let $U_1 := \emptyset$, $U_2 := \emptyset$.

While $V \setminus (U_1 \cup U_2) \neq \emptyset$, do

pick $v \in V \setminus (U_1 \cup U_2)$,

if $\displaystyle\sum_{u\in U_2} w_{uv} > \sum_{u\in U_1} w_{uv}$

then $U_1 := U_1 \cup \{v\}$,

otherwise, $U_2 := U_2 \cup \{v\}$

end{while}.

The above algorithm can also be seen as a derandomized version of a very simple randomized algorithm (in which every node is put into one of the sets $U_1$, $U_2$ with equal, and independently chosen, probability $\frac{1}{2}$).

Until 1994, the above was the best polynomial-time approximation algorithm (in terms of having a small approximation ratio). There had been some improvements to the above algorithm, but none was proven to be significantly better than $(2 - \epsilon)$-approximation.

One of the reasons can be explained by a much older result of Erdős from 1967 [116]. Researchers studying the polynomial-time approximation algorithms for MAX CUT would be tempted to consider linear programming relaxations of the problem that can be solved in polynomial-time. A very well-known integer programming formulation is by utilizing the so-called *triangle inequalities*. We assign a variable $x_e$ for every edge $e$ in the complete graph with the intended meaning that $x_e$ will be 1 if edge $e$ is *cut*, zero otherwise. Then the integer programming formulation is:

$$\begin{array}{lrcl} \max & \sum_{e\in E} w_e x_e & & \\ & \text{subject to:} & & \text{for every triangle } (e,f,g) \text{ in the complete graph:} \\ & x_e + x_f + x_g & \le & 2, \\ & x_e - x_f - x_g & \le & 0, \\ & -x_e + x_f - x_g & \le & 0, \\ & -x_e - x_f + x_g & \le & 0, \\ & x_e & \in & [0,1], \forall e, \\ & x_e & \text{is} & \text{integer}, \forall e. \end{array}$$

The nontrivial inequalities simply state that (for integral $x$) a cut can have only 0 or 2 edges cut from any triangle. They are called the *triangle inequalities.*

Notice that the number of variables and inequalities are bounded by a polynomial in $|V|$. Also, the feasible region of the IP or the LP relaxation does not change, as long as $|V|$ stays the same. We handle any graph by assigning $w_e := 0$ for all $e \notin E$.

The projection of the LP relaxation onto the affine subspace spanned by the variables $\{x_e : e \in E\}$ gives the *odd-cycle polytope* for MAX CUT. The main inequalities are:

$$\sum_{e\in F} x_e - \sum_{e\in C\setminus F} x_e \le |F| - 1,$$

for every cycle $C$ in $G$ and for every $F \subseteq C$ such that $|F|$ is odd. For many graphs the number of these odd cycle inequalities (which define facets of the MAX CUT polytope) is exponential in $|V|$. However, we can optimize any linear function over the odd-cycle polytope of any graph by utilizing the triangle inequality formulation in polynomial-time. (There are also polynomial-time separation oracles for these odd-cycle inequalities). This discussion illustrates the fact that some polytopes with exponentially many facets (in relation to their dimension) can be represented as a projection of a higher dimensional polytope whose dimension as well as the number of facets are bounded by a polynomial of the dimension of the original polytope. We call such representations *lifted representations*.

All these seem promising; however, it easily follows from a result of Erdős [116] that there exist graphs for which the ratio of the optimal objective value of the LP relaxation (given by the odd-cycle polytope or equivalently by the lifted triangle inequality formulation) is about twice the value of the MAX CUT. Indeed, this observation knocks out the above IP formulation from further consideration. Nevertheless, it turns out that an SDP formulation is very helpful.

Let

$$u_i := \begin{cases} 1 & \text{if } i \in U, \\ -1 & \text{if } i \notin U. \end{cases}$$

Let's also set $w_{ij} := 0$ for every $\{i,j\} \notin E$. Then the Max Cut Problem is equivalently written as

$$\begin{array}{ll} \max & \frac{1}{4}\sum_{i,j} w_{ij}(1 - u_i u_j) \\ \text{subject to:} & u \in \{-1,1\}^V. \end{array}$$

Now, we can represent all the weights in a symmetric matrix $W \in \Sigma^V$ such that $W_{ij} := w_{ij}$ for every $i, j \in V$. The last optimization problem can be expressed as

$$\begin{array}{llcl} \max & \frac{1}{4}\langle W, \bar{e}\bar{e}^T\rangle - \frac{1}{4}\langle W, X\rangle & & \\ \text{subject to:} & \text{diag}(X) & = & \bar{e}, \\ & X & \succeq & 0, \\ & \text{rank}(X) & = & 1. \end{array}$$

The equivalence of the last two optimization problems goes through the correspondence

$$X \leftrightarrow uu^T.$$

Let $\bar{u}$ be an optimal solution of the former problem. Let $\bar{X} := \bar{u}\bar{u}^T$. Then $\bar{X}$ is feasible in the latter problem and the objective values of $\bar{u}$ in the first problem and $\bar{X}$ in the second problem are the same. Now, let $\bar{X}$ be a feasible solution of the latter problem. Then the last two constraints imply $\bar{X} = \bar{u}\bar{u}^T$ for some $\bar{u} \in \mathbb{R}^V \setminus \{0\}$. The constraint $\text{diag}(X) = \bar{e}$ implies $\bar{u} \in \{-1, 1\}^V$. Therefore, $\bar{u}$ is a feasible solution of the former optimization problem. Since the objective values of $\bar{X}$ and $\bar{u}$ in their respective problems again coincide, the equivalence is proven.

The last problem, even though it looks more like an SDP is still (very) nonconvex. To obtain a tractable relaxation, we remove the nonconvex constraint "$\text{rank}(X) = 1$" and obtain the SDP relaxation:

$$\begin{array}{lllcll} (P) & \max & -\frac{1}{4}\langle W, X\rangle & & & (+\frac{1}{4}\langle W, \bar{e}\bar{e}^T\rangle) \\ & \text{subject to:} & \text{diag}(X) & = & \bar{e}, & \\ & & X & \succeq & 0, & \end{array}$$

$$\begin{array}{lllcll} (D) & \min & \bar{e}^T y & & & (+\frac{1}{4}\langle W, \bar{e}\bar{e}^T\rangle) \\ & \text{subject to:} & \text{Diag}(y) - S & = & -\frac{1}{4}W, & \\ & & S & \succeq & 0. & \end{array}$$

We wrote "max" and "min" in the above problems, because both $(P)$ and $(D)$ have Slater points. $\bar{X} := I$ works in the primal. Letting $\eta := \frac{1}{4}\sum_{i,j} |W_{ij}| + 1$, we see that $\bar{y} := \eta\bar{e}$ works in the dual (then $\bar{S} = \frac{1}{4}W + \eta I$ which is positive definite by diagonal dominance).

For these Slater points, we compute (to use Theorems 4.5 and 4.13)

$$\begin{aligned} \Psi(\bar{X}, \bar{S}) &= |V| \ln\left(\frac{\langle \bar{X}, \bar{S}\rangle}{|V|}\right) - \ln(\det(\bar{X})) - \ln(\det(\bar{S})) \\ &= |V| \ln(\eta) - \ln\det\left(\eta I + \frac{1}{4}W\right) \\ &\leq |V| \ln(\eta). \end{aligned}$$

We used that facts that $\text{Tr}(W) = 0$ and the smallest eigenvalue of $\eta I + \frac{1}{4}W$ is at least 1 (by definition and diagonal dominance, $\eta I + \frac{1}{4}W \succeq I$) and hence, so is $\det\left(\eta I + \frac{1}{4}W\right)$. Thus, when $W$ is rational, we have

$$\Psi(\bar{X}, \bar{S}) \leq |V|\text{size}(W).$$

This yields a complexity bound that is polynomial in the input size and $1/\epsilon$. However, if we make the pair $(\bar{X}, \bar{S})$ more "central" (closer to the central path) we can obtain a better complexity bound out of Theorems 4.5 and 4.13. Let us keep $\bar{X} := I$; but redefine

$$\bar{y} := \frac{1}{4}\begin{bmatrix} \sum_j |W_{1j}| \\ \sum_j |W_{2j}| \\ \vdots \\ \sum_j |W_{|V|j}| \end{bmatrix} + \frac{\eta}{\sqrt{|V|}}\bar{e} \text{ and } \bar{S} := \frac{1}{4}W + \text{Diag}(\bar{y}).$$

Then, the new $\bar{S}$ is also a Slater point for $(D)$. Moreover,

$$\langle \bar{X}, \bar{S} \rangle = \left(\sqrt{|V|} + 1\right)\eta - 1,$$

and

$$\begin{aligned} \Psi(\bar{X}, \bar{S}) &= |V| \ln\left(\frac{\left(\sqrt{|V|}+1\right)\eta - 1}{|V|}\right) - \ln(\det(\bar{S})) \\ &\leq |V| \ln\left(\frac{\left(\sqrt{|V|}+1\right)\eta}{|V|}\right) - |V| \ln\left(\frac{\eta}{\sqrt{|V|}}\right) \\ &= |V| \ln\left(1 + \frac{1}{\sqrt{|V|}}\right) \\ &\leq \sqrt{|V|}. \end{aligned}$$

For a straightforward application of Theorems 4.5 and 4.13, we may choose $\epsilon := \frac{1}{3\left(\sqrt{|V|}+1\right)\eta}$ yielding the iteration complexity bound

$$O\left(\sqrt{|V|}\ln(|V|\eta)\right)$$

to obtain $X^{(k)}$, $S^{(k)}$ feasible with $\langle X^{(k)}, S^{(k)} \rangle < 1$ (in fact,

$$\langle X^{(k)}, S^{(k)} \rangle < \left(\frac{1}{3\left(\sqrt{|V|}+1\right)\eta}\right)^{O(1)}$$

for any absolute constant $O(1)$). One can improve this bound by exploiting the special structure of $W$ and/or the underlying graph $G$.

If we find an optimal solution $X^*$ of $(P)$ such that $\text{rank}(X^*) = 1$, then we have solved the MAX CUT problem exactly. Otherwise, write

$$X^* =: BB^T, \text{ where } B^T =: \left[v^{(1)}, v^{(2)}, \ldots, v^{(n)}\right], \quad v^{(i)} \in \mathbb{R}^d, d \leq n := |V|.$$

Note that with these definitions, $X^*_{ij} = \langle v^{(i)}, v^{(j)} \rangle$ and $1 = X^*_{ii} = \|v^{(i)}\|_2^2$.

**5.1.1 Random Hyperplane Technique.** Let us generate $r \in \mathbb{R}^d$ on the unit hypersphere randomly. Then setting

$$U := \left\{i \in V : r^T v^{(i)} \geq 0\right\}$$

defines a cut in $G$. If $X^*$ is rank-1, then we have $d = 1$ and $r \in \{-1, 1\}$. Therefore, in this very special case, the random hyperplane technique would behave properly and identify the Max Cut given by $X^*$ correctly.

Note that in the above decomposition $B$ is not unique. If $B$ satisfies the definition then for any $d \times d$ orthogonal matrix $Q$, $BQ^T$ also satisfies the definition. But this is equivalent to applying the orthonormal transformation $Q$ to every vector $v^{(i)}$. Then, it is easy to see that the Random Hyperplane Technique is invariant under orthonormal transformations. In fact, $BB^T = \bar{B}\bar{B}^T$ iff there exists an orthogonal matrix $Q$ such that $\bar{B} = BQ^T$. To prove the nontrivial direction of this fact, use the singular value decomposition of $\bar{B}^T B$:

$$Q_1 D Q_2^T = \bar{B} B^T$$

(where $Q_1$ and $Q_2$ are orthogonal and $D$ is diagonal), and let $Q := Q_2 Q_1^T$.

**Lemma 5.1** *Let $v^{(i)}$ and $r$ be as above. Then*

$$\Pr\left\{\operatorname{sign}\left(r^T v^{(i)}\right) \neq \operatorname{sign}\left(r^T v^{(j)}\right)\right\} = \frac{\theta}{\pi},$$

*where $\theta := \arccos\langle v^{(i)}, v^{(j)}\rangle$.*

Note that in the above lemma, $\theta$ is the angle between $v^{(i)}$ and $v^{(j)}$ and for $v \in \mathbb{R}^n$, $\operatorname{sign}(v) \in \{-1,1\}^n$ is defined by

$$[\operatorname{sign}(v)]_j := \begin{cases} 1, & \text{if } v_j \geq 0; \\ -1, & \text{otherwise} \end{cases}$$

so that we are compatible with the definition of the Random Hyperplane Technique.

**5.1.2 The Analysis of the Goemans-Williamson Algorithm.** For the rest of our discussion, we will reverse the roles of $\rho$ and $\frac{1}{\rho}$ to conform to the original presentation of Goemans and Williamson [145].

**Lemma 5.2** *For every $u \in [-1,1]$, we have*

$$\frac{1}{\pi}\arccos(u) \geq \frac{\rho}{2}(1-u),$$

*and*

$$1 - \frac{1}{\pi}\arccos(u) \geq \frac{\rho}{2}(1+u),$$

*where $\rho := 0.87856$.*

**Proof** Use the following series representation of $\arccos(u)$ (valid for every $u \in [-1,1]$):

$$\arccos(u) = \frac{\pi}{2} - u - \frac{1}{6}u^3 - \frac{3}{40}u^5 - \frac{15}{336}u^7 - \cdots = \frac{\pi}{2} - \sum_{k=0}^{\infty} \frac{(2k)!}{2^{2k}(k!)^2(2k+1)} u^{2k+1}.$$

For the second part, substitute $u := -u$ and note that the domain of $u$ for the first result is invariant under this transformation. $\square$

**Theorem 5.3** *The expected value of the above mentioned cut is at least*

$$\frac{\rho}{4}\sum_{i,j} w_{ij}\left(1 - \langle v^{(i)}, v^{(j)}\rangle\right) = \rho\,(\text{opt. obj. value of } (P)).$$

**Proof** Let $X^*$ be an optimal solution of the SDP relaxation of the MAX CUT problem. Then the optimal objective function value of the SDP relaxation is

$$\frac{1}{4}\sum_{i,j} W_{ij}\left(1 - X^*_{ij}\right) = \frac{1}{4}\sum_{i,j} w_{ij}\left(1 - \langle v^{(i)}, v^{(j)}\rangle\right).$$

The expected value of the cut given by the random hyperplane technique (using Lemma 5.1) is

$$\begin{aligned}\sum_{i,j} w_{ij}\frac{\arccos\langle v^{(i)}, v^{(j)}\rangle}{2\pi} &\geq \frac{\rho}{4}\sum_{i,j} w_{ij}\left(1-\langle v^{(i)}, v^{(j)}\rangle\right)\\ &= \rho\,(\text{opt. obj. value of } (P)),\end{aligned}$$

as desired (where the inequality used the fact that $w_{ij} \geq 0$ for every $i, j$ and Lemma 5.2). □

The above theorem implies that for every graph $G$, there exists a cut in $G$ whose weight is at least $\rho$ times the optimal objective function value of the SDP relaxation. Moreover, we can *derandomize* the above algorithm to obtain a deterministic polynomial-time algorithm which always delivers a cut of objective value at least $\rho$ times the max. cut value.

**Theorem 5.4** *Let $G = (V, E)$ with $w \in \mathbb{Q}_+^E$ be given. Then a cut in $G$ of value at least $\rho$.(weight of the max. cut) can be computed in polynomial-time.*

Note that the $\rho$ in the above theorem corresponds to $1/\rho$ in the Introduction for this chapter. Two more substantial remarks about the above theorem are in order:

**Remark 5.5** We proved that the SDP relaxation and its dual satisfy the Slater condition. Therefore, they both attain their optimal values and the optimal objective values of both the primal and the dual problems are the same. Moreover, we can use the Ellipsoid Method or a polynomial-time interior-point method to solve the SDP relaxation. In particular, we can use the primal-dual interior-point algorithm that we studied. As we mentioned earlier, we may use $X^{(0)} := I$ and $S^{(0)} := \frac{1}{4}W + \frac{1}{4}\text{Diag}\begin{bmatrix}\sum_j |W_{1j}|\\ \sum_j |W_{2j}|\\ \vdots\\ \sum_j |W_{nj}|\end{bmatrix} + \frac{\eta}{\sqrt{n}}I$. Then,

$$\langle X^{(0)}, S^{(0)}\rangle = \left(\sqrt{n}+1\right)\eta - 1.$$

Moreover, $\Psi(X^{(0)}, S^{(0)})$ is also bounded from above by $\sqrt{n}$. Say, we run the primal-dual interior-point algorithm until

$$\langle X^{(k)}, S^{(k)}\rangle \leq 10^{-10}.$$

Note that we had some room (say $10^{-8}$) in the proof of Lemma 5.2 in the definition of $\rho$. Therefore, after polynomially many iterations (polynomial in $(n, \text{size}(W))$), we will have the feasible solution $X$ of $(P)$ with the desired accuracy. Also note that all entries of all the iterates $X^{(k)}, S^{(k)}$ stay bounded.

**Remark 5.6** So far, we really only proved that there is a polynomial-time randomized algorithm (because of the usage of the Random Hyperplane in the algorithm) that attains the claimed $\rho$-approximation. As we mentioned above, the algorithm has been derandomized. Hence, yielding the deterministic algorithm promised in the last theorem. However, derandomization is outside the scope of these lectures. There are some general techniques for derandomization. We will briefly discuss the most basic form when we discuss the problem Max 2Sat.

## 5.2 Karloff's Worst-Case Analysis

The above analysis by Goemans and Williamson was extremely influential in capturing the attention of many combinatorial optimizers and many researchers in theoretical computer science (in general many discrete mathematicians) and drawing it to semidefinite optimization (as well as to the random hyperplane-like techniques and the analysis).

One of the first questions at hand was whether the analysis can be improved. Karloff [190] proved that for certain families of graphs, the above algorithm will give exactly the approximation ratio $\rho$ above.

Let $p$ and $q$ be positive integers such that $q \leq (p/12)$. Define an undirected graph $G_{p,q} = (V_{p,q}, E_{p,q})$ as follows. $V_{p,q}$ is all $\binom{p}{p/2}$ subsets of $\{1, 2, \ldots, p\}$ with cardinality $(p/2)$. $E_{p,q}$ is all pairs $\{S, T\}$ such that $S, T \subset \{1, 2, \ldots, p\}$ and $|S \cap T| = q$. These graphs are closely related to *Kneser Graphs*.

Now, for all $S \in V_{p,q}$ define $v_S \in \mathbb{R}^p$ by

$$(v_S)_i := \begin{cases} \frac{1}{\sqrt{p}}, & \text{if } i \in S; \\ -\frac{1}{\sqrt{p}}, & \text{if } i \notin S. \end{cases}$$

Also define $B \in \mathbb{R}^{p \times V_{p,q}}$, where each column of $B$ is one of the $v_S$ defined above. Finally,

$$\bar{X} := B^T B.$$

Karloff proved that $\bar{X}$ is an optimal solution of the SDP relaxation for the MAX CUT problem. He then proved that the Random Hyperplane Technique applied to $\bar{X}$ delivers a cut whose value is no better than the worst-case guarantee of Goemans-Williamson.

This establishes the fact that the performance guarantee of $\rho$ cannot be improved. Nevertheless, we can ask whether this optimal solution $\bar{X}$ is an extreme point of the feasible region of $(P)$. We observe that

$$\operatorname{rank}\left(\bar{X}\right) = \operatorname{rank}(B) = p - 1,$$

since we can get the incidence vectors of all of the arcs on the complete digraph on $p$ nodes in $B$.

It is not difficult to show that an upper-bound on the rank of an extreme point of $\{X \in \Sigma_+^n : \mathcal{A}(X) = b\} \neq \emptyset$, where $\mathcal{A} : \Sigma^n \to \mathbb{R}^m$ surjective, $n \geq 3$, $m \geq 3$, is

$$\sqrt{2m},$$

see Friedland and Loewy [131], Barvinok [26], Pataki [287]. The above given bound can be decreased by one if we use a result of Barvinok [24] (this is justified since the feasible region is bounded). Note that in our case, $m = \binom{p}{p/2}$. Therefore, to conclude that $\bar{X}$ is not extreme, the above general result does not suffice. However, we can directly argue that $\bar{X}$ is not an extreme point of the feasible region by exhibiting a small nonzero $D$ such that $\bar{X} \pm D$ lie in the feasible region.

So, what does it mean that $\bar{X}$ is not an extreme point of

$$\left\{X \in \Sigma_+^{V_{p,q}} : \operatorname{diag}(X) = \bar{e}\right\}?$$

Well, as we discussed before, given any feasible solution of an SDP with a bounded feasible region, it is easy (in poly. time) to find an extreme point of the feasible region whose objective value is at least as good as the original non-extreme solution. Therefore, a reasonable implementation of the Goemans-Williamson algorithm can first find an approximately optimal, extreme point solution of $(P)$, and then apply the random hyperplane technique etc. to it. Therefore, it seems that Karloff's worst-case analysis does not apply to this variant of the Goemans-Williamson algorithm.

On the other hand, if we do not insist on using extreme point solutions, Karloff's results are even stronger: even adding *all* the triangle inequalities to our SDP relaxation still does not improve the worst-case guarantee of $\rho \approx 0.87856$.

## 5.3 MAX 2SAT

Another combinatorial optimization problem that is closely related to Max Cut is the Max 2Sat problem. First, we define the Satisfiability Problem (SAT).

Let $x_1, x_2, \ldots, x_n$ denote Boolean variables. So, $x_j \in \{0,1\}$ and for $x_j = 0$ the Boolean variable has value "False" and for $x_j = 1$ the Boolean variable has value "True."

A *literal* is either a variable $x_j$ or its complement $\bar{x}_j$. A *clause* is a disjunction of a finite collection of literals (e.g., clause $C_j$ can be $(x_5 \vee \bar{x}_3 \vee \bar{x}_2 \vee x_{10})$). A *formula* is a conjunction of a finite collection of clauses. For example,

$$(x_5 \vee \bar{x}_3 \vee \bar{x}_2 \vee x_{10}) \wedge (\bar{x}_1) \wedge (x_2 \vee \bar{x}_4 \vee x_8 \vee x_9 \vee x_7).$$

The *Satisfiability Problem (SAT)* is: given a Boolean formula to decide if there exists an assignment of values to the variables which makes the formula "True." For instance, in the above example, assigning $x_1 := 0$, $x_2 := 1$, $x_3 := 0$ and all other variables to any values yields an assignment evaluating "True." Hence, it is a *satisfying assignment.*

One of the reasons for the fame of SAT is that it was the model decision problem that was proven to be $\mathcal{NP}$-complete in Cook's seminal theorem [90]:

**Theorem 5.7** *SAT is $\mathcal{NP}$-complete.*

A closely related optimization problem is *Max Sat*: given a Boolean formula and nonnegative weights for the clauses, find an assignment which maximizes the total weight of satisfied clauses.

Clearly, Max Sat is $\mathcal{NP}$-hard (by assigning unit weights to clauses, the existence of a polynomial-time algorithm for even the cardinality version of the Max Sat implies $\mathcal{P} = \mathcal{NP}$).

To bring some more order to our study, let's look into special cases of Max Sat where the number of literals in each clause is restricted. Given and integer $k \geq 2$, the problem *kSat* is a SAT in which every clause has at most $k$ literals. Moreover, the related Max Sat problem on $k$Sat with given weights is called Max $k$Sat.

We can formulate SAT with clauses $C_1, C_2, \ldots, C_m$ and variables $x_1, x_2, \ldots, x_n$ as follows:

$$\sum_{j: x_j \in C_i} x_j + \sum_{j: \bar{x}_j \in C_i} (1 - x_j) \geq 1, \forall i \in \{1, 2, \ldots, m\};$$

$$x \in \{0,1\}^n.$$

Similarly, given weights $w_1, w_2, \ldots, w_m$, the underlying Max Sat problem can be formulated as an Integer Programming problem:

$$\begin{array}{lrcl} \max & \multicolumn{3}{l}{\sum_{i=1}^{m} w_i z_i} \\ \text{subject to:} & z_i & \leq & \sum_{j:x_j \in C_i} x_j + \sum_{j:\bar{x}_j \in C_i} (1 - x_j), \quad \forall i \in \{1, 2, \ldots, m\}, \\ & x & \in & \{0,1\}^n, \\ & z & \in & \{0,1\}^m. \end{array}$$

The problem 2Sat is extremely easy. Solving 2Sat essentially amounts to checking the strong connectivity of a directed graph whose size is essentially the same as that of the 2Sat formula. Therefore, 2Sat $\in \mathcal{P}$. However, Max 2Sat is $\mathcal{NP}$-hard. Let's see a part of the very close relationship between Max Cut and Max 2Sat. Let $G = (V, E)$ be a given instance of the Max Cut (just the max. cardinality version). Now, we describe the corresponding Max 2Sat instance such that solving Max 2Sat instance determines the size of the max. cardinality cut in $G$. We make a variable $x_v$ for every node $v \in V$. We construct the 2Sat formula on these $|V|$ variables as follows: for every edge $\{u, v\} \in E$, we make the clause $(x_u \vee x_v)$ with weight $w_{uv} := 2$; for every $v \in V$, we make a unit clause $(\bar{x}_v)$ with weight $w_v := |\delta(v)|$.

It is easy to prove that the maximum total weight of satisfied clauses in the Max 2Sat instance is exactly

$$|\text{max. card. cut in } G| + 2|E|.$$

Max 2Sat is also $MAXSNP$-hard and $\mathcal{APX}$-hard.

First, let's construct an approximation algorithm. For each variable $x_j$, pick $p_j \in [0, 1]$ independently and let $p_j$ represent the probability that the variable $x_j$ takes the value 1. Let $u_i$ denote the probability that clause $C_i$ is not satisfied. Then

$$u_i = \prod_{j:\bar{x}_j \in C_i} p_j \prod_{j:x_j \in C_i} (1 - p_j).$$

Thus, the expected total weight of such an assignment is:

$$E[w;p] = \sum_{i=1}^{m} w_i (1 - u_i) = \sum_{i=1}^{m} w_i \left[ 1 - \prod_{j:\bar{x}_j \in C_i} p_j \prod_{j:x_j \in C_i} (1 - p_j) \right].$$

Suppose that every clause has precisely two literals and also assume that we choose $p_j := \frac{1}{2}$, for every $j \in \{1, 2, \ldots, n\}$. Then

$$E[w;p] = \sum_{i=1}^{m} w_i \left(1 - \frac{1}{4}\right) = \frac{3}{4} \sum_{i=1}^{m} w_i \geq \frac{3}{4} OPT.$$

So, we have a randomized $\frac{3}{4}$-approximation algorithm for Max 2Sat when every clause has exactly two literals.

Let's *derandomize* this algorithm to achieve the same ratio with a deterministic algorithm. Let $E[w;p|A]$ denote the conditional expectation of the total weight under the probability distribution $p$, given event $A$. Pick $j$ arbitrarily and focus on the equation

$$E[w;p] = p_j E[w;p|x_j = 1] + (1 - p_j) E[w;p|x_j = 0].$$

Since $p_j \in [0,1]$, we clearly have

$$\max\{E[w;p|x_j=1], E[w;p|x_j=0]\} \geq E[w;p].$$

Note that the conditional expectations above are easy to compute. E.g., for the first one, we plug in $x_j := 1$ in all the clauses and then compute the expectation over the simplified clauses with only $(n-1)$ variables.

$$E[w;p|x_j=1] = \sum_{i:x_j\in C_i} w_i + \sum_{i:\bar{x}_j\in C_i} w_i\left(1-\frac{u_i}{p_j}\right) + E[\text{clauses not involving } x_j],$$

$$E[w;p|x_j=0] = \sum_{i:\bar{x}_j\in C_i} w_i + \sum_{i:x_j\in C_i} w_i\left(1-\frac{u_i}{1-p_j}\right) + E[\text{clauses not involving } x_j].$$

Therefore,

$$E[w;p|x_j=1] \geq E[w;p|x_j=0] \iff p_j \sum_{i:x_j\in C_i} w_i u_i \geq (1-p_j) \sum_{i:\bar{x}_j\in C_i} w_i u_i.$$

This observation suggests a deterministic algorithm. We can pick a variable $x_j$ and compute the above terms. If $E[w;p|x_j=1] \geq E[w;p|x_j=0]$ then we assign $x_j := 1$; otherwise, we assign $x_j := 0$. Then we substitute for $x_j$ the value we assigned and repeat the procedure on the remaining formula with $n-1$ variables... After $n-1$ steps, we will have a complete assignment and the total weight of the assignment will be at least $E[w;p]$.

Therefore, our deterministic algorithm delivers a solution with objective value at least $\frac{3}{4}OPT$ if we choose $p := \frac{1}{2}\bar{e}$ and the formula has exactly two literals per clause.

We can run this algorithm using $x^*$, an optimal solution of the LP relaxation of the integer programming formulation above; i.e., we can set the probabilities $p := x^*$ or define a distribution closely related to the $x_j^*$ values for each $j$. For Max 2Sat, even if there are some clauses with single literals, the above algorithm still delivers an assignment whose objective value is at least $\frac{3}{4}OPT$. See Yannakakis [371]. (Also see Goemans and Williamson [146].)

**5.3.1 SDP Relaxation of Max 2Sat.** Now, we turn to an SDP relaxation and utilizing its optimal solution to improve on the approximation ratio $\frac{3}{4}$.

We will use a new variable $x_0 \in \{-1,1\}$. The value $x_0$ takes in a solution will stand for "True." So, $x_i$ will be set to "True" iff $x_i = x_0$. Now, given the formula $(x_i)$ this evaluates to "True" iff $(1+x_0x_i) = 2$ (and evaluates to "False" iff $(1+x_0x_i) = 0$). So, we can write

$$(x_i) = \frac{1}{2}(1+x_0x_i).$$

Similarly,

$$(\bar{x}_i) = \frac{1}{2}(1-x_0x_i).$$

Finally,

$$\begin{aligned}(x_i \vee x_j) &= 1-(\bar{x}_i)(\bar{x}_j) = 1-\frac{1}{4}(1-x_0x_i)(1-x_0x_j)\\ &= \frac{1}{4}\left[(1+x_0x_i)+(1+x_0x_j)+(1-x_ix_j)\right].\end{aligned}$$

Similarly,

$$(\bar{x}_i \vee x_j) = 1 - \left[\frac{1}{2}(1 + x_0 x_i)\frac{1}{2}(1 - x_0 x_j)\right] = \frac{1}{4}\left[(1 - x_0 x_i) + (1 + x_0 x_j) + (1 + x_i x_j)\right],$$

where we used the fact that $x_0^2 = 1$. Therefore, any Max 2Sat instance can be formulated as an instance of the following $\{-1, 1\}$ quadratic optimization problem:

$$\max \quad \sum_{\{i,j\}\in E^+} w_{ij}\,(1 + u_i u_j) + \sum_{\{i,j\}\in E^-} w_{ij}\,(1 - u_i u_j)$$
$$u \in \{-1, 1\}^{n+1},$$

where $E^+$ and $E^-$ are chosen appropriately. We extract the corresponding $W \in \Sigma^{n+1}$ matrix from the objective function above, replace the variables $x_i$ by vectors $v^{(i)} \in \mathbb{R}^{n+1}$ so that we have the matrix variable $X$ representing $BB^T$, and write the SDP relaxation

$$\begin{array}{llrcl} (P) & \max & -\langle W, X\rangle & & (+\text{constant}) \\ & \text{subject to:} & \operatorname{diag}(X) & = & \bar{e}, \\ & & X & \succeq & 0, \end{array}$$

$$\begin{array}{llrcl} (D) & \min & \bar{e}^T y & & (+\text{constant}) \\ & \text{subject to:} & \operatorname{Diag}(y) - S & = & -W, \\ & & S & \succeq & 0. \end{array}$$

Using arguments similar to our arguments for the Max Cut SDP relaxation (we also have Slater points for both $(P)$ and $(D)$), and by Strong Duality Theorem, we conclude that there is no duality gap. Goemans-Williamson analysis also trivially applies here (as they pointed out in their original paper). Note that the objective function of the primal problem $(P)$, in terms of the vector variables $v^{(i)}$ is

$$2\sum_{\{i,j\}\in E^-} w_{ij}(1 - \langle v^{(i)}, v^{(j)}\rangle) + 2\sum_{\{i,j\}\in E^+} w_{ij}(1 + \langle v^{(i)}, v^{(j)}\rangle).$$

Thus, under the Random Hyperlane Technique, the expected value of the objective function is

$$2\sum_{\{i,j\}\in E^-} w_{ij}\Pr\left\{\operatorname{sign}\langle r, v^{(i)}\rangle \neq \operatorname{sign}\langle r, v^{(j)}\rangle\right\}$$
$$+2\sum_{\{i,j\}\in E^+} w_{ij}\Pr\left\{\operatorname{sign}\langle r, v^{(i)}\rangle = \operatorname{sign}\langle r, v^{(j)}\rangle\right\}.$$

We compute an approximately optimal solution of $(P)$ either using a polynomial-time interior-point method or the ellipsoid method. Let $X^*$ be that solution. Then we compute $B \in \mathbb{R}^{(n+1)\times(n+1)}$ such that $X^* = BB^T$. We apply the Random Hyperplane Technique to obtain an assignment of values to the Boolean variables $x_1, x_2, \ldots, x_n$. Using Lemmas 5.1 and 5.2 we arrive at

**Theorem 5.8** *The expected value of the above mentioned assignment is at least*

$$\rho\,(\text{opt. obj. value of } (P))\,.$$

This algorithm also has been derandomized. Thus, we have

**Theorem 5.9** *Let a 2Sat formula with weights $w \in \mathbb{Q}_+^E$ be given. Then a truth assignment of value at least $\rho$.(weight of the optimal truth assignment) can be computed in polynomial-time.*

This approximation ratio $\rho$ has been further improved to 0.931 by Feige and Goemans [123]. They add the following valid inequalities to the initial formulation of the problem: For every distinct triple $\{i, j, k\}$ from $\{0, 1, 2, \ldots, n\}$,

$$\begin{aligned}
-\langle v^{(i)}, v^{(j)}\rangle - \langle v^{(i)}, v^{(k)}\rangle - \langle v^{(j)}, v^{(k)}\rangle &\leq 1, \\
\langle v^{(i)}, v^{(j)}\rangle + \langle v^{(i)}, v^{(k)}\rangle - \langle v^{(j)}, v^{(k)}\rangle &\leq 1, \\
\langle v^{(i)}, v^{(j)}\rangle - \langle v^{(i)}, v^{(k)}\rangle + \langle v^{(j)}, v^{(k)}\rangle &\leq 1, \\
-\langle v^{(i)}, v^{(j)}\rangle + \langle v^{(i)}, v^{(k)}\rangle + \langle v^{(j)}, v^{(k)}\rangle &\leq 1.
\end{aligned}$$

In terms of the matrix variable $X$, these become **linear inequalities**:

$$\begin{aligned}
-X_{ij} - X_{ik} - X_{jk} &\leq 1, \\
X_{ij} + X_{ik} - X_{jk} &\leq 1, \\
X_{ij} - X_{ik} + X_{jk} &\leq 1, \\
-X_{ij} + X_{ik} + X_{jk} &\leq 1.
\end{aligned}$$

These are the "triangle inequalities expressed in the matrix variable $X$." Analogous inequalities can be added to the Max Cut formulation as well (and they do help in some individual cases). However, unlike the Max 2Sat case, the added triangle inequalities do not improve the worst-case bound for Max Cut. To obtain the improved ratio 0.931, Feige and Goemans also modify the rounding technique.

**5.3.2 Connection to the Boolean Quadric Polytope.** Another way of expressing the Max 2Sat problem is

$$\begin{aligned}
&\max && c^T x + x^T Q x \\
&\text{subject to:} && x \in \{0,1\}^n,
\end{aligned}$$

where $c \in \mathbb{R}^n$ (corresponding to the unit clauses), $Q \in \Sigma^n$ (containing the proper weights for the clauses with exactly two literals per clause). We can assume that $\operatorname{diag}(Q) = 0$.

*Aside:* It is also possible to assume that $Q$ is upper-triangular, instead of symmetric.

We can express the above quadratic 0-1 optimization problem as an instance of integer linear programming problem. Let us use new variables $y_{ij}$ to represent the quadratic term $x_i x_j$ and linearize using the constraints:

$$x_i + x_j - y_{ij} \leq 1, \tag{5.1}$$
$$-x_i + y_{ij} \leq 0, \tag{5.2}$$
$$-x_j + y_{ij} \leq 0, \tag{5.3}$$
$$y_{ij} \geq 0, \tag{5.4}$$
$$x \text{ integer }, y \text{ integer }. \tag{5.5}$$

A well-studied object in the combinatorial optimization literature is the *Boolean Quadric Polytope*

$$BQP_n := \operatorname{conv}\left\{ \begin{pmatrix} x \\ y \end{pmatrix} \in \mathbb{R}^n \oplus \mathbb{R}^{n(n+1)/2} : \begin{pmatrix} x \\ y \end{pmatrix} \text{ satisfies } (5.1) - (5.5) \right\}.$$

Next, we study general quadratic optimization problems with $\{-1, +1\}$ variables. Indeed, the linear transformation

$$x \mapsto 2x - \bar{e}$$

allows us to trivially go from $\{0,1\}$ variables to $\{-1,+1\}$ variables.

## 5.4 Generalizations to Quadratic Optimization Problems

Let $W \in \Sigma^n$. Consider the optimization problems

$$\overline{f}(W) := \max_{x\in\{-1,1\}^n} x^T W x,$$

$$\underline{f}(W) := \min_{x\in\{-1,1\}^n} x^T W x.$$

Both of the above problems are $\mathcal{NP}$-hard. Note that $\overline{f}(W) = -\underline{f}(-W)$. As we did to the Max Cut problem, we can rewrite the above optimization problem in terms of a positive semidefinite matrix variable restricted to be of rank-1.

$$\begin{array}{rlrcl} \overline{f}(W) & = & \max & \langle W, X\rangle & \\ & & \text{subject to:} & \operatorname{diag}(X) & = \bar{e}, \\ & & & X & \succeq 0, \\ & & & \operatorname{rank}(X) & = 1. \end{array}$$

Hence, we obtain an SDP relaxation as follows

$$\begin{array}{lclcl} \max\limits_{x\in\{-1,1\}} x^T W x & \leq & \max & \langle W, X\rangle & =: \overline{F}(W) \\ & & & \operatorname{diag}(X) = \bar{e} & \\ & & & X \succeq 0. & \end{array}$$

The latter value $\overline{F}(W)$ is also equal to the optimal value of

$$\begin{array}{ll} \min & \bar{e}^T y \\ \text{subject to:} & \operatorname{Diag}(y) \succeq W. \end{array}$$

This is justified, since both SDP problems admit Slater points. Similarly, we define and establish

$$\begin{array}{lclclcl} \underline{F}(W) & := & \min & \langle W, X\rangle & = & \max & \bar{e}^T y \\ & & & \operatorname{diag}(X) = \bar{e} & & & \operatorname{Diag}(y) \preceq W. \\ & & & X \succeq 0 & & & \end{array}$$

Again, we used the fact that both problems admit Slater points and utilized the SDP duality theory.

**5.4.1 Approximation Ratios for General Optimization Problems.** For many combinatorial optimization problems with nonnegative weights in the objective function, there is a natural scale of values. E.g., all feasible solutions must have objective values between 0 and $\sum w_j$. For instance, the Max Cut problem with nonnegative weights on the edges lies in such a class. Clearly, the empty cut is the worst with objective value zero and no cut can have value more than the sum of all the weights (attained when the graph is bipartite).

What about Max Cut problems for which some edges have negative weights? What about other nonconvex optimization problems with continuous decision variables? Indeed, we can use a similar reasoning to arrive at a range of values. In our current study of the special class of quadratic optimization problems, we already have the necessary notation. The interval

$$\left[\underline{f}(W), \overline{f}(W)\right] \text{ serves in place of } \left[0, \sum_j w_j\right].$$

Therefore, we can measure the *relative quality* of a feasible solution $\bar{x}$ by the ratio

$$\frac{\overline{f}(W) - \bar{x}^T W \bar{x}}{\overline{f}(W) - \underline{f}(W)}.$$

In our case, both $\overline{f}(W)$ and $\underline{f}(W)$ are $\mathcal{NP}$-hard to compute. To deal with this problem, we will estimate these quantities and their differences by using the tractable quantities $\overline{F}(W)$ and $\underline{F}(W)$. Note that this *relative approximation ratio* (which we will call $\mu$-*approximation*), is similar to what we used with the ellipsoid method when the objective function was an arbitrary convex function given by a subgradient oracle.

**5.4.2 The analysis of the approximation ratio.** First, let us summarize the obvious relationships among $\underline{f}, \overline{f}, \underline{F}$ and $\overline{F}$.

**Proposition 5.10** *For every* $W \in \Sigma^n$, *we have*

$$\underline{f}(W) = -\overline{f}(-W), \quad \underline{F}(W) = -\overline{F}(-W)$$

*and*

$$\underline{F}(W) \leq \underline{f}(W) \leq \overline{f}(W) \leq \overline{F}(W).$$

We will first consider the special case that $W \succeq 0$. This will constitute the crux of the generalization and it will contain some of the most interesting special cases.

5.4.2.1 *Positive Semidefinite* $W$. For this subsection, we assume $W \succeq 0$. Then the fact that the *maximum value of a convex function over a compact convex nonempty set is attained at an extreme point* implies

$$\overline{g}(W) := \max_{x \in [-1,1]^n} x^T W x = \overline{f}(W).$$

In the next lemma, we take the first step to simulate the random hyperplane technique by utilizing a nonlinear optimization problem.

**Lemma 5.11** *For every* $W \in \Sigma^n_+$, *we have*

$$\begin{array}{rcll} \overline{g}(W) & = & \max & \xi^T W \xi \\ & & \text{subject to:} & \xi = \text{sign}(Br), \\ & & & \|r\|_2 = 1, \\ & & & \|B^T e_i\|_2 = 1, \quad \forall i \in \{1, 2, \ldots, n\}, \\ & & & B \in \mathbb{R}^{n \times n}, r \in \mathbb{R}^n. \end{array}$$

**Proof** At an optimal solution of the RHS problem, we have $\xi \in \{-1, 1\}^n$. Therefore, $\overline{g}(W) \geq RHS$. For the other direction, Let $\bar{x} \in \{-1, 1\}^n$ attain $\overline{g}(W)$. Take $r \in \mathbb{R}^n$, $\|r\|_2 = 1$. Define the matrix $B \in \mathbb{R}^{n \times n}$ as follows

$$B^T e_i := \begin{cases} r & \text{if } \bar{x}_i = 1, \\ -r & \text{if } \bar{x}_i = -1. \end{cases}$$

Then we have $\xi = \bar{x}$ in the RHS problem and we obtain $\overline{g}(W) \leq RHS$. □

In the above lemma, $r \in \mathbb{R}^n$ plays the role of the random hyperplane and the columns of $B^T$ play the role of $v^{(i)}$ (from the Max Cut setting). Let's define the expected value of a function $\gamma$ on the hypersphere in $\mathbb{R}^n$:

$$E_r(\gamma(r)) := \frac{\int_{r \in \mathbb{R}^n, \|r\|_2 = 1} \gamma(r) dr}{\int_{r \in \mathbb{R}^n, \|r\|_2 = 1} dr}.$$

Now, we are ready to simulate the random hyperplane technique by utilizing a nonlinear optimization problem.

**Lemma 5.12** *For every $W \in \Sigma_+^n$, we have*

$$\begin{array}{rcll} \overline{g}(W) & = & \max & E_r\left(\xi^T W \xi\right) \\ & & \text{subject to:} & \xi = \operatorname{sign}(Br), \\ & & & \|r\|_2 = 1, \\ & & & \|B^T e_i\|_2 = 1, \quad \forall i \in \{1, 2, \ldots, n\}, \\ & & & B \in \mathbb{R}^{n \times n}, r \in \mathbb{R}^n. \end{array}$$

**Proof** First of all, note that the optimization takes place in terms of the matrix variable $B$ only. Also, by the previous lemma, $\overline{g}(W) \geq RHS$. To see the other direction, let $\bar{x} \in \{-1, 1\}^n$ be an optimal solution determining $\overline{g}(W)$. Define the matrix $B \in \mathbb{R}^{n \times n}$ by

$$B^T e_i := \begin{cases} \frac{1}{\sqrt{n}} \bar{x} & \text{if } \bar{x}_i = 1, \\ -\frac{1}{\sqrt{n}} \bar{x} & \text{if } \bar{x}_i = -1. \end{cases}$$

Then

$$\begin{aligned} E_r\left[\operatorname{sign}\left(r^T B^T e_i\right) \operatorname{sign}\left(r^T B^T e_j\right)\right] &= \begin{cases} 1 & \text{if } \bar{x}_i = \bar{x}_j \\ -1 & \text{if } \bar{x}_i \neq \bar{x}_j \end{cases} \\ &= \bar{x}_i \bar{x}_j. \end{aligned}$$

Note that we used above the fact that the set of vectors $r$ which are perpendicular to $B^T e_i$ and $B^T e_j$ (then we would have $\operatorname{sign}(r^T B^T e_i) = \operatorname{sign}(r^T B^T e_j)$) has measure zero relative to the unit hypersphere. Thus, for this choice of $B$,

$$E_r\left(\xi^T W \xi\right) = \sum_{i,j} W_{ij} \bar{x}_i \bar{x}_j = \bar{x}^T W \bar{x}.$$

Therefore, $\overline{g}(W) \leq RHS$. □

The above lemma puts us into the position of being able to repeat the part of the analysis from Max Cut where we computed the expected weight of the cut by focusing on the probability that a certain edge in the graph was cut. In the current setting, we focus on the expected value of the product $x_i x_j$ which is represented by $\operatorname{sign}\langle r, B^T e_i \rangle \operatorname{sign}\langle r, B^T e_j \rangle$. For $X \in \Sigma^n$, we will write $\arcsin(X)$ to mean the matrix in $\Sigma^n$ with the $ij$th entry $\arcsin X_{ij}$.

**Theorem 5.13** *For every $W \in \Sigma_+^n$, we have*

$$\begin{array}{rcllcl} \overline{g}(W) & = & \max & \frac{2}{\pi}\langle W, \arcsin(X) \rangle & & \\ & & \text{subject to:} & \operatorname{diag}(X) & = & \bar{e}, \\ & & & X & \succeq & 0. \end{array}$$

**Proof** First of all, note that since $X \succeq 0$ and $\operatorname{diag}(X) = \bar{e}$ imply $|X_{ij}| \leq 1$ for every $i, j$, every feasible solution of the optimization problem in the statement of the theorem also have this property and therefore, the optimization problem is well-defined. The maximum is also well-defined and attained since the feasible region is nonempty and compact and the objective function is continuous. We will proceed as in the Max Cut analysis. Let $B$ be an optimal solution of the optimization problem in Lemma 5.12. Then, as before we write $v^{(i)}$ for $B^T e_i$. We have

$$E_r\left[\operatorname{sign}\left(r^T B^T e_i\right) \operatorname{sign}\left(r^T B^T e_j\right)\right]$$

$$
\begin{aligned}
&= \Pr\left\{\text{sign}\langle r, v^{(i)}\rangle = \text{sign}\langle r, v^{(j)}\rangle\right\} - \Pr\left\{\text{sign}\langle r, v^{(i)}\rangle \neq \text{sign}\langle r, v^{(j)}\rangle\right\} \\
&= 1 - 2\Pr\left\{\text{sign}\langle r, v^{(i)}\rangle \neq \text{sign}\langle r, v^{(j)}\rangle\right\} \\
&= 1 - \frac{2}{\pi}\arccos\langle v^{(i)}, v^{(j)}\rangle.
\end{aligned}
$$

The last equation above used Lemma 5.1. Since

$$\arccos\langle v^{(i)}, v^{(j)}\rangle + \arcsin\langle v^{(i)}, v^{(j)}\rangle = \frac{\pi}{2},$$

using Lemma 5.12, we obtain

$$E_r\left(\xi^T W \xi\right) = \frac{2}{\pi}\langle W, \arcsin\left(BB^T\right)\rangle,$$

where $\xi := \text{sign}(Br)$. Since the symmetric positive semidefinite matrix $BB^T$ satisfies all the constraints of the $RHS$ problem, we conclude $\overline{g}(W) \leq RHS$. Now, let $\bar{X}$ be an optimal solution of the problem on the $RHS$ of the statement of the theorem. Then find $B$ such that $BB^T := \bar{X}$. We again have the equation

$$\frac{2}{\pi}\langle W, \arcsin\left(BB^T\right)\rangle = E_r\left(\xi^T W \xi\right)$$

this time for the $B$ defined by $\bar{X}$. Using Lemma 5.12, we obtain the remaining inequality $\overline{g}(W) \geq RHS$. □

The last theorem established that we can compute the exact value of $\overline{f}(W)$ by computing the maximum value of the nonlinear function $\frac{2}{\pi}\langle W, \arcsin(X)\rangle$ over the feasible region of the SDP relaxation. Even though the feasible region of this nonlinear optimization problem is identical to that of the SDP relaxation, computing the optimal value is no easier than optimizing $\langle W, X\rangle$ over the feasible region intersected by $\text{rank}(X) = 1$. However, the formulation of Theorem 5.13 easily lends itself to a tractable estimation via the next lemma.

**Lemma 5.14** *Let $X \in \Sigma_+^n$ such that $|X_{ij}| \leq 1$, for every $i, j$. Then*

$$\arcsin(X) \succeq X.$$

**Proof** Taylor's Theorem applied to $\arcsin(u)$ at $u := 0$ yields the matrix equation

$$\arcsin(X) = X + \frac{1}{2}\frac{X \odot X \odot X}{3} + \frac{1}{2}\frac{3}{4}\frac{X \odot X \odot X \odot X \odot X}{5} + \cdots.$$

Also see the proof of Lemma 5.2. Therefore, using the observation from the first chapter that the Hadamard product of two positive semidefinite symmetric matrices is positive semidefinite, we have $\arcsin(X) \succeq X$ as desired. □

Now, we are ready to estimate the quality of the SDP relaxation using Theorem 5.13 and Lemma 5.14.

**Theorem 5.15** *For every $W \in \Sigma_+^n$, we have*

$$
\begin{array}{rcll}
\overline{f}(W) & \geq & \max & \frac{2}{\pi}\langle W, X\rangle \\
& & \text{subject to:} & \text{diag}(X) = \bar{e}, \\
& & & X \succeq 0.
\end{array}
$$

**Proof** Since $X \succeq 0$ and $\mathrm{diag}(X) = \bar{e}$ imply $|X_{ij}| \leq 1$ for every $i, j$, the above lemma applies to every feasible solution of the SDP problem on the RHS. Therefore, the inequality follows from Theorem 5.13 and Lemma 5.14. $\square$

This result guarantees an approximation ratio of $\approx 0.58$:

$$\frac{\text{value of the relax.}}{\overline{f}(W)} \leq \frac{\pi}{2}.$$

The assumption $W \succeq 0$ is rich enough to include the Max Cut problem with nonnegative weights as a special case. Even though the ratio is not as good as the Goemans-Williamson result, it still beats the previous best ratio of $(1/2)$ that was in place for 20 years as the best until the Goemans-Williamson result. Let $w_{ij} : \{i,j\} \in E$ be the given weights. Consider the (Max Cut) objective function

$$\max \frac{1}{4} \sum_{i,j} w_{ij} \left(1 - u_i u_j\right).$$

Treat the term "1" as $x_i^2$. Then it is easy to see that defining

$$4W_{ij} := \begin{cases} -w_{ij} & \text{if } \{i,j\} \in E, \\ \sum_{k:\{i,k\} \in E} w_{ik} & \text{if } i = j, \\ 0 & \text{otherwise,} \end{cases}$$

we find that the matrix $W$ is diagonally dominant and hence is positive semidefinite.

Note that Goemans-Williamson analyzed the quantity

$$\langle W, \arcsin(X) \rangle$$

by bounding the individual terms in the above inner-product

$$\sum_{i,j} W_{ij} \arcsin(X_{ij}).$$

In contrast, Nesterov's approach uses the order induced by the positive semidefinite cone.

5.4.2.2 *Arbitrary* $W$. In this subsection, we will see how the results extend to arbitrary weight matrices $W$. The set-up behaves very well under diagonal perturbations of the weight matrix $W$. Let $y \in \mathbb{R}^n$ denote the diagonal perturbation. Then for every such $y$ we have

$$\begin{aligned} \overline{f}\left(W + \mathrm{Diag}(y)\right) &= \overline{f}(W) + \bar{e}^T y, \\ \underline{f}\left(W + \mathrm{Diag}(y)\right) &= \underline{f}(W) + \bar{e}^T y. \end{aligned}$$

Also for every such $y$, we have the analogous equations for the SDP relaxations:

$$\begin{aligned} \overline{F}\left(W + \mathrm{Diag}(y)\right) &= \overline{F}(W) + \bar{e}^T y, \\ \underline{F}\left(W + \mathrm{Diag}(y)\right) &= \underline{F}(W) + \bar{e}^T y. \end{aligned}$$

The first two equations are obvious. For the last two, we used

$$\langle \mathrm{Diag}(y), X \rangle = y^T \mathrm{diag}(X) = \bar{e}^T y.$$

Therefore, all values are shifted exactly by $\bar{e}^T y$.

These observations and Theorem 5.15 are sufficient to obtain nontrivial estimates of $\underline{f}(W)$ and $\overline{f}(W)$ for arbitrary $W \in \Sigma^n$. The key idea is to modify $W$ by a suitable diagonal matrix to make the resulting matrix positive semidefinite so that we can apply Theorem 5.15. Recall that the duals of the SDP relaxations determining $\overline{F}(W)$ and $\underline{F}(W)$ have the constraints $(\mathrm{Diag}(y) - W) \succeq 0$ and

$(W - \text{Diag}(y)) \succeq 0$ respectively. So in this respect, we already have access to some "optimal" diagonal perturbations.

**Theorem 5.16** *Let $W \in \Sigma^n$. Then*

$$\underline{f}(W) \leq \frac{2}{\pi}\underline{F}(W) + \left(1 - \frac{2}{\pi}\right)\overline{F}(W) \leq \left(1 - \frac{2}{\pi}\right)\underline{F}(W) + \frac{2}{\pi}\overline{F}(W) \leq \overline{f}(W).$$

**Proof** Let $\overline{y}$ be an optimal solution of the dual of the SDP problem determining $\overline{F}(W)$:

$$\min\left\{\bar{e}^T y : \text{Diag}(y) \succeq W\right\}.$$

Then, we have

$$\begin{aligned}
\overline{F}(W) - \underline{f}(W) &= \bar{e}^T\overline{y} - \underline{f}(W) \\
&= \bar{e}^T\overline{y} + \overline{f}(-W) \\
&= \overline{f}\left(\text{Diag}(\overline{y}) - W\right) \\
&\geq \frac{2}{\pi}\overline{F}\left(\text{Diag}(\overline{y}) - W\right) \\
&= \frac{2}{\pi}\left[\bar{e}^T\overline{y} - \underline{F}(W)\right] \\
&= \frac{2}{\pi}\left[\overline{F}(W) - \underline{F}(W)\right].
\end{aligned}$$

To get the inequality, we used the fact that $\overline{y}$ is feasible in its SDP and hence $(\text{Diag}(\overline{y}) - W)$ is positive semidefinite; therefore, Theorem 5.15 applies. Similarly, we can derive

$$\overline{f}(W) - \underline{F}(W) \geq \frac{2}{\pi}\left[\overline{F}(W) - \underline{F}(W)\right].$$

□

Using convex combinations of $\overline{F}(W)$ and $\underline{F}(W)$, we can approximate $\overline{f}(W)$ within a constant fraction of $\left[\overline{f}(W) - \underline{f}(W)\right]$.

**Corollary 5.17** *For every $W \in \Sigma^n$, the value*

$$c' := \left(1 - \frac{2}{\pi}\right)\underline{F}(W) + \frac{2}{\pi}\overline{F}(W)$$

*satisfies*

$$\frac{\overline{f}(W) - c'}{\overline{f}(W) - \underline{f}(W)} < \frac{4}{7}.$$

Note that we can also handle the linear terms in the objective function. Suppose that the quadratic optimization problem is given as:

$$\max_{x \in \{-1,1\}^n}\left\{x^T W x + w^T x\right\}.$$

We add a new variable $x_0 \in \{-1, 1\}$ and maximize

$$(x_0, x^T)\begin{pmatrix} 0 & \frac{1}{2}w^T \\ \frac{1}{2}w & W \end{pmatrix}\begin{pmatrix} x_0 \\ x \end{pmatrix}$$

subject to $x_0, x_1, \ldots, x_n \in \{-1, 1\}$. Hence we eliminated the linear term at the cost of increasing the dimension by one.

This generalization of Goemans-Williamson result that we have been discussing, leading all the way to Corollary 5.17 is due to Nesterov [274]. Similar problems with some simple linear constraints are $\mathcal{NP}$-hard to approximate within a constant.

For example, Bellare and Rogaway [32] proved that optimizing a quadratic function over a convex polytope (even if it is defined by $Ax = \bar{e}, 0 \leq x \leq \bar{e}$, where $A$ has only $\{0,1\}$ entries) is hard to approximate in polynomial-time unless $\mathcal{P} = \mathcal{NP}$. It follows from their results that in terms of $\mu$-approximation, no $\mu$-approximation algorithm is possible for $\mu \leq 1/3$ unless $\mathcal{P} = \mathcal{NP}$.

Other extensions are due to Ye [372], Nesterov [273] (so-called *Matrix Cube Theorem*), and Ben-Tal, Roos, Nemirovski [36] (*Extended Matrix Cube Theorem*). Also see, Nemirovski [265]. *Matrix Cube* is simply the image of a hypercube under an affine mapping with given matrix coefficients. Then the question is "what is the largest cube whose image is contained in $\Sigma^n_+$?" I.e., given $A_0, A_1, \ldots, A_d \in \Sigma^n$, find the smallest $r \in \mathbb{R}_+$ such that

$$\left\{ A_0 + \sum_{i=1}^{d} y_i A_i : \|y\|_\infty \leq r \right\} \subseteq \Sigma^n_+.$$

We can write down an SDP (using $d$ matrix variables $X^{(i)} \in \Sigma^n$) whose feasibility guarantees the above inclusion. We simply need

$$\left(X^{(i)} \pm r A_i\right) \in \Sigma^n_+, \text{ and } \sum_{i=1}^{d} X^{(i)} \preceq A_0.$$

The above-mentioned references work with the nontrivial direction. That is "if the above SDP is infeasible, then for what value of $r$ can we say that the matrix cube is not contained in $\Sigma^n_+$?" For some related results, see Luo, Sidiropoulos, Tseng and Zhang [247] and Nie and Sturmfels [279].

On a somewhat related front, Bomze and DeKlerk [53] gave a polynomial-time approximation scheme (an algorithm whose complexity, to obtain a $(1 \pm \epsilon)$-approximate solution, can be bounded by a polynomial function of the size of the input and $\frac{1}{\epsilon}$) for the $\mathcal{NP}$-hard quadratic optimization problem over a simplex ($W \in \Sigma^n$ is given):

$$\begin{array}{ll} \min & x^T W x \\ \text{subject to:} & \bar{e}^T x = 1, \\ & x \geq 0. \end{array}$$

Motivated by dynamical system analysis and design, Megretski [254] considers quadratic optimization problems, their SDP relaxations and proves approximations bounds. Another related result is due to Alon and Naor [8]. Given $W \in \mathbb{R}^{m \times n}$, the *cut norm of* $W$ is

$$\|W\|_{\text{cut}} := \max \left\{ \left| \sum_{i \in I, j \in J} W_{ij} \right| : I \subseteq \{1, 2, \ldots, m\}, J \subseteq \{1, 2, \ldots, n\} \right\}.$$

They prove that computing the Cut Norm is MAXSNP-hard. They give an approximation algorithm with performance ratio 0.56. They utilize Semidefinite Optimization, a rounding technique and a classical inequality of Grothendieck. Consider the problem

$$\begin{array}{ll} \max & v^T W x \\ \text{subject to:} & x \in \{-1, +1\}^n, \\ & v \in \{-1, +1\}^m. \end{array}$$

We have the SDP relaxation

$$\begin{array}{rrcl} \max & \langle \tilde{W}, X \rangle & & \\ \text{subject to:} & \operatorname{diag}(X) & = & \bar{e}, \\ & X & \in & \Sigma_+^{m+n}, \end{array}$$

where, we represented each of the $m+n$ variables $v_1, v_2, \ldots, v_m, x_1, x_2, \ldots, x_n$ by vectors in $\mathbb{R}^{m+n}$ and the $(i,j)$ component of our matrix variable $X$ represents the inner product of the $i$th and the $j$th vectors from that list. Then, in the objective function, we want $\sum_{i,j} W_{ij} \langle v^{(i)}, v^{(j)} \rangle$; we choose $\tilde{W} \in \Sigma^{m+n}$ such that

$$\langle \tilde{W}, X \rangle = \sum_{i,j} W_{ij} \langle v^{(i)}, v^{(j)} \rangle.$$

E.g.,

$$\tilde{W} := \frac{1}{2} \begin{bmatrix} 0 & W \\ W^T & 0 \end{bmatrix}.$$

Grothendieck's inequality implies that the optimal objective value of the SDP relaxation is at most a constant times the optimal objective value of the previous quadratic $\{-1, +1\}$ optimization problem. The constant is called the Grothendieck's constant and is known to be in the interval

$$\left[ \frac{\pi}{2}, \frac{\pi}{2 \ln(\sqrt{2}+1)} \right].$$

## 5.5 Further Bibliographical Notes

Shor pioneered a lot of the convex quadratic relaxation work [331, 330]. The SDP relaxation for Max Cut was stated and studied before the work of Goemans and Williamson (see for instance, Delorme and Poljak [104], Poljak and Rendl [292], Poljak, Rendl and Wolkowicz [293]). However, no worst-case approximation ratio better than $1/2$ was known before the GW work.

For a vast amount of information on Max Cut and closely related problems, see the book by Deza and Laurent [106].

It is $\mathcal{NP}$-hard to approximate Max Cut with an approximation ratio $\frac{16}{17} + \epsilon$, for any $\epsilon > 0$. For hardness of approximation results, see Håstad [163].

Max 3-Cut was studied by Frieze and Jerrum [132] who gave a 0.800217 approximation algorithm (based on SDP). Later the approximation ratio was improved by Goemans and Williamson [144] as well as de Klerk, Pasechnik, Warners [103] (independently of GW) to 0.836 (again based on SDP). In fact, GW use Semidefinite Optimization over Hermitian positive semidefinite matrices over the complex numbers (the related cone is a symmetric cone, our results about SDPs and the positive semidefinite cone easily apply to that case too). Of course, at the cost of doubling the dimension, we can also represent any convex conic optimization problem over Hermitian positive semidefinite cones over complex numbers as ordinary SDPs over the real numbers.

For some other applications of semidefinite optimization in combinatorial optimization, see Karger, Motwani, Sudan [188] for 3-coloring; see Kleinberg and Goemans [202], Charikar [72] and Hatami, Magen and Markakis [166] for various results related to SDP relaxations of Vertex Cover. It is $\mathcal{NP}$-hard to approximate Max 3Sat with an approximation ratio $\frac{7}{8} + \epsilon$, for any $\epsilon > 0$. Karloff and Zwick [191] gave an approximation algorithm with the approximation ratio $\frac{7}{8}$ (hence the best possible unless $\mathcal{P} = \mathcal{NP}$).

In this section, we kept utilizing the Random-Hyperplane Technique to obtain a feasible solution of the combinatorial optimization problem from an approximately optimal solution $X$ of the SDP relaxation. It is also possible to obtain good approximation guarantees by treating $X$ as the covariance matrix of a distribution and then sampling from this underlying distribution or its close variants. See Bertsimas and Ye [42]. Another connection is as follows: Suppose our matrix variable is of the form

$$\begin{bmatrix} 1 & x^T \\ x & X \end{bmatrix}$$

and $X = xx^T$ leads to an exact formulation of the problem at hand. Then our SDP relaxation replaces $X = xx^T$ by the condition that the matrix variable is positive semidefinite. The latter is equivalent to (by the Schur Complement lemma), $X \succeq xx^T$. So, for an approximately optimal solution of the SDP relaxation we have a measure of closeness to solutions of $X = xx^T$. Namely, the positive semidefinite matrix $\left(X - xx^T\right)$. We can treat this *gap* matrix $\left(X - xx^T\right)$ as a covariance matrix which could be used in generating "confidence intervals" for the feasible solutions we generate by randomized algorithms (that are similar to the Random Hyperplane Technique).

We emphasized many times that if we can obtain a rank-1 optimal solution to our SDP relaxations, we would solve the original problem exactly. So, Ye and Zhang [334] take a different but related viewpoint by relaxing the linear equality constraints $\mathcal{A}(X) = b$ and requiring $X$ to be symmetric, positive semidefinite as well as rank-1. They obtain some bounds on $\|b - \mathcal{A}(X)\|$ for their rank-1 solutions. Their work also has some connection to the approximate geometric embedding lemma of Johnson and Lindenstrauss (see [185] and [253]).

CHAPTER 6

# Geometric Representations of Graphs

Geometric representation of graphs was born as soon as the first graph was drawn. This area has grown tremendously and made very important, deep links to various areas of mathematics and applications. In this chapter, we will get a glimpse of the area of geometric representations of graphs and see a few branches of the area that have clearly deep connections to Semidefinite Optimization and the chapters to come...

We will first talk about $L_1$ embeddability which has some connections to the SDP relaxations we worked on for Max Cut, Max 2Sat, and general quadratic optimization over binary variables. Then we will study some ways of approximating the sparsest cut in graphs via SDP. Next, we will focus on the unit distance representations, including those on the surface of the unit hypersphere. Such representations will naturally lead us to orthonormal representations of graphs and the underlying linear inequality constraints for the stable set problem. Through this connection, we will study the Lovász $\theta$ number of a graph and also briefly touch on the Shannon Capacity of a communication channel.

## 6.1 $L_1$ embeddability

Recall the formulation we used when we showed that the SDP relaxation for the Max Cut problem can be treated as a special case of the SDP relaxation for the quadratic optimization problem

$$\max_{x \in \{-1,1\}^n} x^T W x,$$

where $W \in \Sigma_+^n$. To prove this claim, we took an arbitrary instance of the Max Cut problem with input $G = (V, E)$ and weights $w_{ij}, \{i,j\} \in E$, and defined

$$W_{ij} := \begin{cases} -w_{ij} & \text{if } \{i,j\} \in E, \\ \sum_{k:\{i,k\} \in E} w_{ik} & \text{if } i = j, \\ 0 & \text{otherwise.} \end{cases}$$

The matrix that we defined is called the *Laplacian matrix of $G$ with weights $w_{ij}$*. For any matrix $M \in \Sigma^n$, we define the *Laplacian of $M$* as

$$[L(M)]_{ij} := \begin{cases} -m_{ij} & \text{if } \{i,j\} \in E, \\ \sum_{k:\{i,k\} \in E} m_{ik} & \text{if } i = j, \\ 0 & \text{otherwise.} \end{cases}$$

Usually, when the Laplacian is used on a matrix $M$, the matrix $M$ has the properties

- $\operatorname{diag}(M) = 0$, and
- every entry of $M$ is nonnegative.

Let us denote by $L(G_w)$ the Laplacian of the weighted graph for a Max Cut instance. Then the dual of the SDP relaxation becomes

$$\begin{array}{lrcl} \min & \frac{1}{4}\bar{e}^T y & & \\ \text{subject to:} & \mathrm{Diag}(y) & \succeq & L(G_w). \end{array}$$

Now, we turn to metrics on graphs. Let us consider distance functions $w : V \oplus V \to \mathbb{R}_+$. That is

- $w(i,i) = 0, \forall i \in V$,
- $w(i,k) \leq w(i,j) + w(j,k), \forall i,j,k \in V$.

It is sometimes convenient to represent the distance function $w$ as a symmetric matrix $W \in \Sigma^V$ with zero diagonal.

Let $p \in [1, +\infty]$. We say that a distance function $w$ on graph $G = (V,E)$ is $L_p$-embeddable in $\mathbb{R}^d$ if there exists $\{u^{(i)} \in \mathbb{R}^d : i \in V\}$ such that

$$\|u^{(i)} - u^{(j)}\|_p = w(i,j), \forall i,j \in E.$$

Now, let us represent each cut $S$ in $G$ by the edges of $G$ that it *cuts*, that is $\delta(S)$. Consider the 0,1 incidence vectors of such cuts $\delta(S)$ in $\mathbb{R}^E$. Then the set of all nonnegative linear combinations of these incidence vectors define a polyhedral convex cone in $\mathbb{R}^E$. We denote this cone by $K_{\mathrm{Cut}}(G)$ and call it the *Cut Cone of* $G$.

The following well-known theorem ties the Max Cut problem to $L_1$-embeddability and proves, among other things, that deciding $L_1$-embeddability is $\mathcal{NP}$-hard.

**Theorem 6.1** *Let $G = (V,E)$ be a graph and $w$ be a distance function on $G$. Then, the distance function $w$ on graph $G = (V,E)$ is $L_1$-embeddable in $\mathbb{R}^V$ iff $w \in K_{\mathrm{Cut}}(G)$.*

$L_2$-embeddability is even more closely connected to the SDP relaxation for Max Cut. A distance function $w$ is $L_2$-embeddable (in $\mathbb{R}^V$) iff

$$(W \odot W)e_1\bar{e}^T + \bar{e}e_1^T(W \odot W) \succeq W \odot W,$$

where we could have replaced $e_1$ by any other unit vector $e_j$ (the conclusion is independent of which $j$ we choose). Therefore, in contrast to $L_1$-embeddability, $L_2$-embeddability seems easier (we just determine if the above positive semidefiniteness condition is satisfied).

Let us define the convex cone

$$K_2 := \left\{X \in \Sigma^V : \mathrm{diag}(X) = 0,\ Xe_1\bar{e}^T + \bar{e}e_1^T X - X \succeq 0\right\}.$$

Then, $w$ is $L_2$-embeddable iff $(W \odot W) \in K_2$. The dual of cone $K_2$ is

$$K_2^* = \left\{S \in \Sigma^V : \mathrm{diag}(S) = 0, L(S) \succeq 0\right\}.$$

The way that the Max Cut SDP relaxation approximates the Max Cut value in $G$, the SDP formulation for $L_2$-embeddability can be considered an approximation to $L_1$-embeddability.

## 6.2 Approximating the Sparsest Cuts via SDP

One of the most fundamental tools in graph theory (and its applications) is decomposition. Given a graph, if we can separate it into two roughly equal size pieces with the "interaction" between the pieces "minimized," we would be in a good position to make progress.

One basic notion looks at the sizes of the cuts in the graph and normalizes the cardinality of the set of edges in each cut with the cardinality of the smaller shore of it. Determining the minimum of this ratio is the *Sparsest Cut Problem*:

$$\operatorname{expand}(G) := \min_{U \subset V : |U| \leq \frac{|V|}{2}} \left\{ \frac{|\delta(U)|}{|U|} \right\}.$$

In a seminal work (in 1988), Leighton and Rao [226] gave an $O(\ln(|V|))$-approximation algorithm for the sparsest cut problem. This approximation factor was significantly improved to $O\left(\sqrt{\ln(|V|)}\right)$ by Arora, Rao and Vazirani [13]. This latter work fits into our current discussion very well since it used SDP relaxations and is related to geometric embeddings of graphs (and many other subjects as we describe below).

Let $G = (V, E)$ be given. An *$L_2^2$-representation of $G$* is $v : V \to \mathbb{R}^d$ such that

$$\|v^{(i)} - v^{(j)}\|_2^2 + \|v^{(j)} - v^{(k)}\|_2^2 \geq \|v^{(i)} - v^{(k)}\|_2^2, \quad \forall i, j, k.$$

Up to a factor of $|V|$, the optimal objective function value of the following optimization problem is a valid lower bound on $\operatorname{expand}(G)$:

$$\begin{array}{rrcl} \min & \sum_{\{i,j\} \in E} \|v^{(i)} - v^{(j)}\|_2^2 & & \\ \text{subject to:} & \|v^{(i)} - v^{(j)}\|_2^2 + \|v^{(j)} - v^{(k)}\|_2^2 & \geq & \|v^{(i)} - v^{(k)}\|_2^2, \quad \forall i, j, k, \\ & \sum_{i<j} \|v^{(i)} - v^{(j)}\|_2^2 & = & 1, \\ & v^{(1)}, v^{(2)}, \ldots, v^{(|V|)} & \in & \mathbb{R}^{|V|}. \end{array}$$

Equivalently, we have the SDP:

$$\begin{array}{rrcl} \min & \sum_{\{i,j\} \in E} (X_{ii} + X_{jj} - 2X_{ij}) & & \\ \text{subject to:} & X_{ij} + X_{jk} - X_{ik} & \leq & X_{jj} \quad \forall i, j, k, \\ & \sum_{i<j} (X_{ii} + X_{jj} - 2X_{ij}) & = & 1, \\ & X & \in & \Sigma_+^V. \end{array}$$

This SDP has Slater points. E.g., consider $\bar{X} := \frac{1}{|V|(|V|-1)} I$. Similarly, its dual also has Slater points. To see this, notice that the dual constraints are equivalent to

$$\left(\sum y_{ijk} A^{(i,j,k)}\right) - \eta \left(|V| I - \bar{e}\bar{e}^T\right) \preceq C, \text{ and } y_{ijk} \leq 0, \quad \forall i, j, k,$$

where $A^{(i,j,k)}$ is suitably defined so that $\langle A^{(i,j,k)}, X \rangle = X_{ij} + X_{jk} - X_{ik} - X_{jj}$ (similarly, $C$ is defined). Hence, setting $\bar{y} := 0$, $\bar{\eta}$ large enough (e.g., $\bar{\eta} := \|C\|_2 + 1$),

$$C + \bar{\eta} \left(|V| I - \bar{e}\bar{e}^T\right) \succ 0,$$

we conclude that the dual also has Slater points. Therefore, by a corollary of Strong Duality Theorem, both problems attain their optimal objective values and these values coincide.

Now, let us see how the optimal objective value of the SDP, denoted by $OPT(SDP)$, can be used to give a lower bound on $\operatorname{expand}(G)$. Let $U \subset V$ be a nonempty set that is otherwise arbitrary. Define

$$\bar{x}_v := \frac{1}{2} \begin{cases} \frac{1}{\sqrt{|U|(|V|-|U|)}}, & \text{if } v \in U, \\ -\frac{1}{\sqrt{|U|(|V|-|U|)}}, & \text{otherwise.} \end{cases}$$

Then, let $\bar{X} := \bar{x}\bar{x}^T$. Clearly, $\bar{X} \succeq 0$. Moreover, for every triple,

$$\begin{aligned}\bar{X}_{ij} + \bar{X}_{jk} - \bar{X}_{ik} &\leq \frac{1}{2|U|(|V|-|U|)} - \frac{1}{4|U|(|V|-|U|)} \\ &= \frac{1}{4|U|(|V|-|U|)} \\ &= \bar{X}_{jj},\end{aligned}$$

where we used the fact that either all three terms have the same sign or exactly two of them have the same sign. We also have

$$\sum_{i<j} (\bar{X}_{ii} + \bar{X}_{jj} - 2\bar{X}_{ij}) = \sum_{\{i,j\}\in\delta(U)} \frac{1}{|U|(|V|-|U|)} = \frac{|\delta(U)|}{|U|(|V|-|U|)} \leq 1.$$

Therefore $\bar{X}$ is feasible in our SDP. Its objective value is

$$\frac{|\delta(U)|}{|U|(|V|-|U|)}.$$

Now, let $\bar{U}$ be a cut such that

$$1 \leq |\bar{U}| \leq \frac{|V|}{2} \text{ and } \text{expand}(G) = \frac{|\delta(\bar{U})|}{|\bar{U}|}.$$

Then

$$OPT(SDP) \leq \frac{1}{|V|-|\bar{U}|}\text{expand}(G)$$

and we conclude

$$\frac{|V|}{2} OPT(SDP) \leq \text{expand}(G).$$

We will not get into the rest of the details of Arora-Rao-Vazirani work as it would take us away from our main discussion of the SDP techniques. However, we just mention that an approximately optimal solution $\bar{X}$ of the SDP relaxation can be used to guide a "rounding procedure" reminiscent of the Random Hyperplane Technique after finding $B$ such that $BB^T = \bar{X}$ and defining

$$\left[v^{(1)}, v^{(2)}, \ldots, v^{(|V|)}\right] := B^T.$$

Indeed, $\{v^{(i)} : i \in V\}$ is a $L_2$-representation of $G$.

A closely related problem is the so-called *c-balanced graph-separator problem* (in which given a constant $c \in (0,1)$, we want to find a cut attaining):

$$\text{expand}(G,c) := \min_{U\subset V:|U|\geq c|V| \text{ and } |V\setminus U|\geq c|V|} \left\{\frac{|\delta(U)|}{|U|}\right\}.$$

$c$-balanced graph-separator problem clearly captures the crux of a basic, unsophisticated divide-and-conquer approach on graphs. We want to break the graph into *two pieces* one supported by $U$ and the other supported by $(V \setminus U)$; we want each piece to *substantial* (i.e., $|U| \geq c|V| = \Theta(|V|)$ and $|V \setminus U| \geq c|V| = \Theta(|V|)$); finally, we want the *scaled interaction* between the pieces $\frac{|\delta(U)|}{|U|}$ to be minimum.

A very closely related concept is that of *expander graphs*. These concepts also have fundamental relevance in the theory of Markov Chains. Let $W \in [0,1]^{n\times n}$ be the transition matrix of an ergodic ($W$ is irreducible) Markov Chain. Let $p \in$

$[0,1]^n$ denote the stationary distribution. Assume that the Markov Chain is *time-reversible* that is

$$W_{ij}p_j = W_{ji}p_i, \quad \forall i,j.$$

Define $\gamma_{ij} := p_i W_{ij}$, $\forall i,j$. Then the *conductance* of the Markov Chain is

$$\Phi(W) := \min_{U: 0 < \sum_{i \in U} p_i \leq \frac{1}{2}} \left\{ \frac{\sum_{i \in U, j \notin U} \gamma_{ij}}{\sum_{i \in U} p_i} \right\}.$$

Under these definitions, assume that $\mathrm{diag}(W) > 0$. Then

$$1 - |\lambda_2(W)| = \Omega(\Phi(W)) \text{ and } \Phi(W) = \Omega(1 - |\lambda_2(W)|).$$

In the current context, $\lambda_1(W) = 1$. So,

$$(1 - |\lambda_2(W)|) = \lambda_1(W) - |\lambda_2(W)|$$

is the *spectral gap*. Here, conductance and hence the spectral gap characterize the mixing rate of the random walks etc. on the underlying Markov Chain.

Let us turn to the $c$-balanced graph-separator problem, and $\mathrm{expand}(G,c)$. An SDP relaxation to get a lower bound on $\mathrm{expand}(G,c)$ is (up to a scaling factor of $c|V|$):

$$\begin{array}{lrcl} \min & \frac{1}{2}\sum_{\{i,j\} \in E}(1 - X_{ij}) & & \\ \text{subject to:} & \mathrm{diag}(X) & = & \bar{e}, \\ & X_{ij} + X_{jk} - X_{ik} & \leq & 1, \quad \forall i,j,k, \\ & \sum_{i<j} X_{ij} & = & \frac{|V|(|V|-1)}{2} - 2c(1-c)|V|^2, \\ & X & \in & \Sigma_+^V. \end{array}$$

Note that $\bar{X} := I$ is a Slater point for this SDP. The dual constraints have the form

$$\mathrm{Diag}(z) + \cdots \preceq \cdots.$$

Hence, taking $\bar{z} := -\eta\bar{e}$ for $\eta$ large enough and setting all the other dual variables to zero gives Slater points for the dual. Again the corollary to Strong Duality Theorem applies...

## 6.3 Unit Distance Representations of Graphs

Let $G = (V,E)$ be given. *A unit distance representation of* $G$ is $v : V \to \mathbb{R}^d$ for some $d \geq 1$ such that

$$\|v^{(i)} - v^{(j)}\|_2 = 1, \quad \forall \{i,j\} \in E.$$

There are many interesting questions/problems about such representations and their relatives:

- What is the smallest $d$ such that a unit distance representation of $G$ exists?
- Given "distances" $w_{ij} \in \mathbb{R}_+^{\mathcal{S}}$, for $\mathcal{S} \subseteq E$ find the smallest dimension $d$ such that a representation of $G$ in $\mathbb{R}^d$ with these prescribed distances exists.
- (For any $d$) What is the radius of the smallest Euclidean ball containing such unit distance representation?

The next theorem establishes the fact that we can answer the last question exactly by solving an SDP problem!

**Theorem 6.2** *Let $G = (V, E)$ be a given graph. Then, the optimal objective value of the following semidefinite optimization problem is finite, it is attained and is equal to the square of the radius of the smallest ball containing a unit distance representation of $G$:*

$$\begin{array}{rlrcll} t(G) := & \min & & t & & \\ & \text{subject to:} & \operatorname{diag}(X) - t\bar{e} & \leq & 0, & \\ & & X_{ii} - 2X_{ij} + X_{jj} & = & 1, & \forall \{i,j\} \in E, \\ & & X & \in & \Sigma_+^V. & \end{array}$$

**Proof** Let us define

$$A^{\{i,j\}} := e_i e_i^T + e_j e_j^T - \left(e_i e_j^T + e_j e_i^T\right), \forall \{i,j\} \in E.$$

Then the dual of the SDP problem given in the statement of the theorem is:

$$\begin{array}{llcl} \max & \displaystyle\sum_{\{i,j\} \in E} z_{ij} & & \\ \text{subject to:} & \operatorname{Diag}(y) + \displaystyle\sum_{\{i,j\} \in E} z_{ij} A^{\{i,j\}} & \preceq & 0, \\ & -\bar{e}^T y & = & 1, \\ & y & \leq & 0. \end{array}$$

The dual SDP above has Slater points; e.g., set $\bar{y} := -\frac{1}{n}\bar{e}$, $\bar{z} := 0$. The primal SDP also has Slater points, for instance $\bar{X} := \frac{1}{2}I$, $\bar{t} := \frac{1}{2} + \epsilon$, for any $\epsilon > 0$. Therefore, by the corresponding corollary of the Strong Duality Theorem, both problems have finite optimal objective values which coincide. Moreover, both SDPs attain their optimal values.

We first note that *any* unit distance representation of $G$ in $\mathbb{R}^d$ for *any* $d$ can be represented in $\mathbb{R}^V$ with the same radius Euclidean ball containing the representation of $G$ (for $d \leq |V|$ the claim is obvious; for $d > |V|$, note that the representation of $G$ lies in a $|V|$ dimensional affine subspace of $\mathbb{R}^d$, so this too is obvious).

Let

$$\left\{v^{(i)} \in \mathbb{R}^V : i \in V\right\}$$

denote an optimal (contained in the smallest ball) unit distance representation of $G$ in $\mathbb{R}^V$. (Such representation exists; e.g., we can pose the problem as an optimization problem with a continuous objective function over a nonempty compact set.) Without loss of generality, the minimum radius Euclidean ball containing the unit distance representation is centered at the origin. Therefore,

$$r := \max_{i \in V} \left\{\|v^{(i)}\|_2\right\}$$

is the radius of that ball. We define

$$B^T := \left[v^{(1)}, v^{(2)}, \ldots, v^{(|V|)}\right]$$

as before. We then set $\bar{X} := BB^T$. By construction,

$$\bar{X} \succeq 0 \text{ and } \bar{X}_{ij} = \langle v^{(i)}, v^{(j)} \rangle, \forall i, j.$$

We observe that

$$\operatorname{diag}(\bar{X}) \leq \max_{i \in V} \left\{\|v^{(i)}\|_2^2\right\} \bar{e} = r^2 \bar{e},$$

and

$$\bar{X}_{ii} - 2\bar{X}_{ij} + \bar{X}_{jj} = \|v^{(i)} - v^{(j)}\|_2^2 = 1, \forall \{i,j\} \in E.$$

Thus, $(\bar{X}, r^2)$ is feasible in the primal SDP with objective value $r^2$.

Now, let $(\bar{X}, \bar{t})$ be an optimal solution of the primal SDP. Then there exists $B \in \mathbb{R}^{V \times V}$ such that $BB^T = \bar{X}$. Let $v^{(i)}$ denote the $i$th column of $B^T$. From the constraints of the primal SDP, we immediately obtain

$$\|v^{(i)} - v^{(j)}\|_2 = 1, \forall \{i,j\} \in E,$$

that is, $\left\{v^{(i)} : i \in V\right\}$ form a unit distance representation of $G$ in $\mathbb{R}^V$, and

$$\|v^{(i)}\|_2 \leq \sqrt{\bar{t}},$$

that is, the representation is contained in the Euclidean ball centered at the origin with radius $\sqrt{\bar{t}}$.

Therefore, the optimal objective value of the SDP is precisely the square-root of the smallest Euclidean ball in $\mathbb{R}^V$ containing a unit distance representation of $G$. □

Note that the dual problem also has an interpretation. Each edge $\{i,j\} \in E$ has a weight $z_{ij} \in \mathbb{R}$ to be determined such that for some $y \in \mathbb{R}^V_+$ with $\bar{e}^T y = 1$,

$$\operatorname{Diag}(y) - Z \succeq 0, \text{ and } \sum_{\{i,j\} \in E} z_{ij} \text{ is maximized},$$

where

$$Z_{ij} := \begin{cases} \sum_{k:\{i,k\} \in E} z_{ik}, & \text{if } i = j, \\ -z_{ij}, & \text{if } \{i,j\} \in E, \\ 0, & \text{otherwise.} \end{cases}$$

We observe that the total weight assigned to the edges incident to a node $i$ cannot exceed 1, since each $y_i \leq 1$ and $y_i$s give an upper bound on the total weight of edges incident on a node. So the dual problem could be interpreted in two stages: we want to distribute a unit allowance to the nodes (decide on the values of $y_i \geq 0$) and then assign weights $z_{ij}$ on the edges such that $\operatorname{Diag}(y) - Z \succeq 0$ and the sum of all the edge weights is maximized.

In fact, this problem has connections to the Max Cut SDP relaxation as well. Recall our discussion on $L_1$-embeddability in the previous section. In the notation of the previous section, the dual SDP is

$$\begin{array}{lrcl} \max & \sum_{\{i,j\} \in E} z_{ij} & & \\ \text{subject to:} & \operatorname{Diag}(y) & \succeq & L(G_z), \\ & \bar{e}^T y & = & 1, \\ & y & \in & \mathbb{R}^V_+. \end{array}$$

Note the similarity as well as the dissimilarity (that now the weights $z_{ij}$ are variables, in Max Cut problem they were given).

**6.3.1 Symmetries in Graph Representations.** In many cases, the given graph $G$ has various symmetries. These symmetries can be exploited very nicely in SDP formulations and SDP relaxations (convexity of the relaxations and the invariance of positive semidefiniteness under such symmetries are very helpful here).

Suppose that $G$ is node-symmetric (or vertex-transitive). That is, for every $i, j \in V$ there exists $f : V \to V$, $f \in \operatorname{Aut}(G)$ such that $f(i) = j$. Let $\bar{X} \in \Sigma^V_+$ be an optimal solution of an SDP formulation associated with the graph $G$. Suppose that the objective function and the constraints (other than the semidefiniteness constraint) are invariant under every $f \in \mathrm{Aut}(G)$. The action of $f$ on the matrix variable $X$ can be represented by the linear operator

$$P \cdot P^T : X \mapsto PXP^T,$$

where $P \in \{0,1\}^{V \times V}$ is a permutation matrix. Note that $P \cdot P^T \in \mathrm{Aut}(\Sigma_+^V)$. Let $\Gamma \subseteq \mathrm{Aut}(G)$ denote the set of automorphisms under which all the constraints and the objective function are invariant. Therefore, by the convexity of the set of optimal solutions of the SDP,

$$\hat{X} := \frac{1}{|\Gamma|} \sum_{P \cdot P^T \in \Gamma} P\bar{X}P^T$$

is an optimal solution of the underlying SDP. In the view of this, we can reduce the number of unknowns in our variable matrix. Such a technique has been used in many areas over the years (mostly in theory; however, recently there has been significant interest in using it to speed up computation and to reduce memory requirements as well).

When a given problem has enough symmetries, the given SDP can be solved as an LP. An elegant partial characterization is given in the next theorem. But, first, let's see some nice examples.

Let us focus on $G$ being the 5-cycle: $C_5$. Automorphisms of $C_5$ are given by the *dihedral group*. We can express them as the permutation matrices $P$ (in accordance with our discussion above):

$$I, \begin{bmatrix} 1&0&0&0&0\\0&0&0&0&1\\0&0&0&1&0\\0&0&1&0&0\\0&1&0&0&0 \end{bmatrix}, \begin{bmatrix} 0&1&0&0&0\\1&0&0&0&0\\0&0&0&0&1\\0&0&0&1&0\\0&0&1&0&0 \end{bmatrix}, \begin{bmatrix} 0&0&1&0&0\\0&1&0&0&0\\1&0&0&0&0\\0&0&0&0&1\\0&0&0&1&0 \end{bmatrix},$$

$$\begin{bmatrix} 0&0&0&1&0\\0&0&1&0&0\\0&1&0&0&0\\1&0&0&0&0\\0&0&0&0&1 \end{bmatrix}, \begin{bmatrix} 0&0&0&0&1\\0&0&0&1&0\\0&0&1&0&0\\0&1&0&0&0\\1&0&0&0&0 \end{bmatrix}, \begin{bmatrix} 0&0&0&0&1\\1&0&0&0&0\\0&1&0&0&0\\0&0&1&0&0\\0&0&0&1&0 \end{bmatrix},$$

$$\begin{bmatrix} 0&0&0&1&0\\0&0&0&0&1\\1&0&0&0&0\\0&1&0&0&0\\0&0&1&0&0 \end{bmatrix}, \begin{bmatrix} 0&0&1&0&0\\0&0&0&1&0\\0&0&0&0&1\\1&0&0&0&0\\0&1&0&0&0 \end{bmatrix}, \begin{bmatrix} 0&1&0&0&0\\0&0&1&0&0\\0&0&0&1&0\\0&0&0&0&1\\1&0&0&0&0 \end{bmatrix}.$$

Denoting the above ten matrices by $P_i$, we write

$$\hat{X} := \frac{1}{10} \sum_{i=1}^{10} P_i X P_i^T.$$

We see that $\hat{X}$ has three different entries:

- the same entry on the diagonal (call it $t$),
- the same entry on $\hat{X}_{ij}$ when $\{i,j\}$ is an edge in $C_5$ (call its value $z$),
- the same entry on $\hat{X}_{ij}$ when $\{i,j\}$ is not an edge in $C_5$ (call its value $\bar{z}$).

Then denoting by $A(C_5) \in \{0,1\}^{5\times 5}$ the node-node incidence matrix of $C_5$, we observe that

$$\hat{X} = (z - \bar{z})A(C_5) + (t - \bar{z})I + \bar{z}\left(\bar{e}\bar{e}^T\right).$$

The eigenvalues of $A(C_5)$ are 2 (corresponding to $\bar{e}$ and with multiplicity one), $\frac{\sqrt{5}-1}{2}$ (with multiplicity 2), and $-\frac{\sqrt{5}+1}{2}$ (with multiplicity 2). Since the matrices $A(C_5)$, $I$ and $\bar{e}\bar{e}^T$ can be simultaneously diagonalized (using the eigenspace decomposition given by $A(C_5)$), positive semidefiniteness condition on $\hat{X}$ becomes equivalent to linear inequalities involving the variables $t, z, \bar{z}$. Unit norm constraint becomes a single linear equation: $z = t - \frac{1}{2}$. Using this equation, we eliminate the variable $z$ from the formulation and thus the SDP reduces to the following two-variable LP with variables $t$ and $\bar{z}$:

$$\begin{array}{lrcl} \min & t & & \\ \text{subject to:} & 3t + 2\bar{z} & \geq & 1, \\ & t - \bar{z} & \geq & \frac{3-\sqrt{5}}{4}, \\ & -t + \bar{z} & \geq & -\frac{3+\sqrt{5}}{4}. \end{array}$$

The optimal solution of the LP is given by $\hat{t} := \frac{1}{2} - \frac{\sqrt{5}}{10}$, $\hat{\bar{z}} := \frac{3\sqrt{5}-5}{20}$. Together with the implied value $\hat{z} := -\frac{\sqrt{5}}{10}$, they describe all of the entries of an optimal solution $\hat{X}$ of the SDP problem. Note that $\hat{X}$ has rank two. The dual LP

$$\begin{array}{lrcl} \max & y_1 + \frac{3-\sqrt{5}}{4}y_2 - \frac{3+\sqrt{5}}{4}y_3 & & \\ \text{subject to:} & 3y_1 + y_2 - y_3 & = & 1, \\ & 2y_1 - y_2 + y_3 & = & 0, \\ & y_1, y_2, y_3 & \geq & 0 \end{array}$$

has the optimal solution $\hat{y}_1 := 1/5$, $\hat{y}_2 := 2/5$, $\hat{y}_3 := 0$.

As the next example, let us take as $G$ the Petersen graph which has a very rich automorphism group. For every pair of nodes $i, j$ there exists $f \in \text{Aut}(G)$ such that $f(i) = j$ and $f(j) = i$. For every pair of edges $\{i,j\}$ and $\{k,\ell\} \in E$, there exists $f \in \text{Aut}(G)$ such that $f$ maps $\{i,j\}$ to $\{k,\ell\}$ and $\{k,\ell\}$ to $\{i,j\}$. Finally, for every pair of non-edges $\{i,j\}$ and $\{k,\ell\} \in \bar{E}$, there exists $f \in \text{Aut}(G)$ such that $f$ maps $\{i,j\}$ to $\{k,\ell\}$ and $\{k,\ell\}$ to $\{i,j\}$.

Then applying the above symmetry averaging scheme, we conclude that there exists an optimal solution $\hat{X}$ of the SDP formulation of the unit distance representation problem for the Petersen graph such that

- $\hat{X}_{ii} = t, \forall i \in V$,
- $\hat{X}_{ij} = z, \forall \{i,j\} \in E$,
- $\hat{X}_{ij} = \bar{z}, \forall \{i,j\} \in \bar{E}$.

Let $A(G) \in \{0,1\}^{V\times V}$ denote the node-node incidence matrix of the Petersen graph. Then

$$\hat{X} = (z - \bar{z})A(G) + (t - \bar{z})I + \bar{z}\left(\bar{e}\bar{e}^T\right).$$

Also, the eigenvalue decomposition of $A(G)$ is very well-known. The eigenvalues of $A(G)$ are:

- $\lambda_1 = 3$ with the corresponding eigenvector $\bar{e}$,
- $\lambda_2 = \lambda_3 = \lambda_4 = \lambda_5 = \lambda_6 = 1$,
- $\lambda_7 = \lambda_8 = \lambda_9 = \lambda_{10} = -2$.

In the expression for $\hat{X}$ (using the fact that $\bar{e}$ is an eigenvector of $A(G)$), we see that the orthonormal matrix diagonalizing $A(G)$ also diagonalizes $\hat{X}$. Therefore, the eigenvalues of $\hat{X}$ are:

$(3z+6\bar{z}+t), (z-2\bar{z}+t)$ with multiplicity 5 , and $(-2z+\bar{z}+t)$ with multiplicity 4.

So, the SDP formulation can be solved by solving the 3-variable LP:

$$\begin{array}{lrcl} \min & & & t \\ \text{subject to:} & -3z-6\bar{z} & \le & t, \\ & -z+2\bar{z} & \le & t, \\ & 2z-\bar{z} & \le & t, \\ & 2t-2z & = & 1. \end{array}$$

Since $t$ and $z$ are free variables, using the last equation, we can eliminate one of them and the equation from the LP. Then we are left with a 2-variable LP to solve to compute $t(G)$, instead of an SDP with 55 unknowns. It turns out that for the Petersen graph, $t(G) = \frac{3}{10}$. Note that the underlying $\hat{X}$ generated from the optimal solution of the above LP problem has rank four. As in the 5-cycle example, we can eliminate the variable $z$, obtain an equivalent 2-variable LP and consider its dual. For the Petersen graph, we have the following dual LP:

$$\begin{array}{lrcl} \max & 3y_1+y_2-y_3 & & \\ \text{subject to:} & 8y_1+4y_2-y_3 & = & 1, \\ & 12y_1-4y_2+y_3 & = & 0, \\ & y_1, y_2, y_3 & \ge & 0. \end{array}$$

The optimal solution of this dual LP is $\hat{y}_1 := \frac{1}{20}$, $\hat{y}_2 := \frac{3}{20}$, and $\hat{y}_3 := 0$. From these computations, we should be able to conclude a lot information on an optimal solution of the dual SDP as well. Indeed this is possible. Moreover, the structures arising in the above two examples can be generalized. The first immediate generalization is to consider the node-node incidence matrices of $k$-regular graphs. For such graphs $G = (V, E)$, $\bar{e}$ is an eigenvector of $A(G)$ corresponding to $\lambda_1(A(G)) = k$. Thus, if we have access to the eigenvalues of $A(G)$, then for every $z \in \mathbb{R}^3$, we can reduce the semidefiniteness constraint

$$z_1 A(G) + z_2 I + z_3 \bar{e}\bar{e}^T \succeq 0$$

to (at most) $|V|$ linear inequalities on $z_1, z_2$ and $z_3$. However, another key ingredient we needed was a very rich automorphism group for $G$ to ensure that an optimal solution $\hat{X}$ could be expressed as a linear combination of $A(G)$, $I$, and $\bar{e}\bar{e}^T$. The second generalization we consider backs off from this latter extremely rich symmetry property (which reduces the underlying SDP problems to 2- or 3-variable LP problems) and focuses on the first property ($A(G)$, $I$, and $\bar{e}\bar{e}^T$ admitting a simultaneous diagonalization). The next section points to such a generalization.

### 6.3.2 Symmetries in SDPs, Reducing SDPs to LPs.

**Theorem 6.3** *If $A_1, A_2, \ldots, A_m, C \in \Sigma^n$ pairwise commute, then the SDP problems (P) and (D) are equivalent to a pair of primal-dual LP problems. Moreover, for such SDPs, the duality theory of linear programming applies (for example, there is no need for using Slater points or similar constraint qualifications).*

**Proof** Suppose $A_1, A_2, \ldots, A_m, C \in \Sigma^n$ pairwise commute. Then there exists an orthonormal matrix $Q \in \mathbb{R}^{n \times n}$ which mutually diagonalizes all the matrices $A_1, A_2, \ldots, A_m, C$ in our data for $(P)$ and $(D)$. Then

$$\langle C, X\rangle = \langle Q\mathrm{Diag}(\lambda(C))Q^T, X\rangle = \langle \mathrm{Diag}(\lambda(C)), Q^T XQ\rangle = [\lambda(C)]^T \operatorname{diag}(Q^T XQ).$$

Similarly, for every $i \in \{1, 2, \ldots, m\}$, we have

$$\langle A_i, X\rangle = [\lambda(A_i)]^T \operatorname{diag}(Q^T XQ).$$

Since $Q^T \cdot Q \in \mathrm{Aut}(\Sigma^n_+)$, $X \succeq 0$ iff $Q^T XQ \succeq 0$. Let us represent $\operatorname{diag}(Q^T XQ)$ by $\tilde{x} \in \mathbb{R}^n$. Consider the LP problem

$$\begin{array}{llcl} \min & [\lambda(C)]^T \tilde{x} & & \\ \text{subject to:} & [\lambda(A_i)]^T \tilde{x} & = & b_i, \quad \forall i \in \{1, 2, \ldots, m\} \\ & \tilde{x} & \geq & 0, \end{array}$$

which is equivalent to the primal SDP problem $(P)$.

Now, let us turn to the dual problems. We can do our transformation on the dual as follows. The dual SDP constraint

$$\sum_{i=1}^{m} y_i A_i \preceq C,$$

is equivalent to (using the automorphism $Q \cdot Q^T$),

$$\sum_{i=1}^{m} y_i \mathrm{Diag}\left(\lambda(A_i)\right) \preceq \mathrm{Diag}\left(\lambda(C)\right).$$

Now, both the RHS and LHS are diagonal matrices. Thus, the partial order "$\preceq$" decomposes and becomes the partial order "$\leq$" induced by the nonnegative orthant $\mathbb{R}^n_+$. Therefore, the dual SDP is equivalent to the LP:

$$\begin{array}{llcl} \max & b^T y & & \\ \text{subject to:} & \displaystyle\sum_{i=1}^{m} y_i \lambda(A_i) & \leq & \lambda(C). \end{array}$$

Note that the latter LP is the dual of the former LP. Therefore, the duality theory for LP applies to the original pair of SDP problems with pairwise commuting data matrices. □

Note that, in the above proof, if $x^* \in \mathbb{R}^n$ is an optimal solution of the LP, then $Q\mathrm{Diag}(x^*)Q^T$ is an optimal solution of the SDP. Moreover, $\lambda(A_i)$, $\lambda(C)$ are not necessarily sorted.

## 6.4 Hypersphere Representations of Graphs

Let us turn to graph representations. Last, we discussed representing graphs inside the smallest radius Euclidean balls. All of our work trivially extends to the case where we want all the nodes of the graph to lie exactly on the hypersphere (the boundary of the smallest radius ball). All we have to do is to replace the constraint

$$\operatorname{diag}(X) \leq t\bar{e} \text{ by the constraint } \operatorname{diag}(X) = t\bar{e}$$

in our SDP formulation. Let us call such representations *smallest hypersphere representations*. We have

**Theorem 6.4** *Let $G = (V, E)$ be a given graph. Then, the optimal objective value of the following semidefinite optimization problem is finite, it is attained and is equal to the square of the radius of the smallest ball containing a smallest hypersphere representation of $G$:*

$$\begin{array}{lrcll} \min & t & & & \\ \text{subject to:} & \operatorname{diag}(X) - t\bar{e} & = & 0, & \\ & X_{ii} - 2X_{ij} + X_{jj} & = & 1, & \forall\{i,j\} \in E, \\ & X & \in & \Sigma_+^V. & \end{array}$$

The proof of Theorem 6.2 (with trivial changes) can be adapted to give a proof of the above theorem. These representations are connected to so-called *orthonormal representations of graphs.*

Some related work on graph representations goes back at least to Koebe and it is even connected to the topology of three dimensional manifolds.

**Theorem 6.5** *(Koebe-Andreev-Thurston's Circle Packing Theorem) Every planar graph can be represented in such a way that its nodes correspond to disjoint disks which touch iff the corresponding nodes are adjacent. Moreover, if the graph is 3-connected then such a circle representation is unique up to Möbius transformations of the plane (which maps all circles into circles).*

In fact, there exists a corresponding representation of the dual graph $G^*$ by disks such that intersecting edges of $G$ and $G^*$ are represented by disks whose boundaries intersect orthogonally.

## 6.5 Orthonormal Representations of Graphs

Let graph $G = (V, E)$ be given. Then $\{u^{(i)} \in \mathbb{R}^d : i \in V\}$ is called an *orthonormal representation of* $G$ if

- $\|u^{(i)}\|_2 = 1, \ \forall i \in V$, and
- $\langle u^{(i)}, u^{(j)} \rangle = 0, \ \forall\{i,j\} \in \bar{E}$.

Suppose $\{v^{(i)} \in \mathbb{R}^d : i \in V\}$ is a smallest hypersphere representation of $\bar{G}$ with radius $t$. Then we have

$$\|v^{(i)}\|_2^2 = t, \ \forall i \in V \text{ and } \langle v^{(i)}, v^{(j)} \rangle = \frac{2t-1}{2}, \forall\{i,j\} \in \bar{E}.$$

Note that setting $\bar{X} := \frac{1}{2}I$, $\bar{t} := \frac{1}{2}$ give a feasible solution for the SDP formulation in Theorem 6.4. Therefore, we can assume $t \le \frac{1}{2}$. Now, using these $v^{(i)}$s from the hypersphere representation of $\bar{G}$, we can give an orthonormal representation of $G$ in $\mathbb{R}^{d+1}$. Let 0 denote the index of the new coordinate. For the orthonormal representation vectors $u^{(i)}$, we set the zeroth coordinate to $\sqrt{\frac{1}{2} - t}$. More specifically, we embed the $d$-dimensional hypersphere representation of $\bar{G}$ in $\mathbb{R}^{d+1}$, on the hyperplane $\left\{x \in \mathbb{R}^{d+1} : x_0 = \sqrt{\frac{1}{2} - t}\right\}$. Thus, after an appropriate scaling, we obtain

$$u^{(i)} := \sqrt{2}\begin{bmatrix} \sqrt{\frac{1}{2} - t} \\ v^{(i)} \end{bmatrix}, \quad \forall i \in V.$$

It is clear that

$$\|u^{(i)}\|_2 = 1, \ \forall i \in V \text{ and } \langle u^{(i)}, u^{(j)} \rangle = 0, \ \forall\{i,j\} \in \bar{E}.$$

Therefore, $\{u^{(i)} \in \mathbb{R}^{d+1} : i \in V\}$ gives an orthonormal representation of $G$.

Clearly, every orthonormal representation of $\bar{G}$ gives a hypersphere representation of $G$ as well: Let $\{u^{(i)} \in \mathbb{R}^{d+1} : i \in V\}$ be an orthonormal representation of $\bar{G}$. Define

$$v^{(i)} := \frac{1}{\sqrt{2}} u^{(i)}, \forall i \in V.$$

Now, we clearly have

$$\left\| v^{(i)} - v^{(j)} \right\|_2 = 1, \forall \{i, j\} \in E,$$

and $\|v^{(i)}\|_2 = \frac{1}{\sqrt{2}}$, for every $i \in V$. Therefore, $\{v^{(i)} : i \in V\}$ is a hypersphere representation of $G$ contained in $B(0, 1/\sqrt{2})$. (The size of the ball is not particularly good, it is the same as the objective value guaranteed by our trivial Slater point for the SDP formulation.)

A very important application of orthonormal representations is to the *Stable Set Problem*. Recall our definition from Chapter 1: A subset $\mathcal{S} \subseteq V$ is a *stable set of* $G$ if for every $\{i, j\} \in E$ at most one of $i$, $j$ is in $\mathcal{S}$. The *stability number of* $G$ is the cardinality of the largest cardinality stable set in $G$ and is denoted by $\alpha(G)$. Computing $\alpha(G)$ is $\mathcal{NP}$-hard (and it is also hard to approximate $\alpha(G)$ in polynomial time —no constant ratio, polynomial time approximation algorithm can exist unless $\mathcal{P} = \mathcal{NP}$). If we are given weights $w \in \mathbb{R}^V_+$, we can define the weight of a stable set as the total weight of the nodes that are in that stable set. Then the problem of finding a stable set of the largest weight in $G$ is also of interest to us.

We associate each stable set in $G$ with a 0,1 incidence vector $x \in \{0, 1\}^V$ and define the *Stable Set Polytope*:

$$\text{STAB}(G) := \text{conv}\left\{x \in \{0, 1\}^V : x \text{ is an incidence vector of a stable set in } G\right\}.$$

Let us define STAB($G$) in terms of a linear integer programming formulation:

$$\text{FRAC}(G) := \left\{x \in [0, 1]^V : x_i + x_j \leq 1, \forall \{i, j\} \in E\right\}.$$

Then we have

$$\text{STAB}(G) = \text{conv}\left(\text{FRAC}(G) \cap \{0, 1\}^V\right).$$

We have another interesting formulation based on the clique inequalities. For any clique $\mathcal{C}$ in $G$ *the clique inequality*

$$\sum_{i \in \mathcal{C}} x_i \leq 1$$

is a valid inequality for STAB($G$). Let $A_{\text{clq}}(G)$ denote the matrix whose rows are the incidence vectors of cliques in $G$. So, $A_{\text{clq}}(G)$ is a 0,1 matrix with $|V|$ columns and as many rows as the number of cliques in $G$ (ignoring the empty clique, but including the singletons and edges, ...). Then the *clique polytope of* $G$ is

$$\text{CLQ}(G) := \left\{x \in \mathbb{R}^V_+ : A_{\text{clq}}(G)x \leq \bar{e}\right\}.$$

Note that including a row for each inclusion-wise maximal clique suffices.

Now, let $\{u^{(i)} \in \mathbb{R}^d : i \in V\}$ be an orthonormal representation of $G$. Then for every $c \in \mathbb{R}^d$ such that $\|c\|_2 = 1$, the inequality

$$\sum_{i \in V} \left(c^T u^{(i)}\right)^2 x_i \leq 1$$

is called an *orthonormal representation constraint*. We now define a compact, convex set with nonempty interior, called the *theta body for* $G$:

$$\mathrm{TH}(G) := \left\{x \in \mathbb{R}^V_+ : x \text{ satisfies all orthonormal representation constraints for } G\right\}.$$

Note that there are uncountably many orthonormal representations of $G$ and for every such representation, there are uncountably many $c$ that we can choose to make a single orthonormal representation constraint. Thus, $\mathrm{TH}(G)$ is defined as the intersection of uncountably many half-spaces. Since the intersection of any collection of closed convex sets is a closed, convex set, $\mathrm{TH}(G)$ is closed and convex. Clearly $0 \in \mathrm{TH}(G)$; in fact $e_i \in \mathrm{TH}(G)$ for every $i \in V$, but this will be trivially clear from the next theorem.

**Theorem 6.6** *For every graph* $G = (V, E)$, *we have*

$$\mathrm{STAB}(G) \subseteq \mathrm{TH}(G) \subseteq \mathrm{CLQ}(G) \subseteq \mathrm{FRAC}(G).$$

**Proof** First, we show $\mathrm{STAB}(G) \subseteq \mathrm{TH}(G)$. Let $\mathcal{S} \subseteq V$ be a stable set in $G$. Let $\{u^{(i)} : i \in V\}$ be an arbitrary orthonormal representation of $G$; also let $c$ such that $\|c\|_2 = 1$ be otherwise arbitrary. Let $\mathcal{U}_{\mathcal{S}}^T$ denote the $|V| \times |\mathcal{S}|$ matrix with columns $u^{(i)}, i \in \mathcal{S}$. Then

$$\sum_{i \in \mathcal{S}} \left(c^T u^{(i)}\right)^2 = \|\mathcal{U}_{\mathcal{S}} c\|_2^2 \leq \|c\|_2^2 = 1.$$

Therefore, for every stable set in $G$ its incidence vector $x$ satisfies

$$\sum_{i \in V} \left(c^T u^{(i)}\right)^2 x_i \leq 1,$$

for every orthonormal representation $\{u^{(i)} : i \in V\}$ of $G$ and every unit vector $c$. Since $\mathrm{STAB}(G)$ is the smallest convex set containing the incidence vectors of all stable sets in $G$ and $\mathrm{TH}(G)$ is a convex set containing the same vectors, we proved $\mathrm{STAB}(G) \subseteq \mathrm{TH}(G)$.

Secondly, we prove $\mathrm{TH}(G) \subseteq \mathrm{CLQ}(G)$. Let $\mathcal{C} \subseteq V$ be a clique in $G$. Let $\{u^{(i)} : i \in V \setminus \mathcal{C}\} \cup \{c\}$ be a set of orthonormal vectors. We set

$$u^{(i)} := c, \ \forall i \in \mathcal{C}.$$

Then clearly, $\{u^{(i)} : i \in V\}$ form an orthonormal representation of $G$. The corresponding orthonormal representation constraint is

$$\begin{aligned} \sum_{i \in V} \left(c^T u^{(i)}\right)^2 x_i \leq 1 \quad &\iff \quad \sum_{i \in \mathcal{C}} \left(c^T c\right)^2 x_i \leq 1 \\ &\iff \quad \sum_{i \in \mathcal{C}} x_i \leq 1. \end{aligned}$$

Thus, every clique inequality is a consequence of an orthonormal representation constraint. Therefore, $\mathrm{TH}(G) \subseteq \mathrm{CLQ}(G)$.

Finally, $\mathrm{CLQ}(G) \subseteq \mathrm{FRAC}(G)$ is obvious, since $\mathrm{FRAC}(G)$ is defined only by the nonnegativity constraints and clique constraints on one or two nodes. □

Note that $0, e_1, e_2, \ldots, e_{|V|} \in \mathrm{STAB}(G)$; whence, all the sets in the above theorem have nonempty interior. The theorem implies that all three convex sets $\mathrm{FRAC}(G)$, $\mathrm{CLQ}(G)$ and $\mathrm{TH}(G)$ provide a convex relaxation of $\mathrm{STAB}(G)$. Clearly we can optimize any linear function over $\mathrm{FRAC}(G)$ in polynomial time. However, this relaxation is usually extremely weak. It is $\mathcal{NP}$-hard to find the largest clique in a graph and linear optimization problem over CLQ($G$) is also $\mathcal{NP}$-hard. Nevertheless, as the next theorem shows, optimizing a linear function over TH($G$) is tractable.

We are given a graph $G = (V, E)$ and a weight vector $w \in \mathbb{R}^V_+$. We define

$$\theta(G, w) := \max\{w^T x : x \in \mathrm{TH}(G)\}.$$

Then, $W \in \Sigma^V$ is defined componentwise as $W_{ij} := \sqrt{w_i w_j}$, $\forall i, j \in V$. The next central result and the proof are due to Grötschel, Lovász and Schrijver (see [153]) showing how to connect $\theta(G, w)$ to Semidefinite Optimization and how to connect optimal solutions of those underlying SDP problems to geometric representations of graphs.

**Theorem 6.7** *Let $G = (V, E)$ be a given graph together with a weight vector $w \in \mathbb{R}^V_+$. Using the above definitions, the following are all equal:*

(i)

$$\theta(G, w)$$

(ii)

$$\min_{\textit{all orth. representations}} \max_{i \in V} \left\{ \frac{w_i}{\left(c^T u^{(i)}\right)^2} \right\}$$

(iii)

$$\min\left\{\eta : \mathrm{diag}(S) = 0; S_{ij} = 0, \forall \{i, j\} \in \bar{E}; \eta I - S \succeq W\right\}$$

(iv)

$$\max\left\{\langle W, X\rangle :\ X_{ij} = 0, \forall \{i, j\} \in E;\ \langle I, X\rangle = 1;\ X \succeq 0\right\}.$$

**Proof** Note that if $w = 0$, then we trivially have (i)=(ii)=(iii)=(iv). So, we may assume $w \neq 0$. First, let us prove (iii) = (iv). Call the semidefinite optimization problems in (iii) and (iv), $(D)$ and $(P)$ respectively. Then, $(P)$ and $(D)$ are dual to each other. $\bar{X} := \frac{1}{|V|} I$ is a Slater point for $(P)$ and $\bar{S} := 0$, $\bar{\eta} := \lambda_1(W) + 1$ yield a Slater point for $(D)$. Therefore, by the corresponding corollary of the Strong Duality Theorem, both $(P)$ and $(D)$ attain their optimum values and these values are the same (iii) = (iv).

Secondly, we prove (i) $\leq$ (ii): Let $\bar{x} \in \mathrm{TH}(G)$ be a maximizer of $w^T x$ over TH($G$) (since TH($G$) is nonempty and compact and $w^T x$ is continuous over TH($G$), such maximizer exists). Let $\{\bar{u}^{(i)} : i \in V\}$ be an orthonormal representation of $G$ and $\bar{c}$ be the corresponding unit vector such that $\bar{u}^{(i)}$, $\bar{c}$ attain the min-max in (ii) (this min-max is also attained; moreover, for every $i$, $\bar{c}^T \bar{u}^{(i)} \neq 0$). Then

$$\begin{aligned}
\text{(i)} &= \theta(G, w) = \sum_{i \in V} w_i \bar{x}_i = \sum_{i \in V} \frac{w_i}{\left(\bar{c}^T \bar{u}^{(i)}\right)^2} \left(\bar{c}^T \bar{u}^{(i)}\right)^2 \bar{x}_i \\
&\leq \left(\max_{i \in V} \left\{ \frac{w_i}{\left(\bar{c}^T \bar{u}^{(i)}\right)^2} \right\}\right) \underbrace{\sum_{i \in V} \left(\bar{c}^T \bar{u}^{(i)}\right)^2 \bar{x}_i}_{\substack{\leq 1 \\ \text{since } \bar{x} \in \mathrm{TH}(G)}} \leq \max_{i \in V} \left\{ \frac{w_i}{\left(\bar{c}^T \bar{u}^{(i)}\right)^2} \right\} \\
&= \text{(ii)}.
\end{aligned}$$

Thirdly, we prove (ii) $\leq$ (iii): Let $\bar{S}$ and $\bar{\eta}$ make up an optimal solution of $(D)$ (so that $\bar{\eta} =$ (iii) ). Then

$$\bar{\eta} \geq \langle W, \bar{X} \rangle = \frac{1}{|V|} \bar{e}^T w > 0.$$

(We used the weak duality inequality and the facts that $w \geq 0$, $w \neq 0$.) Let $Y \in \mathbb{R}^{V \times V}$ such that $YY^T = \bar{\eta} I - \bar{S} - W \succeq 0$. Let $v^{(i)}$ denote the $i$th column of $Y^T$. Then

$$\langle v^{(i)}, v^{(i)} \rangle = \bar{\eta} - w_i, \ \forall i \in V,$$

and

$$\langle v^{(i)}, v^{(j)} \rangle = -\sqrt{w_i w_j}, \ \forall \{i, j\} \in \bar{E}.$$

We claim that $\text{rank}(Y) < |V|$. (Otherwise, we would have $\bar{\eta} I - \bar{S} - W \succ 0$ and there would exist a small enough positive $\epsilon$ such that the pair $\bar{S}, (\bar{\eta} - \epsilon)$ would be feasible in $(D)$, contradicting the optimality of $\bar{\eta}$.) Whence, there exists $c \in \mathbb{R}^V$ such that $\|c\|_2 = 1$ and $Yc = 0$. Let us define some vectors that will form an orthonormal representation for $G$:

$$u^{(i)} := \frac{1}{\sqrt{\bar{\eta}}} \left( \sqrt{w_i} c + v^{(i)} \right), \ \forall i \in V.$$

Then, we have

$$\|u^{(i)}\|_2^2 = 1, \ \forall i \in V$$

and

$$\langle u^{(i)}, u^{(j)} \rangle = 0, \ \forall \{i, j\} \in \bar{E}.$$

The next inequality follows from the fact that $\{u^{(i)} : i \in V\}$ is an orthonormal representation for $G$ and $c$ is a unit vector:

$$\text{(ii)} \leq \max_{i \in V : w_i > 0} \left\{ \frac{w_i}{\left(c^T u^{(i)}\right)^2} \right\} = \frac{w_i}{\frac{w_i}{\bar{\eta}}} = \bar{\eta} = \text{(iii)}.$$

Finally, we prove (iv) $\leq$ (i): Let $\bar{X}$ be an optimal solution of $(P)$. Then there exists $Y \in \mathbb{R}^{V \times V}$ such that

$$YY^T = \bar{X}.$$

Let $y^{(i)}$ denote the $i$th column of $Y^T$ and

$$u^{(i)} := \frac{y^{(i)}}{\|y^{(i)}\|_2}, \ \forall i \in V \text{ such that } y^{(i)} \neq 0.$$

For the remaining $i \in V$, define $\{u^{(i)}\}$ from an orthonormal basis for $\left(\text{span}\left\{u^{(i)} : y^{(i)} \neq 0\right\}\right)^{\perp}$. Now, $\{u^{(i)} : i \in V\}$ is an orthonormal representation for $\bar{G}$. Define

$$c := \frac{1}{\sqrt{\langle W, \bar{X} \rangle}} Y^T \begin{bmatrix} \sqrt{w_1} \\ \sqrt{w_2} \\ \vdots \\ \sqrt{w_{|V|}} \end{bmatrix}.$$

Then $\|c\|_2^2 = \frac{\langle W, \bar{X} \rangle}{\langle W, \bar{X} \rangle} = 1$. Let

$$U^T := \left[ u^{(1)}, u^{(2)}, \ldots, u^{(|V|)} \right].$$

**Claim 1** $Uc \odot Uc \in \text{TH}(G)$.

**Proof.** To prove the claim, we show that the vector $Uc \odot Uc$ satisfies all orthonormal representation constraints for graph $G$. To this end, let $\{v^{(i)} : i \in V\}$ be an orthonormal representation of $G$ and let $d$ be an arbitrary unit vector of suitable size. Note that

$$\langle cd^T, cd^T \rangle = (c^T c)(d^T d) = 1,$$

and

$$\langle u^{(i)} \left(v^{(i)}\right)^T, u^{(j)} \left(v^{(j)}\right)^T \rangle = \langle u^{(i)}, u^{(j)} \rangle \langle v^{(i)}, v^{(j)} \rangle = \begin{cases} 1, & \text{if } i = j; \\ 0, & \text{otherwise.} \end{cases}$$

Now, we are ready to check the orthonormal representation constraint for $Uc \odot Uc$:

$$\sum_{i \in V} \left(d^T v^{(i)}\right)^2 \left(c^T u^{(i)}\right)^2 = \sum_{i \in V} \left(\langle cd^T, u^{(i)} \left(v^{(i)}\right)^T \rangle\right)^2 \leq \langle cd^T, cd^T \rangle = 1.$$

The above inequality needs some explanation: we established that $cd^T$ has unit norm; we also established that $u^{(i)} \left(v^{(i)}\right)^T$ are orthonormal vectors. (If we completed them to an orthonormal basis then we would have an equality above.) Whence, the claim is proved. ◊

Note that the proof of the above claim is applicable to all orthonormal representations of $\bar{G}$ and all unit vectors $c$.

**Claim 2** *With the above definitions, we have*

$$\langle W, \bar{X} \rangle = \left[\sum_{i \in V} \sqrt{w_i} \|y^{(i)}\|_2 \left(c^T u^{(i)}\right)\right]^2.$$

**Proof.** Direct calculation. ◊

Thus,

$$\begin{aligned} \langle W, \bar{X} \rangle &\leq \underbrace{\left(\sum_{i \in V} \|y^{(i)}\|_2^2\right)}_{=\text{Tr}(\bar{X})} \left[\sum_{i \in V} w_i \left(c^T u^{(i)}\right)^2\right] \\ &= \sum_{i \in V} w_i \left(c^T u^{(i)}\right)^2 \\ &\leq \theta(G, w), \end{aligned}$$

where the first inequality is the Cauchy-Schwarz inequality applied to the RHS of Claim 2, and the last inequality follows from the fact that $Uc \odot Uc \in \text{TH}(G)$ (Claim 1). This completes the proof. □

An important consequence of this theorem is that $\theta(G, w)$ can be approximated to any precision efficiently. Also this theorem when combined with the next theorem shows that to solve the stable set problem on *perfect graphs* it suffices to solve an SDP problem.

Recall that a graph $G = (V, E)$ is *perfect* if for every node induced subgraph $H$ of $G$, the clique number of $H$ and the chromatic number of $H$ coincide:

$$\omega(H) = \chi(H).$$

An *odd-hole* is a chordless odd cycle of length at least 5. An *odd-anti-hole* is the complement of an odd-hole. The next theorem summarizes some important equivalent characterizations of perfect graphs.

**Theorem 6.8** *Let $G$ be a graph. Then TFAE*

(i) *$G$ is perfect;*
(ii) $\mathrm{STAB}(G) = \mathrm{CLQ}(G)$*;*
(iii) $\mathrm{TH}(G)$ *is a polytope;*
(iv) $\mathrm{STAB}(G) = \mathrm{TH}(G)$*;*
(v) *$G$ does not contain an odd-hole or its complement.*

Note that if $G$ is not perfect, then we know that $\mathrm{STAB}(G) \subset \mathrm{TH}(G)$. However, for some given weight function $w$, we might get lucky and it may still be true that

$$\max\left\{w^T x : x \in \mathrm{STAB}(G)\right\} = \max\left\{w^T x : x \in \mathrm{TH}(G)\right\}.$$

(We only know that if $G$ is not perfect, then there exists some $w \in \mathbb{R}^V_+$ such that

$$\max\left\{w^T x : x \in \mathrm{STAB}(G)\right\} < \max\left\{w^T x : x \in \mathrm{TH}(G)\right\}.)$$

Let us consider the proof of Theorem 6.7. It seems that $\mathrm{TH}(\cdot)$ behaves well under duality and graph complementation. Indeed, this observation can be made more precise:

**Theorem 6.9** *For every graph $G = (V, E)$,*

$$[\mathrm{TH}(G)]^o \cap \mathbb{R}^V_+ = \mathrm{TH}(\overline{G}).$$

That is, the polar of $\mathrm{TH}(G)$ when restricted to the nonnegative orthant, coincides with the $\mathrm{TH}(\cdot)$ set of the complement of $G$. (Polar of a set intersected with the nonnegative orthant is called *the antiblocker* of the set— another notion of duality.)

The theta body $\mathrm{TH}(G)$ also admits a representation as the projection of the feasible region of a tractable SDP problem:

**Theorem 6.10** *Let $G = (V, E)$ be an undirected graph. Then*

$$\mathrm{TH}(G) = \left\{x \in \mathbb{R}^V : \begin{pmatrix} 1 \\ x \end{pmatrix} = Ye_0; Y_{ij} = 0, \forall\{i,j\} \in E; Ye_0 = \mathrm{diag}(Y); Y \in \Sigma_+^{\{0\}\cup V}\right\}.$$

This theorem and some other considerations in combinatorial optimization (such as the *disjunctive programming technique*) led to a general procedure for combinatorial optimization problems based on SDP relaxations. We will study those beginning in Chapter 7.

## 6.6 Products of Graphs, Kronecker Products

Let $G = (V, E)$, $H = (W, F)$ be given undirected graphs. Then *the strong product of $G$ and $H$* is

$$G \otimes H := (V(G \otimes H), E(G \otimes H)),$$

where

$$V(G \otimes H) := V \times W,$$

$$E(G \otimes H) := \left\{ \{(i,u),(j,v)\} : \begin{array}{l} \{i,j\} \in E \text{ and } \{u,v\} \in F \\ \text{or } \{i,j\} \in E \text{ and } u = v \\ \text{or } i = j \text{ and } \{u,v\} \in F \end{array} \right\}.$$

We also define

$$G^k := (\underbrace{G \otimes G \otimes \cdots \otimes G}_{k \text{ times}}).$$

As we will see shortly, the strong product on graphs is closely related to the Kronecker Product on the matrices representing algebraic entities related to the underlying graph.

## 6.7 Stable Set Problem and Shannon Capacity

As we discussed, many different geometric representations of graphs have connections to Semidefinite Optimization. Some have been studied more than 100 years ago. One of these considerations that led to Orthonormal Representations is that of the *Shannon Capacity*.

We are sending messages through a noisy channel. As a result, some of the characters (letters) may be confused with others. Our objective is to select as many words of length $k$ as possible such that no two can be confused. This number grows as $\Theta^k$ for some $\Theta \geq 1$. $\Theta$ is called *the Shannon zero-error capacity of the channel.*

Let's associate the given alphabet with $V$, the node set of a graph. Put an edge between nodes $i$ and $j$ iff the letters corresponding to the nodes $i$ and $j$ may be confused with each other. Denote the corresponding graph by $G = (V, E)$.

When $k = 1$, we would like to compute the maximum number of letters that cannot be confused. This corresponds to $\alpha(G)$, the maximum cardinality of a stable set in $G$. Consequently, the Shannon Capacity of $G$ is

$$\Theta(G) := \lim_{k\to\infty} \left[\alpha\left(G^k\right)\right]^{1/k}.$$

If $\mathcal{S}_G$ is a stable set in $G$ and $\mathcal{S}_H$ is a stable set in $H$ then $\mathcal{S}_G \times \mathcal{S}_H$ is a stable set in $G \otimes H$. Thus,

$$\alpha\left(G^k\right) \geq [\alpha(G)]^k, \forall k \geq 1, \forall G.$$

Therefore,

$$\Theta(G) \geq \alpha(G).$$

Note that $\alpha\left(G^k\right)$ is the maximum number of words of length $k$ so that for every pair of words, there exist at least one $i$ (among those $k$ positions) such that the $i$th letters are different and cannot be confused (i.e., the letters are not adjacent in $G$).

Let $C_j$ denote the cycle on $j$ nodes. Then $C_5^2$ has a stable set of size 5 which implies $\alpha(C_5^{2k}) \geq 5^k$. Therefore, $\Theta(C_5) \geq \sqrt{5}$. Lovász (1979) proved $\Theta(C_5) = \sqrt{5}$. Graham offered \$1000 for finding $\Theta(C_7)$.

Let $\theta(G) := \theta(G, \bar{e})$. Then we can show that

$$\Theta(G) \leq \theta(G).$$

**Lemma 6.11** *For every pair of graphs $G, H$,*

$$\theta(G \otimes H) \leq \theta(G)\theta(H).$$

**Proof** We will use Theorem 6.7. Let $\left\{u^{(i)} : i \in V(G)\right\}$, $\left\{v^{(i)} : i \in V(H)\right\}$ denote orthonormal representations of $G, H$ respectively such that together with the unit vectors $c$ and $d$ they satisfy

$$\theta(G) = \min_{\text{all orth. representations}} \max_{i\in V(G)} \left\{\frac{1}{(c^T u^{(i)})^2}\right\}$$

$$\theta(H) = \min_{\text{all orth. representations}} \max_{k\in V(H)} \left\{\frac{1}{(c^T v^{(k)})^2}\right\}.$$

We define

$$w^{(i,k)} := u^{(i)} \otimes v^{(k)}, \forall (i,k) \in V(G) \times V(H),$$

and claim that $\{w^{(i,k)}\}$ is an orthonormal representation of $G \otimes H$. Just note that

$$\langle u^{(i)} \otimes v^{(k)}, u^{(j)} \otimes v^{(\ell)}\rangle = \langle u^{(i)}, u^{(j)}\rangle\langle v^{(k)}, v^{(\ell)}\rangle.$$

Finally, for the choice of $h := c \otimes d$, we have $\|h\|_2^2 = \|c\|_2^2\|d\|_2^2 = 1$ and

$$\max_{\{i,k\}\in V(G\otimes H)} \left\{\frac{1}{(h^T w^{(i,k)})^2}\right\} = \max_{i\in V(G), k\in V(H)} \left\{\frac{1}{(c^T u^{(i)})^2 (d^T v^{(k)})^2}\right\} \le \theta(G)\theta(H).$$

For the last inequality, we used Theorem 6.7. □

We can strengthen the inequality in the statement of the last lemma to an equation. In fact, analogous to this graph product construction, we can define a Kronecker product operation for pairs of SDPs. Under mild conditions, it suffices to take the Kronecker products of optimal solutions to the smaller SDP problems to make an optimal solution of the big SDP problem. So, the result has been further generalized in that direction (with equality).

**Corollary 6.12** *For every graph $G$, and every $k \in \mathbb{Z}_{++}$, we have*

$$\theta\left(G^k\right) = [\theta(G)]^k.$$

We conclude

$$\Theta(G) \le \theta(G).$$

**Theorem 6.13** *For every graph $G$,*

$$\begin{array}{rclcrcl} \theta(G) & = & \max\langle \bar{e}\bar{e}^T, X\rangle & & = & & \min t \\ & & X_{ij} = 0, \forall\{i,j\} \in E & & \operatorname{diag}(Z) & = & (t-1)\bar{e} \\ & & \langle I, X\rangle = 1 & & Z_{ij} & = & -1, \forall\{i,j\} \in \bar{E} \\ & & X \succeq 0 & & Z & \succeq & 0. \end{array}$$

*Moreover,*

$$\alpha(G) \le \Theta(G) \le \theta(G) \le \chi(\overline{G}), \text{ for all graphs } G.$$

*Finally, we have equality all the way through if $G$ is a perfect graph.*

The second statement (with a string of inequalities) of the above theorem is also known as *the Sandwich Theorem.*

**Proof** The first equation follows from the definition of $\theta(G)$ and Theorem 6.7. The dual of the first problem is

$$\begin{array}{lrcl} \min & \eta & & \\ \text{subject to:} & S_{ij} & = & 0, \forall\{i,j\} \notin E \\ & \eta I + S & \succeq & \bar{e}\bar{e}^T. \end{array}$$

Letting $Z := \eta I + S - \bar{e}\bar{e}^T$, we see that the above dual is same as the second problem mentioned in the theorem. Since both problems have Slater points, the equality of their optimal objective values follows from the corresponding corollary of the Strong Duality Theorem.

We already noted earlier that $\alpha(G) \le \Theta(G)$ and more recently $\Theta(G) \le \theta(G)$. However, we will have to prove $\theta(G) \le \chi(\overline{G})$ here. One of our proofs of this latter case will be given a colouring of $\overline{G}$ with $\chi(\overline{G})$ colours, to construct a feasible solution to the SDP whose objective value is $\chi(\overline{G})$. To have a counterpart to this upcoming argument we show next that $\alpha(G) \leq \theta(G)$ in an analogous way (even though we do not need it). Let $\mathcal{S} \neq \emptyset$ be any stable set in $G$.

$$\bar{X}_{ij} := \begin{cases} \frac{1}{|\mathcal{S}|} & \text{if } i \text{ and } j \in \mathcal{S}; \\ 0 & \text{otherwise.} \end{cases}$$

We have $\bar{X}_{ij} = 0$, for all $\{i,j\} \in E$ and $\langle I, \bar{X} \rangle = \frac{|\mathcal{S}|}{|\mathcal{S}|} = 1$. Also, under a suitable permutation, we can express $\bar{X}$ as

$$\frac{1}{|\mathcal{S}|} \begin{pmatrix} \bar{e}\bar{e}^T & 0 \\ 0 & 0 \end{pmatrix}$$

which is clearly positive semidefinite. Therefore, $\bar{X} \succeq 0$. So, we see once more that $\alpha(G) \leq \theta(G)$.

**Claim 1** *For every graph $G$, $\theta(G) \leq \chi(\overline{G})$.*

**Proof.** We give two proofs of this fact. Since $\text{TH}(G) \subseteq \text{CLQ}(G)$ for every graph $G$, we have the next inequality:

$$\begin{aligned} \theta(G) &\leq \max\left\{\bar{e}^T x : x \in \text{CLQ}(G)\right\} \\ &\leq \min\left\{\bar{e}^T y : [A_{\text{clq}}(G)]^T y \geq \bar{e}, y \geq 0\right\} \\ &\leq \min\left\{\bar{e}^T y : [A_{\text{clq}}(G)]^T y \geq \bar{e}, y \geq 0, y \text{ integral}\right\}. \end{aligned}$$

In the above, the second inequality is by the Weak Duality Relation of LP, third one is obvious (the feasible region of the last problem is a subset of the previous one). The last term is the card. of a minimum cardinality clique cover in $G$, which is equal to the chromatic number of $\overline{G}$. This concludes the first proof of the claim.

The second proof uses the SDP formulation above. Now, suppose we have a colouring of the nodes of $\overline{G}$ with $k$ colours. Then we define $\bar{t} := k$ and

$$\bar{Z}_{ij} := \begin{cases} -1, & \text{if colour}(i) \neq \text{colour}(j); \\ (k-1), & \text{if colour}(i) = \text{colour}(j). \end{cases}$$

Notice that $(\bar{Z}, \bar{t})$ satisfies all the equations in the dual SDP above. If we can prove that $\bar{Z} \succeq 0$, then we are done. Under a suitable permutation (group the nodes in the same colour class together), $\bar{Z}$ is equal to

$$\begin{bmatrix} (k-1) & \cdots & (k-1) & (-1) & \cdots & (-1) & \cdots & (-1) & \cdots & (-1) \\ \vdots & \ddots & \vdots & \vdots & \ddots & \vdots & \ddots & \vdots & \ddots & \vdots \\ (k-1) & \cdots & (k-1) & (-1) & \cdots & (-1) & \cdots & (-1) & \cdots & (-1) \\ (-1) & \cdots & (-1) & (k-1) & \cdots & (k-1) & \cdots & (-1) & \cdots & (-1) \\ \vdots & \ddots & \vdots & \vdots & \ddots & \vdots & \ddots & \vdots & \ddots & \vdots \\ (-1) & \cdots & (-1) & (k-1) & \cdots & (k-1) & \cdots & (-1) & \cdots & (-1) \\ \vdots & \ddots & \vdots & \vdots & \ddots & \vdots & \ddots & \vdots & \ddots & \vdots \\ (-1) & \cdots & (-1) & (-1) & \cdots & (-1) & \cdots & (k-1) & \cdots & (k-1) \\ \vdots & \ddots & \vdots & \vdots & \ddots & \vdots & \ddots & \vdots & \ddots & \vdots \\ (-1) & \cdots & (-1) & (-1) & \cdots & (-1) & \cdots & (k-1) & \cdots & (k-1) \end{bmatrix},$$

where the number of diagonal blocks is exactly $k$ (since there are $k$ colour classes). Any symmetric minor of this matrix that has nonzero determinant can take at most one column (and hence at most one row) from each block. Therefore, $\bar{Z} \succeq 0$ iff the following $k$-by-$k$ matrix is:

$$\begin{bmatrix} (k-1) & (-1) & \cdots & (-1) \\ (-1) & (k-1) & \cdots & (-1) \\ \vdots & \vdots & \ddots & \vdots \\ (-1) & (-1) & \cdots & (k-1) \end{bmatrix}.$$

The latter is positive semidefinite by diagonal dominance. Now, it is clear that $(\bar{Z}, \bar{t})$ is feasible in the second optimization problem and its objective value is $k$. Therefore, $\theta(G) \leq \chi(\overline{G})$. $\Diamond$

By the last lemma we have $\theta(G^k) \leq (\theta(G))^k$, for every $k \geq 1$. Since $\alpha(G^k) \leq \theta(G^k)$, $\Theta(G) \leq \theta(G)$ follows.

If $G$ is perfect, then so is $\overline{G}$ and $\alpha(G) = \omega(\overline{G}) = \chi(\overline{G})$. Thus, the last statement follows. $\square$

Let us finish this section by pointing out some nice results about the theta function on graphs with rich automorphism groups. Recall that $G$ *is node-symmetric* (vertex-transitive) if the automorphism group of $G$ acts transitively on $V$. We have,

**Theorem 6.14** *If $G$ is node-symmetric then $\theta(G)\theta(\overline{G}) = |V|$.*

It is fun to verify the above theorem for the Petersen graph.

## 6.8 Realizability of Graphs

Let us go back to a variant of one of the graph representation questions we discussed at the beginning of this chapter: Given a simple undirected graph $G = (V, E)$, distances $w \in \mathbb{R}^E_{++}$, and a positive integer $d$, find a drawing of the graph $G$ in $\mathbb{R}^d$ obeying all the given the distance relations. There are very many applications of this problem and its variants in areas as wide as from sensor network localization to design of drugs to overcome various illnesses.

We know how to solve this problem efficiently if we are free to choose $d$; however, in general (e.g., for the smallest possible $d$) this problem is $\mathcal{NP}$-hard.

In some variants, the nodes of the graph are partitioned into two sets: *pinned* (those that cannot move) and *unpinned* (those that are allowed to move). Also, the edges are divided into three groups: *cables* (can allow compression), *struts* (can allow stretching) and *bars* (no movement is allowed, we must realize the exact distance for this edge). This kind of set up allows lower and upper bounds on the realized distances and leads to more flexibility in modeling. See, Barvinok [26], Belk [30], Beck and Connelly [31], So and Ye [333], Al-Homidan and Wolkowicz [4].

Rigidity problems on graphs attracted the attention of many application oriented researchers long before other applications of graph theory became widely known. This was indeed partly due to engineering applications in structural engineering and related areas. We will not get into rigidity here. See, Connelly [89].

## 6.9 Bibliographical Notes

This chapter is partly based on explaining some of the details of the material in Lovász' talk at the Fields Institute, during the 1999-2000 special year on Combinatorial Optimization and Graph Theory. A very nice introduction to geometric representations of graphs is given in Lovász' survey [238] and monograph [237];

we followed the basic structure of Lovász' presentation in [238] and his examples in this chapter. Some of the classical results on the *theta function* and the *theta body* go back to Lovász' classical 1979 paper [243]. After this seminal paper, a more detailed and comprehensive work of Grötschel, Lovász and Schrijver followed [155, 154, 153]. For results like Theorem 6.3 and extensions see Goemans and Rendl [142]. For more recent results on exploiting symmetry in SDPs, see [138, 102]. For one of the earliest, see Bochner [50]. For results on Kronecker products of SDP problems as they relate to our discussion, see [285]. For an analysis of $\theta(G)$ for random graphs see Juhász [186]. For an analysis of $\theta(G)$ under a randomized version of graph products, see Feige [122].

Some other recent work on graph representations (which has some relation to Semidefinite Optimization) involves the so-called *Colin de Verdière number* of a graph (see for instance, Kotlov, Lovász, Vempala [214] and Lovász [239]).

For some hardness results on the sparsest cut and related problems, see Chawla et al.[73].

CHAPTER 7

# Lift-and-Project Procedures for Combinatorial Optimization Problems

One of the most fruitful (in theory, practice and algorithms) ways of approaching combinatorial optimization problem is by the *polyhedral method.* We are given a set $F \subseteq \{0,1\}^d$ (e.g., incidence vectors of stable sets in a graph, incidence vectors of Hamiltonian cycles in a graph, incidence vectors of matchings in a graph). We are also given a vector $c \in \mathbb{R}^d$. We would like to find an $x \in F$ maximizing (or minimizing) $c^T x$. Usually, it is very easy to describe (in terms of its facets) a polytope $P \subseteq [0,1]^d$ such that

$$P \cap \{0,1\}^d = F.$$

Now, if we can "find" or "compute"

$$P_I := \operatorname{conv}(F) = \operatorname{conv}\left(P \cap \{0,1\}^d\right)$$

then we can solve the linear programming problem

$$\max\left\{c^T x : x \in P_I\right\}$$

and obtain an optimal solution of the original problem (assuming we can compute an extreme point solution that is optimal). In some cases this is possible (even if it means we employ the ellipsoid method). However, in many other cases we do settle for "partial descriptions" (partial set of facets or strong valid inequalities) of $P_I$ (e.g., if the original problem is $\mathcal{NP}$-hard). Even these "partial descriptions" turn out to be very helpful in theory and in applications.

Many classes of interesting facets for combinatorial optimization problems are exponential in number, in the size of the input. (For instance, consider the number of odd-hole inequalities in a given graph $G$, in the worst-case.) We know that if we can design a polynomial-time weak separation oracle for a class of inequalities, then (under very mild assumptions) we can solve the linear optimization problem over them in polynomial time. However, we may have to utilize an algorithm like the ellipsoid method (or an algorithm for which we do not have a polynomial time convergence proof). In some cases, it is possible to have a *compact representation* of exponentially many inequalities provided that we do not mind increasing the number of variables and the number of inequalities polynomially. For instance, the odd-hole inequalities for a graph on $d$ nodes can be represented in $\mathbb{R}^{d(d+1)/2}$ with $O(d^3)$ linear constraints.

Once we have such a compact representation (we refer to it as a *lifting* as well, since it is in a higher dimensional space), our options increase. We can use polynomial-time interior-point methods; we can use further techniques to strengthen this lifting etc.

We have Theorem 1.46 which guarantees that after a finite number of applications of Gomory-Chvátal procedure we will have a description of conv$(F)$ in terms of linear inequalities. However, in the worst-case, even a single iteration of this procedure could be difficult. For example, for some $P$, optimizing a linear function over $GC(P)$ (the first Gomory-Chvátal closure) is $\mathcal{NP}$-hard.

So, *can we construct operators such that they produce a description of* conv$(F)$ *in finitely many steps and at the same time, a few applications of the operator does not disturb the tractability of the relaxation?*

Certain *liftings* to $\Sigma^d$ lead to "tight" SDP relaxations for some hard problems. Lift and project methods of this chapter automize this idea. They start from $P$ and recursively generate tighter and tighter relaxations of $P_I$ and eventually converge to $P_I$. These relaxations are first generated in the *lifted space* and then projected back onto $\mathbb{R}^d$. However, one can equivalently compute with the lifted objects.

The cone generated by all 0-1 vectors $x \in \mathbb{R}^{d+1}$ with $x_0 = 1$ is called $Q$. Let $K \subset Q$ denote a convex cone; for example, $K$ could be a polyhedral cone obtained from a polytope $P$ in $[0,1]^d$ via homogenization using a new variable $x_0$. That is, if

$$P = \{x \in \mathbb{R}^d : \; Ax \leq b, \; 0 \leq x \leq \bar{e}\},$$

then

$$K := \left\{ \begin{bmatrix} x_0 \\ x \end{bmatrix} \in \mathbb{R}^{d+1} : \; Ax \leq x_0 b, \; 0 \leq x \leq x_0 \bar{e} \right\},$$

see Figure 1.

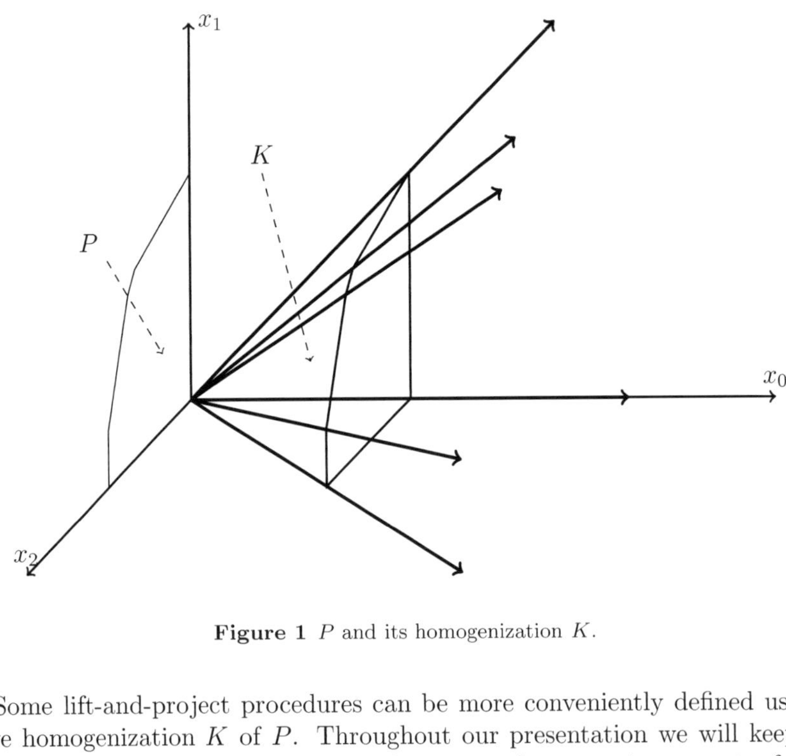


**Figure 1** $P$ and its homogenization $K$.

Some lift-and-project procedures can be more conveniently defined using the above homogenization $K$ of $P$. Throughout our presentation we will keep going back-and-forth between these two equivalent ways of representing our sets of interest in $\mathbb{R}^d$ and $\mathbb{R}^{d+1}$.

Let $K \in \mathbb{R}^{d+1}$ be a convex cone (the components of the vectors are indexed by $0, 1, \ldots, d$, and the elements of $K$ are denoted by $y = (y_0, y_1, \ldots, y_d)^T =: \begin{bmatrix} x_0 \\ x \end{bmatrix}$, where $x \in \mathbb{R}^d$) such that $\left\{x \in \mathbb{R}^d : \begin{bmatrix} 1 \\ x \end{bmatrix} \in K\right\} \subseteq [0,1]^d$.

Similarly, $K_I \subset \mathbb{R}^{d+1}$ denotes the convex cone which is a homogenization of $P_I$.

## 7.1 Lovász-Schrijver Procedures

Recall the theorem from the last section involving an SDP representation of TH($G$):

**Theorem 7.1** *Let $G = (V, E)$ be an undirected graph. Then*

$$TH(G) = \left\{x \in \mathbb{R}^V : \begin{pmatrix} 1 \\ x \end{pmatrix} = Ye_0 \quad ; \quad Y_{ij} = 0, \forall \{i,j\} \in E; \right.$$
$$\left. Ye_0 = \text{diag}(Y); Y \in \Sigma_+^{\{0\} \cup V}\right\}.$$

This is a nice fact showing that linear optimization over TH($G$) is no harder than solving a well-behaved SDP (and hence is a tractable problem). It would be great if we could generalize these types of SDP representations to other combinatorial optimization problems or in general, other problems in discrete mathematics.

In the above representation, all is quite general except the constraints that

$$Y_{ij} = 0, \forall \{i,j\} \in E.$$

If we can replace these conditions by those depending on a geometric object, say a polytope like FRAC($G$) we might have some hope for further generalization. Let's consider the cone of FRAC($G$) in $[0,1]^{\{0\} \cup V}$:

$$\text{cone}\,(\text{FRAC}(G)) := \left\{\begin{bmatrix} x_0 \\ x \end{bmatrix} \in \mathbb{R}^{\{0\} \cup V} : x_i + x_j \leq x_0, \forall \{i,j\} \in E; 0 \leq x \leq x_0 \bar{e}\right\}.$$

Now, we clearly have

$$TH(G) \subseteq \left\{ x \in \mathbb{R}^V : \begin{array}{c} \begin{pmatrix} 1 \\ x \end{pmatrix} = Ye_0; ; Ye_0 = \text{diag}(Y); \\ Ye_i \in \text{cone}\,(\text{FRAC}(G)), \forall i \in \{0\} \cup V; Y \in \Sigma_+^{\{0\} \cup V} \end{array} \right\}.$$

Now, consider the set in the RHS. There, we can easily replace cone (FRAC($G$)) by the cone of a polytope $P$ which approximates $P_I$ for some other combinatorial problem. Since TH($G$) is a lot better of an approximation of STAB($G$) when compared with FRAC($G$), we might get a relaxation of $P_I$ that is better than $P$. Moreover, we can strengthen our approach by requiring that $Y(e_0 - e_i)$ also belong to the cone. (When $Y$ is a rank-one matrix made up from $0, 1$ vectors in $P_I$, these new restrictions are valid!)

This is what we do next. Let $P$ and $K$ be as in the introduction of this chapter. Let us begin by defining three versions of lift-and-project operators proposed by Lovász and Schrijver. We will mostly focus on the one operator that involves the positive semidefiniteness constraints.

$$M_0(K) := \Big\{Y \in \mathbb{R}^{(d+1)\times(d+1)} : \quad \begin{array}{l} Ye_0 = Y^T e_0 = \operatorname{diag}(Y), \\ Ye_i \in K, \forall i \in \{1,2,\ldots,d\}, \\ Y(e_0 - e_i) \in K, \forall i \in \{1,2,\ldots,d\}\} \end{array}$$

defines a lifting of the next relaxation (potentially tighter than $K$) of the cone of all 0-1 vectors in $K$. We project it back onto the space of $K$ to get

$$N_0(K) := \{Ye_0 : Y \in M_0(K)\}.$$

We can now restrict the above lifting to symmetric matrices:

$$M(K) := \{Y \in \Sigma^{d+1} : \quad \begin{array}{l} Ye_0 = \operatorname{diag}(Y), \\ Ye_i \in K, \forall i \in \{1,2,\ldots,d\}, \\ Y(e_0 - e_i) \in K, \forall i \in \{1,2,\ldots,d\}\}. \end{array}$$

Projecting this lifting back results in

$$N(K) := \{Ye_0 : Y \in M(K)\}.$$

Finally, to obtain tighter relaxations, we can restrict the matrix $Y$ to be positive semidefinite. Defining

$$M_+(K) := \{Y \in \Sigma_+^{d+1} : \quad \begin{array}{l} Ye_0 = \operatorname{diag}(Y), \\ Ye_i \in K, \forall i \in \{1,2,\ldots,d\}, \\ Y(e_0 - e_i) \in K, \forall i \in \{1,2,\ldots,d\}\} \end{array}$$

and projecting it back yields the relaxation

$$N_+(K) := \{Ye_0 : Y \in M_+(K)\}.$$

Our main interest lies in the sets

$$\left\{x \in \mathbb{R}^d : \begin{bmatrix} 1 \\ x \end{bmatrix} \in N_0(K)\right\}$$

(and similarly for $N(K)$ and $N_+(K)$). For simplicity, when we say that we are *applying* the $N_0$, $N$, or $N_+$ operator to some convex set $P \subseteq [0,1]^d$, we mean that we consider the cone corresponding to this convex set, apply the corresponding lifting-projecting procedure, then take the convex subset of $[0,1]^d$ defined by the intersection of this new cone with $x_0 = 1$. $N_0(P)$, $N(P)$, or $N_+(P)$, resp., will denote this final subset of $[0,1]^d$, and we will use $N_0^k(P)$, $N^k(P)$, and $N_+^k(P)$ to indicate that we applied the corresponding operator $k$ times in succession ($k = 0$ will refer to the original polytope, so $N^0(P) = P$, etc.).

Note that by definition, the sets $N_0^k(K)$ and $N^k(K)$ are polyhedral cones for every polyhedral cone $K$ and for every $k \in \mathbb{Z}_+$. $N_+^k(K)$ can be nonpolyhedral; however, it is always a convex cone for every convex cone $K$ and for every $k \in \mathbb{Z}_+$.

**Lemma 7.2** *Let $H$ be any face of $[0,1]^d$ and $P \subseteq [0,1]^d$. Then*

$$N(P \cap H) \subseteq N(P) \cap H.$$

*Similarly for $N_+$ and $N_0$.*

Applying Lemma 7.2 finitely many times we obtain the general version of the above fact.

**Corollary 7.3** *Let $H$ be any face of $[0,1]^d$ and $P \subseteq [0,1]^d$. Then for every $k \in \mathbb{Z}_+$,*

$$N^k(P \cap H) \subseteq N^k(P) \cap H.$$

*Similarly for $N_+$ and $N_0$.*

By definition, $M(K)$ depends only on the sets $K \cap \{y : x_i = x_0\}$ and $K \cap \{y : x_i = 0\}$ for all $i \in \{1, 2, \ldots, d\}$. This property also holds for $M_0$ as well as $M_+$. Therefore, we have:

**Lemma 7.4** *Let $K$ and $K'$ be such that $K \cap \{y : x_i = x_0\} = K' \cap \{y : x_i = x_0\}$ and $K \cap \{y : x_i = 0\} = K' \cap \{y : x_i = 0\}$ for all $i \in \{1, 2, \ldots, d\}$. Then $M(K) = M(K')$, $M_0(K) = M_0(K')$ and $M_+(K) = M_+(K')$.*

Since $N_0(K)$, $N(K)$ and $N_+(K)$ are obtained as projections of $M_0(K)$, $M(K)$ and $M_+(K)$, we immediately have

**Corollary 7.5** *Let $K$ and $K'$ be as above. Then for every $k \in \mathbb{Z}_{++}$,*

$$N^k(K) = N^k(K').$$

*Similarly for $N_0$ and $N_+$.*

For example, if we let

$$P := \left\{x \in \mathbb{R}^2 : ||x - 0.5e||_2 \leq \frac{1}{2}\right\}$$

and

$$P' := \left\{x \in \mathbb{R}^2 : ||x - 0.5e||_1 \leq 0.5\right\}$$

then $N^k(P) = N^k(P')$, for every $k \in \mathbb{Z}_{++}$.

The definitions of $M$, $M_0$, $M_+$ and $N$, $N_0$, $N_+$ are invariant under various operations including flipping coordinates $x_i \to (1 - x_i)$ for any subset of the indices $\{1, 2, \ldots, d\}$. More formally,

**Proposition 7.6** *Let $L \in \mathrm{Aut}(Q)$. Then*

$$N(LK) = LN(K).$$

*Similarly for $N_0$ and $N_+$.*

This proposition is just the tip of the iceberg, we will get into some more detail in Chapter 8 involving similar operations that have similar invariance properties with the lift-and-project operators.

Now, let us begin studying the convergence properties of these procedures.

**Lemma 7.7** *Let $K$ be as above. Then*

$$K_I \subseteq N_+(K) \subseteq N(K) \subseteq N_0(K) \subseteq K.$$

**Proof** By definitions of $M_0(K)$, $M(K)$ and $M_+(K)$, we immediately have $N_+(K) \subseteq N(K) \subseteq N_0(K)$. Next, we establish $K_I \subseteq N_+(K)$. If $K_I = \emptyset$, then the inclusion $K_I \subseteq N_+(K)$ is clear. So, we may assume $K_I \neq \emptyset$. Let $y \in K_I \cap \{0,1\}^{d+1}$. If $y_0 = 0$, then $y = 0$. Consider $Y := 0 \in \Sigma^{d+1}$. Then $Y \in M_+(K)$. Thus, $y \in N_+(K)$. If $y_0 \neq 0$, then $y_0 = 1$. Let

$$Y := yy^T.$$

Clearly, $Y \in \Sigma_+^{d+1}$ and $\mathrm{diag}(Y) = Ye_0$ since $y \in \{0,1\}^{d+1}$. Also, for every $i \in \{1, 2, \ldots, d\}$, $Ye_i$ and $Y(e_0 - e_i)$ are in the set $\{0, y\}$ and hence in $K$. We verified the conditions for $Y \in M_+(K)$. Therefore, $Ye_0 \in N_+(K)$ and so is any positive multiple including $y$. Whence, $K_I \subseteq N_+(K)$.

If $N_0(K) = \emptyset$, then the inclusion $N_0(K) \subseteq K$ is clear; otherwise, let $y \in N_0(K)$. So, there exists $Y \in M_0(K)$ such that $Ye_0 = y$. Since $Ye_i$ and $Y(e_0 - e_i)$ both lie in $K$ and $K$ is a convex cone,

$$Ye_i + Y(e_0 - e_i) = Ye_0$$

must be in $K$. Thus, $N_0(K) \subseteq K$. □

**Lemma 7.8** *Let $P$ be as above. Then*

$$N_+(P) \subseteq N(P) \subseteq N_0(P) \subseteq \operatorname{conv}\left[(P \cap \{x : x_i = 0\}) \cup (P \cap \{x : x_i = 1\})\right],$$

*for every $i \in \{1, 2, \ldots, d\}$.*

**Proof** By definition, and by the previous lemma, we have $N_+(P) \subseteq N(P) \subseteq N_0(P)$. So, let $x \in N_0(P)$. In the space of $K$, we have $\begin{bmatrix} 1 \\ x \end{bmatrix} \in N_0(K)$. Thus, there exists $Y \in M_0(K)$ such that

$$Ye_0 = \begin{bmatrix} 1 \\ x \end{bmatrix}.$$

We also have

$$\begin{bmatrix} 1 \\ x \end{bmatrix} = Ye_0 = \underbrace{Ye_i}_{\in K \cap \{y : x_i = x_0\}} + \underbrace{Y(e_0 - e_i)}_{\in K \cap \{y : x_i = 0\}}.$$

Hence, we established

$$N_0(K) \subseteq (K \cap \{y : y_i = 0\}) + (K \cap \{y : y_i = y_0\})$$

which is equivalent to the desired result in the space of $P$. □

Since the index $i$ is arbitrary in the above lemma, we also have

**Corollary 7.9** *Let $P$ be as above. Then*

$$N_+(P) \subseteq N(P) \subseteq N_0(P) \subseteq \bigcap_{i=1}^{d} \operatorname{conv}\left[(P \cap \{x : x_i = 0\}) \cup (P \cap \{x : x_i = 1\})\right].$$

Lemma 7.8 immediately yields the following convergence theorem:

**Theorem 7.10** *Let $P$ be as above. Then*

$$P \supseteq N_0(P) \supseteq N_0^2(P) \supseteq \cdots \supseteq N_0^d(P) = P_I.$$

*Similarly for $N$ as well as $N_+$.*

Therefore, the lift-and-project procedures described above will always terminate with the convex hull $P_I$ in at most $d$ steps.

Moreover, it is easy to establish from the definitions of these procedures that the relaxations obtained after a few iterations are still tractable if the original relaxation $P$ is.

**Theorem 7.11** *Let $P$ be as above. If we have a polynomial time weak separation oracle for $P$ then we can optimize any linear function over any of $N_0^k(P)$, $N^k(P)$, $N_+^k(P)$ in polynomial time, provided $k = O(1)$.*

## 7.2 Balas-Ceria-Cornuéjols Procedure

The general idea of lift-and-project in combinatorial optimization goes back at least to late 1960's, early 1970's. Among other related works, Balas' work [19] in particular established many of the fundamentals.

In this section, it is more convenient not to use homogenization. As in the previous section, we have $A$ and $b$ given, describing a polytope

$$P := \left\{x \in \mathbb{R}^d : Ax \leq b,\ 0 \leq x \leq \bar{e}\right\}.$$

We are interested in integer (that is 0, 1 in this case) vectors in $P$:

$$P_I := \operatorname{conv}\left(P \cap \{0,1\}^d\right).$$

Recall Lemma 7.8. This simple but beautiful lemma was the key to securing the convergence of $N_0$, $N$, and $N_+$ procedures. To have the same convergence result, we could have defined an operator weaker than $N_0$ as follows:

$$N_{(j)}(P) := \operatorname{conv}\left\{(P \cap \{x : x_j = 0\}) \cup (P \cap \{x : x_j = 1\})\right\},$$

where $j \in \{1, 2, \ldots, d\}$. Since the inclusions

$$P_I \subseteq N_{(j)}(P) \subseteq P$$

are clear for every $j$, it makes sense to consider applying this operator iteratively, each time for a new index $j$. Let us define

$$J := \{j_1, j_2, \ldots, j_k\} \subseteq \{1, 2, \ldots, d\}.$$

Let us denote

$$N_{(J)}(P) := N_{(j_k)}\left(N_{(j_{k-1})}\left(\cdots N_{(j_1)}(P)\cdots\right)\right).$$

It is easy to check that in the above context, the operators $N_{(j)}$ commute with each other. Therefore, the notation $N_{(J)}(\cdot)$ is justified.

It is easy to show that

**Lemma 7.12** *For every $J \subseteq \{1, 2, \ldots, d\}$, we have*

$$N_{(J)}(P) = \operatorname{conv}\left(P \cap \{x : x_j \in \{0,1\}, \forall j \in J\}\right).$$

The lemma directly leads to the convergence theorem.

**Theorem 7.13** *Let $P$ be as above. Then*

$$N_{(\{1,2,\ldots,d\})}(P) = P_I.$$

We just described a simple lift-and-project procedure that requires at most $d$ iterations, matching our upper bounds for $N$ and $N_+$. Let us call this procedure *Balas-Ceria-Cornuéjols procedure*. A slightly different version of the last theorem can be obtained by similar ease.

**Proposition 7.14** *Let $P$ and $J$ be as above. Then $N_{(J)}(P) = P_I$ iff*

$$\{x \in P : x_j = z_j,\ \forall j \in J\} \text{ is integral for every } z \in \{0,1\}^J.$$

The above proposition shows that we can apply the lift-and-project procedures to 0,1 mixed integer programming problems (those with 0,1 variables and continuous variables) and guarantee convergence in $|J|$ iterations (under suitable assumptions on the data), where $J$ is the index set of the 0,1 variables.

We can connect the Balas-Ceria-Cornuéjols procedure to the Lovász-Schrijver procedures by the following result.

**Proposition 7.15** *Let $P$ be as above. Then*

$$N_0(P) = \bigcap_{j=1}^{d} N_{(j)}(P).$$

### 7.3 Sherali-Adams (Reformulation-Linearization) Procedure

Consider the Balas-Ceria-Cornuéjols procedure of the last section. As in the previous section, we have $A$ and $b$ given, describing a polytope

$$P := \left\{x \in \mathbb{R}^d : Ax \leq b,\ 0 \leq x \leq \bar{e}\right\}.$$

We are interested in integer (that is 0, 1 in this case) vectors in $P$:

$$P_I := \text{conv}\left(P \cap \{0,1\}^d\right).$$

Note that we have the valid inequalities, $x_1 \geq 0$ and $(1 - x_1) \geq 0$. We can use these inequalities together with the original inequalities

$$Ax \leq b, \quad 0 \leq x \leq \bar{e}$$

to derive valid quadratic inequalities. Namely,

$$x_1 Ax \leq x_1 b, \quad 0 \leq x_1 x \leq x_1 \bar{e}, \quad (1 - x_1)Ax \leq (1 - x_1)b, \quad 0 \leq (1 - x_1)x \leq (1 - x_1)\bar{e}.$$

Now, we want to get back to a system of *linear* inequalities so that we can utilize polyhedral theory (and techniques). In the new quadratic system of inequalities, we have terms like $x_1^2$. Since $x_1 \in \{0,1\}$, $x_1^2 = x_1$. So, we only have to worry about terms like $x_1 x_j$, where $j \neq 1$. We collect all these terms (noting that the scalar multiplication over reals is commutative) and replace them by new variables $y_{1j}$. Thus, we obtain a linear system of inequalities of the form:

$$\tilde{A}x + By \leq \tilde{b}.$$

To go back to the original space of the polytope $P$, we project away the $y$ variables. That is, for every $u$ nonnegative such that $B^T u = 0$, we write the valid inequality

$$u^T \tilde{A} x \leq u^T \tilde{b}.$$

Clearly, each of these inequalities is valid for $P_I$. Let us call the convex set defined by these valid inequalities $N_{(1)}(P)$. Moreover, every facet of $P$ can be derived from this construction. Consider an original inequality

$$a^T x \leq \beta$$

defining the polytope $P$. Multiplying both sides of this inequality by $x_1$ and $(1-x_1)$ and summing up, we obtain

$$x_1(a^T x) + (1 - x_1)(a^T x) \leq x_1 \beta + (1 - x_1)\beta$$

which is the original inequality. The above real-algebraic lift-and-project construction inherits this property. Multiplying by $x_1$ leads to the linear inequality, in the lifted $(x, y)$-space,

$$(a_1 - \beta)x_1 + a_2 y_{12} + \cdots + a_d y_{1d} \leq 0;$$

multiplying by $(1 - x_1)$ yields

$$a^T x + (\beta - a_1)x_1 - a_2 y_{12} - \cdots - a_d y_{1d} \leq \beta.$$

Now, let $\bar{u}$ be the vector with ones corresponding to these two inequalities (in the system $\tilde{A}x + By \leq \tilde{b}$) and with zeroes everywhere else. Then, the valid inequality $\bar{u}^T \tilde{A} x \leq \bar{u}^T \tilde{b}$ is precisely $a^T x \leq \beta$. So, we proved

**Lemma 7.16** *In the above described lift-and-project procedure,*

$$P_I \subseteq N_{(1)}(P) \subseteq P.$$

We described uncountably many inequalities above. However, it suffices to consider only those $u$ that lie in

$$\text{ext}\left\{u : B^T u = 0, \ u \geq 0\right\}.$$

Thus, the set $N_{(1)}(P)$ is finitely constrained and therefore, is a polytope. Our choice of notation for this lift-and-project operator is no accident. It is easy to show that what we described above is a step of Balas-Ceria-Cornuéjols procedure.

The above real-algebraic derivation is the most suitable one to describe the next operator. Now that we have a nice way of generating LP relaxations of $P_I$ using special quadratic inequalities, why not try higher order polynomials for valid inequalities? Let $J_0, J_1 \subseteq \{1, 2, \ldots, d\}$ such that $J_0 \cap J_1 = \emptyset$. We define

$$f(J_0, J_1) := \prod_{j \in J_0} x_j \prod_{j \in J_1} (1 - x_j).$$

Note that $f(J_0, J_1) \geq 0$, for every $x \in [0, 1]^d$. For every such $J_0$ and $J_1$ with the additional property that $|J_0 \cup J_1| = k$, we multiply both sides of the inequalities $Ax \leq b$ and $0 \leq x \leq \bar{e}$ by $f(J_0, J_1)$, and include in this polynomial system of inequalities, the following

$$f(J_0, J_1) \geq 0, \text{ where } |J_0 \cup J_1| = \min\{k + 1, d\}.$$

As before, we replace

$$x_j^2 \text{ by } x_j \text{ (since } x_j \in \{0, 1\}\text{), and}$$

we replace

$$\prod_{j \in J} x_j \text{ by a new variable } y_J.$$

Now, we have a linear system of inequalities

$$\tilde{A}x + \tilde{B}y \leq \tilde{b}.$$

As before, we can project this polyhedron onto the $x$-space using the polyhedral cone

$$\left\{u : \tilde{B}^T u = 0, \ u \geq 0\right\}.$$

Let us call the resulting polytope $SA^k(P)$. The following theorem is easy to establish (especially after our discussion in the next section).

**Theorem 7.17** *For every $P$ as above and every $k \geq 0$,*

$$SA^k(P) \subseteq N^k(P).$$

Among other implications, the above theorem establishes that the Sherali-Adams (Reformulation-Linearization) procedure also terminates in at most $d$ steps.

## 7.4 Optimization over Subset Lattices Interpretation

We have seen many lift-and-project operators and many definitions, interpretations for them. For instance, we gave a *geometric* definition for Balas-Ceria-Cornuéjols operator and then using this definition, we gave a geometric definition of the $N_0$ operator. Our original definition of $N_0$ as well as those of $N$ and $N_+$ operators were *matrix-geometric* in the lifted space. Finally, we gave a *real-algebraic* definition of the Balas-Ceria-Cornuéjols operator and used it to motivate the Sherali-Adams procedure (which was originally proposed in this real-algebraic form, which is still the easiest way to define it). However, when it comes to understanding the deep relationships among these lift-and-project procedures, the *subset lattice* definition/interpretation seems to be the most suitable. This is what we describe next.

Let $F \subseteq \{0,1\}^d$ denote the feasible points (i.e., $F = P \cap \{0,1\}^d$). Let us map $v \in F$ into $\hat{v} \in \{0,1\}^{2^d}$. Entries of $\hat{v}$ are indexed by the subsets of $\{1,2,\ldots,d\}$. So, for every $J \subseteq \{1,2,\ldots,d\}$. We put $\hat{v}_J = 1$ iff $v_j = 1$, for every $j \in J$. So, $\hat{v}$ encodes the information about this feasible solution $v$ in a very very redundant way.

Each $v \in F$ gets mapped into a distinct column of the *zeta matrix* of the *subset lattice* of $\{1,2,\ldots,d\}$. *Zeta matrix* is the $2^d \times 2^d$ matrix whose columns and rows are indexed by the subsets of $\{1,2,\ldots,d\}$ and

$$Z_{I,J} := \begin{cases} 1, & \text{if } I \subseteq J; \\ 0, & \text{otherwise.} \end{cases}$$

E.g., for $d = 3$, the zeta matrix of the 3-element set is

$$Z = \begin{bmatrix} 1 & 1 & 1 & 1 & 1 & 1 & 1 & 1 \\ 0 & 1 & 0 & 0 & 1 & 1 & 0 & 1 \\ 0 & 0 & 1 & 0 & 1 & 0 & 1 & 1 \\ 0 & 0 & 0 & 1 & 0 & 1 & 1 & 1 \\ 0 & 0 & 0 & 0 & 1 & 0 & 0 & 1 \\ 0 & 0 & 0 & 0 & 0 & 1 & 0 & 1 \\ 0 & 0 & 0 & 0 & 0 & 0 & 1 & 1 \\ 0 & 0 & 0 & 0 & 0 & 0 & 0 & 1 \end{bmatrix},$$

where the columns and rows are indexed by the sets

$$\emptyset, \{1\}, \{2\}, \{3\}, \{1,2\}, \{1,3\}, \{2,3\}, \{1,2,3\}$$

respectively. In the above example, the 0,1 zeta matrix $Z$ ending up an upper-triangular matrix with $\text{diag}(Z) = \bar{e}$ was no accident.

**Lemma 7.18** *For every zeta matrix $Z$, there exists an appropriate ordering of the rows and the columns such that $Z$ is upper-triangular and* $\text{diag}(Z) = \bar{e}$.

The above observation immediately implies that $Z$ is nonsingular and $Z^{-1}$ has only integer entries. In fact, all entries of $Z^{-1}$ are from $\{-1,0,1\}$. In particular, it is easy to show that

$$\left[Z^{-1}\right]_{I,J} = \begin{cases} (-1)^{|J \setminus I|}, & \text{if } I \subseteq J; \\ 0, & \text{otherwise.} \end{cases}$$

The underlying matrix is called the *Möbius matrix*. In our 3-element set example

$$Z^{-1} = \begin{bmatrix} 1 & -1 & -1 & -1 & 1 & 1 & 1 & -1 \\ 0 & 1 & 0 & 0 & -1 & -1 & 0 & 1 \\ 0 & 0 & 1 & 0 & -1 & 0 & -1 & 1 \\ 0 & 0 & 0 & 1 & 0 & -1 & -1 & 1 \\ 0 & 0 & 0 & 0 & 1 & 0 & 0 & -1 \\ 0 & 0 & 0 & 0 & 0 & 1 & 0 & -1 \\ 0 & 0 & 0 & 0 & 0 & 0 & 1 & -1 \\ 0 & 0 & 0 & 0 & 0 & 0 & 0 & 1 \end{bmatrix}.$$

Whence, the Möbius matrix is just a signing of the zeta matrix.

Let us try to formulate $K_I$, using the zeta and Möbius matrices. Suppose we have some $\hat{v} \in [0,1]^{2^d}$ such that

$$\hat{v} = \sum \lambda_J \zeta^{(J)},$$

where $\lambda_J \geq 0$, for every $J \subset \{1, 2, \ldots, d\}$ and $\zeta^{(J)}$ denotes the column of $Z$ indexed by the subset $J$ (so we are expressing $v \in [0,1]^d$ as a convex combination of 0,1 vectors, but we are doing so in the homogenized space—the space of $K$—and keeping track of this very very redundant description in the subset lattice). Then

$$\lambda = Z^{-1}\hat{v}.$$

So, now we can formulate $K_I$ using the following constraints on the variable vector $\hat{v}$

$$\begin{aligned} Z^{-1}\hat{v} &\geq 0, \\ \left[Z^{-1}\hat{v}\right]_J &= 0, \text{ for } J \notin \mathcal{J}, \end{aligned}$$

and projecting onto the first $(d+1)$ components. Here, the 0th component is the one corresponding to the empty set (homogenizing index in $K$, $K_I$), $\mathcal{J}$ encodes the elements of $F$ in the subset lattice.

Let $f$ be a function on the subset lattice of $\{0,1\}^{2^d}$. Then $Z^{-1}f$ is called the *upper Möbius inverse of* $f$. Define a $2^d \times 2^d$ matrix $W^f$ as follows:

$$\left[W^f\right]_{I,J} := f(I \cup J).$$

**Lemma 7.19** *Let $\tilde{f}$ be the upper Möbius inverse of $f$. Then*

$$W^f = Z\left[\mathrm{Diag}(\tilde{f})\right]Z^T.$$

*Thus, $W^f$ is positive semidefinite iff $\tilde{f} \geq 0$. Moreover, $W^f$ is positive definite iff $\tilde{f} > 0$.*

**Proof** First, by definitions, we have

$$W^f = Z\left[\mathrm{Diag}(Z^{-1}f\right]Z^T.$$

Secondly,

$$\begin{aligned} \left[W^f\right]_{I,J} &= \sum_{S \supseteq I \cup J} \left[Z^{-1}f\right]_S \\ &= \sum_{S \supseteq I \cup J} \left[\sum_{\tilde{S} \supseteq S} (-1)^{|\tilde{S} \setminus S|} f(\tilde{S})\right] \\ &= f(I \cup J), \end{aligned}$$

as desired. Since $Z$ is a square, nonsingular matrix and $W^f = Z\left[\text{Diag}(Z^{-1}f\right] Z^T$ we have

$$W^f \succeq 0 \text{ iff } Z^{-1}f \geq 0$$

and

$$W^f \succ 0 \text{ iff } Z^{-1}f > 0.$$

□

Next, we point out a connection with *supermodularity*.

**Lemma 7.20** *Let $f$ be as above such that $Z^{-1}f \geq 0$. Then, $f \geq 0$ and $f$ is supermodular.*

**Proof** Assuming $Z^{-1}f \geq 0$, it is easy to deduce $f \geq 0$ (use an upper-triangular representation of $Z^{-1}$, arranged like our example for $\{1,2,3\}$, focus on the implied inequalities on $f(J)$ by the linear system $Z^{-1}f \geq 0$ starting with $f(\{1,2,\ldots,d\})$).

For supermodularity, we will establish

$$f(I \cup J) + f(I \cap J) \geq f(I) + f(J), \quad \forall I, J \subseteq \{1,2,\ldots,d\}.$$

Since $Z^{-1}f \geq 0$, by Lemma 7.19, $W^f$ is positive semidefinite. Thus, for every $h$, $h^T W^f h \geq 0$. Let us take

$$h := e_{I\cup J} + e_{I\cap J} - e_I - e_J.$$

Then, after a lot of cancellations on the RHS below,

$$0 \leq h^T W^f h = f(I \cup J) + f(I \cap J) - f(I) - f(J)$$

we deduce the desired inequality. □

Hence, the nonnegative curvature of the quadratic form $h^T W^f h$ along the directions

$$e_{I\cup J} + e_{I\cap J} - e_I - e_J, \text{ for every } I, J \subseteq \{1,2,\ldots,d\}$$

is equivalent to the supermodularity of $f$. Note that this condition we just mentioned is defined by finitely many linear inequalities on $W^f$ and thus is polyhedral.

Another important connection is that $W^f$ is also known as the *moment matrix of $f$*. This is a hint for connections to algebraic geometry which we will explain further in the later chapters.

We can easily (and implicitly) encode the feasible points, those in $\mathcal{J}$, using linear or nonlinear constraints. These linear or nonlinear constraints would be used to describe an initial relaxation.

To make a better transition to the later chapters, let us assume that the initial formulation is given to us by the polynomials $g_1, \ldots, g_m$. We use the notation $g_\ell \star y$ to denote the underlying function on (or the vector in) $\mathbb{R}^{2^d}$. $J$th component of $g_\ell \star y$ is

$$(g_\ell \star y)_J = \sum_{S\subseteq\{1,2,\ldots,d\}} g_\ell(S)\left(\prod_{i\in S} x_i\right) y(J \cup S).$$

**Theorem 7.21** *The convex cone of $\{0,1\}^{2^d}$ vectors lying in the initial relaxation is*

$$\left\{y \in \mathbb{R}^{2^d} : W^y \succeq 0, W^{(g_\ell \star y)} \succeq 0, \forall \ell \in \{1,2,\ldots,m\}\right\}.$$

The above theorem shows that enforcing the positive semidefiniteness of these huge matrices gives an exact characterization. (Moreover, in our case, these condition become polyhedral!) Instead of enforcing the positive semidefiniteness of the whole matrix, as a relaxation, $N_+$ operator corresponds to enforcing the condition that the $(d+1) \times (d+1)$ leading principle minor be positive semidefinite.

Lasserre [221] proposed requiring positive semidefiniteness of some other submatrices of $W^y$ and $W^{(g_\ell \star y)}$. We will see that Lasserre's procedure is further connected to many other, interesting areas of mathematics in Chapter 10.

## 7.5 Bibliographical Notes

The above subset lattice interpretation was originally given by [245] (and [325] implicitly). More recently it was used by Laurent [224] to compare various lift-and-project operators. Cornuéjols and Li [96] also compare various operators and the cuts that they generate.

In terms of how close the relaxations get to $P_I$ (per iteration), Balas-Ceria-Cornuéjols procedure [20] is the slowest. However, it is the simplest one to present, easiest to analyze, and in the worst-case, it requires no more than $d$ iterations (this is the best bound we proved for all the other operators). Since all the other operators lead to tighter relaxations, the $d$-iteration convergence result for the Balas-Ceria-Cornuéjols operator is the strongest convergence result we proved so far. Also, from a computational point-of-view, amount of work required per iteration for the BCC operator is expected to be much less than all the others. For various properties of the BCC operator, see Ceria [71, 70].

We established some trade-offs for various features of lift-and-project procedures above. However, from a computational complexity point-of-view, we want the strongest possible operator that we can compute with in polynomial time. Indeed $N_+$ seems promising from this point-of-view (we will see more on this in Chapter 8). Along this direction, and using the subset lattice interpretation of lift-and-project operators, Bienstock and Zuckerberg [45] proposed an operator that is significantly stronger than $N_+$ on many packing-covering type problems. They used a doubly exponentially large lattice with $2^{2^d}$ elements to construct their procedures while keeping the desired property that the resulting lift-and-project procedure is tractable for $O(1)$ iterations.

See the books by Sherali and Adams [324] and Tawarmalani and Sahinidis [343] for many applications and related algorithms. For applications of RLT in more general settings, see Sherali and Alameddine [326] and Sherali and Tuncbilek [328].

CHAPTER 8

# Lift-and-Project Ranks for Combinatorial Optimization

We established in the last chapter that the lift-and-project procedures never require more than $d$ iterations. While this result is very nice, it is not very informative about the performance of the lift-and-project procedures on different combinatorial optimization problems. Also, this simple convergence result does not distinguish between those operators utilizing semidefiniteness constraints (such as $N_+$) and those solely based on polyhedral constraints (such as $N$). As a result, it does not help clarify the relative strength of the positive semidefiniteness constraint.

For a more detailed study, we define the notion of a *rank*. The *$N$-rank of $P_I$ relative to $P$* is the smallest $k$ for which $N^k(P) = P_I$. *$N_0$-rank of $P_I$ relative to $P$* as well as the *$N_+$-rank of $P_I$ relative to $P$* are defined analogously.

Whenever the initial polytope $P$ is clear from the context, we drop the "relative to $P$" part of the phrase. Analogously, we define the ranks of valid inequalities. Let $a^T x \leq \beta$ be a valid inequality for $P_I$. Since for some $k \leq d$, $N^k(P) = P_I$, there exists a smallest $k$ for which $a^T x \leq \beta$ becomes valid for $N^k(P)$. This smallest $k$ is called the *$N$-rank of the inequality $a^T x \leq \beta$ relative to $P$*. $N_+$-rank of the valid inequalities are defined similarly.

Next two results tell us ways of obtaining valid inequalities for $N_0(P)$, $N(K)$ and $N_+(K)$ by using the properties of $K \cap \{y : y_i = 0\}$ and $K \cap \{y : y_i = y_0\}$.

**Proposition 8.1** *Let $P$ be as above. Suppose that the linear inequality $a^T x \leq \beta$ is valid for $P \cap \{x : x_j = 0\}$ and for $P \cap \{x : x_j = 1\}$ for some $j \in \{1, 2, \ldots, d\}$. Then $a^T x \leq \beta$ is valid for $N_0(P)$ and hence $a^T x \leq \beta x_0$ is valid for $N(K)$ as well as $N_+(K)$.*

**Proof** Directly follows form the results of Chapter 7. □

**Lemma 8.2** *Let $K$ be as above and let $a \in \mathbb{R}^d_+$. If $a^T x \leq \beta x_0$ is valid for $(K \cap \{y : x_i = x_0\})$ for all $i$ such that $a_i > 0$, then $a^T x \leq \beta x_0$ is valid for $N_+(K)$.*

**Proof** Since $a^T x \leq \beta x_0$ is valid for $(K \cap \{y : x_i = x_0\})$ for all $i$ such that $a_i > 0$, for every such $i$, and every $Y \in M_+(K)$, we have

$$[\beta, -a^T] \, Y e_i \geq 0.$$

Now, consider an arbitrary $y \in N_+(K)$. There exists $Y \in M_+(K)$ such that $Y e_0 = y$. Note that since $Y \succeq 0$, on the one hand we have

$$[\beta, -a^T] \, Y \begin{bmatrix} \beta \\ -a \end{bmatrix} \geq 0.$$

On the other hand,

$$[\beta, -a^T]\, Y \begin{bmatrix} \beta \\ -a \end{bmatrix} = [\beta, -a^T]\, Y \left( \beta e_0 - \sum_{i:a_i>0} a_i e_i \right).$$

Therefore,

$$\beta \left[\beta, -a^T\right] Y e_0 \geq 0,$$

meaning that $a^T x \leq \beta x_0$ is valid for $N_+(K)$ provided $\beta > 0$. If not, then $\beta = 0$. In this case, as we explain below, the inequality $a^T x \leq 0$ is valid for $N_0(K)$ by Proposition 8.1, whence also valid for $N_+(K)$. Let $J := \{i : a_i > 0\}$. So, in the space of $P$, we have $a^T x \leq 0$ valid for $P \cap \{x : x_i = 1\}$ for every $i \in J$. Thus, $P \cap \{x : x_i = 1\} = \emptyset$, for every $i \in J$. Now, Proposition 8.1 implies that $N_0(P) \subseteq \{x : x_i = 0, \forall i \in J\}$. Therefore, $a^T x \leq 0$ is valid for $N_0(K)$ and hence for $N_+(K)$. $\square$

Lemma 8.2 can be extended to derive conditions under which the $N_+$-rank of an inequality is at most $r$.

**Theorem 8.3** *Let $a \geq 0$ and let $I_+ := \{i : a_i > 0\}$. If $a^T x \leq \beta x_0$ is valid for $(K \cap \{y : x_i = x_0,$ for all $i \in I\})$ for all sets $I \subseteq I_+$ satisfying either of the following two conditions*

1. *$|I| = r$,*
2. *$|I| \leq (r-1)$ and $\sum_{i \in I} a_i > \beta$,*

*then $a^T x \leq \beta x_0$ is valid for $N_+^r(K)$.*

Note that the most powerful applications of the above theorem require a lot of symmetry.

## 8.1 Lower Bounds on the $N_0$-Rank, $N$-Rank and $N_+$-Rank

In this section, we provide some lower bounds on the $N_0$-rank, $N$-rank and $N_+$-rank. We also briefly discuss a situation in which the positive semidefiniteness constraints do not help at all (in the context of these lift-and-project operators).

First we give a sufficient condition for a point $\bar{x} \in P$ to remain in $N_+(P)$.

**Theorem 8.4** *Let $\bar{x} \in P$ such that*

$$(\bar{x} - \bar{x}_j e_j), [\bar{x} + (1 - \bar{x}_j) e_j] \in P, \quad \forall j \in \{1, 2, \ldots, d\} \text{ such that } 0 < \bar{x}_j < 1.$$

*Then $\bar{x} \in N_+(P)$.*

Indeed, easy conclusions from such a theorem can be drawn by utilizing polytopes where

$$(P \cap \{x : x_j = 0\}) + e_j = P \cap \{x : x_j = 1\}, \quad \forall j \in \{1, 2, \ldots, d\}.$$

Collecting together all the points $x$ satisfying the sufficient condition of the above theorem on such a polytope $P$, we deduce

**Corollary 8.5** *Let $P$ be such that $(P \cap \{x : x_j = 0\}) + e_j = P \cap \{x : x_j = 1\}$ for every $j \in \{1, 2, \ldots, d\}$. Then*

$$N_+(P) = N(P) = N_0(P) = \bigcap_{j \in \{1, \cdots, d\}} \left\{x \in [0,1]^d : (x - x_j e_j) \in P\right\}.$$

To prove Theorem 8.4, for certain $x \in P$, we can construct a $Y \in M_+(P)$ such that $x$ would also lie in $N_+(P)$.

Let's define for $x \in P$,

$$Y(x) := \begin{bmatrix} 1 \\ x \end{bmatrix} [1, x^T] + \operatorname{Diag} \begin{bmatrix} 0 \\ x_1 - x_1^2 \\ \vdots \\ x_d - x_d^2 \end{bmatrix} + \begin{bmatrix} 0 & 0^T \\ 0 & B(x) \end{bmatrix},$$

where $B(x) \in \Sigma^d$ and $\operatorname{diag}(B(x)) = 0$. Then clearly we have $Y(x) \in \Sigma^{d+1}$, $Y(x)e_0 = \operatorname{diag}(Y(x))$. Moreover, using the Schur complement of $[Y(x)]_{00}$ in $Y(x)$, we have

$$Y(x) \succeq 0 \text{ iff } B(x) + \operatorname{Diag} \begin{bmatrix} x_1 - x_1^2 \\ \vdots \\ x_d - x_d^2 \end{bmatrix} \succeq 0.$$

The latter may be assured in many simple ways, for example by diagonal dominance: It suffices to choose $[B(x)]_{ij}$ such that

$$\left| [B(x)]_{ij} \right| \leq \min \left\{ \frac{x_i - x_i^2}{\text{\# of nonzeros in column } i}, \frac{x_j - x_j^2}{\text{\# of nonzeros in column } j} \right\}.$$

To satisfy $Y(x) \in M(P)$, the entries of such a $B(x)$ will be further restricted by the condition $Y(x)e_i \in K$ for every $i \in \{1, 2, \ldots, d\}$ and $Y(x)(e_0 - e_i) \in K$ for every $i \in \{1, 2, \ldots, d\}$. If this condition is verified for some $B(x)$ then the above argument directly implies $x \in N_+(P)$. In the case of Theorem 8.4, diagonal dominance was utilized; because of the special structure of $P$, we can simply choose $B(x) := 0$ and satisfy all the conditions for $Y(x) \in M_+(P)$ and whence for $x \in N_+(P)$.

**8.1.1 Some General Conditions.** Next, we derive general conditions under which the positive semidefiniteness constraint is not useful. This can be expressed in several ways as

- $M(K) = M_+(K)$, or as
- $N(K) = N_+(K)$ or even as
- $\max\{c^T x : x \in N(K)\} = \max\{c^T x : x \in N_+(K)\}$ for some given $c$.

First, we rewrite the condition

$$Ye_i, Y(e_0 - e_i) \in K.$$

Since $Y$ is symmetric,

$$u^T Y v \geq 0, \ \forall\, u \in Q^*, v \in K^* \iff u^T Y v + v^T Y u \geq 0, \ \forall\, u \in Q^*, v \in K^*.$$

Using the fact that $u^T Y v + v^T Y u = \operatorname{tr}\left(Y(uv^T + vu^T)\right)$, we see that the condition is also equivalent to

$$Y \in [T(K)]^*,$$

where

$$\begin{aligned} T(K) &:= \operatorname{cone}\left\{uv^T + vu^T : u \in Q^*, v \in K^*\right\} \\ &= \operatorname{cone}\left\{uv^T + vu^T : u \in \operatorname{ext}(Q^*), v \in \operatorname{ext}(K^*)\right\}. \end{aligned}$$

Let's define

$$D := \left\{Y \in \Sigma^{d+1} : \operatorname{diag}(Y) = Ye_0\right\}.$$

Note that the cone (more specifically, the subspace in this case) dual to $D$ in the space $\Sigma^{d+1}$ is the orthogonal complement of $D$.

$$D^* = D^\perp = \left\{ \sum_{i=1}^{d} \alpha_i (E_{ii} - E_{0i}) : \alpha \in \mathbb{R}^d \right\},$$

where $E_{ij} := e_i e_j^T + e_j e_i^T$.

**Theorem 8.6** *Let $K \subseteq Q$ be a convex cone. Then we have*

1. $M_+(K) = M(K)$ *if and only if* $T(K) + D^\perp \supseteq \Sigma_{++}^{d+1}$.
2. $N_+(K) = N(K)$ *if and only if for every* $s \in \mathbb{R}^{d+1}$,
   $\mathrm{Diag}(s) \in T(K) + D^\perp + \Sigma_+^{d+1}$ *implies* $\mathrm{Diag}(s) \in \mathrm{cl}\left(T(K) + D^\perp\right)$.

The above theorem is based on duality theory. Indeed, when $K$ is polyhedral, the situation becomes nicer. Then, in particular, in part 1, $\Sigma_{++}^{d+1}$ can be replaced by $\Sigma_+^{d+1}$ and in part 2, $\mathrm{cl}\left(T(K) + D^\perp\right)$ can be replaced by $\left(T(K) + D^\perp\right)$.

Now, let us go back to Corollary 8.5. We can use this corollary recursively, to obtain examples for which $N^k(P) = N_+^k(P)$, for every $k$, and $N_+^{d-1}(P) \neq P_I$. For instance, consider

$$P(k) := \left\{ x \in \mathbb{R}^d : \left\| x - \frac{1}{2}\bar{e} \right\|_1 \leq \frac{k}{2} \right\}.$$

**Theorem 8.7** *For every $k \in \{1, 2, \ldots, d-1\}$, $N_+(P(k)) \supseteq P(k-1)$. Furthermore, $P(0) = \left\{\frac{1}{2}\bar{e}\right\} \neq \emptyset$ and $P_I(d-1) = \emptyset$. Thus, the $N_+$ procedure requires $d$ iterations to prove $P_I(d-1) = \emptyset$.*

**Proof** Corollary 8.5 implies that $N_+(P(k)) = P(k-1)$ for every $k \in \{1, 2, \ldots, d-1\}$. Therefore, the claim follows immediately. □

Note that $P(d-1)$ is a polytope in $\mathbb{R}^d$ with $2^d$ facets. $P(d-1) \cap \mathbb{Z}^d = \emptyset$; however, the facets of $P(d-1)$ are minimal in the sense that if we remove any one of them, the resulting polytope contains some integral points. Doignon's theorem [107] establishes that any system of $m$ linear inequalities in $d$ variables that does not have an integral solution has a subsystem of size at most $2^d$ linear inequalities with the same property (no integral solution). Polytopes like $P(d-1)$ are extreme in this sense and prove that the constant $2^d$ in Doignon's theorem cannot be improved.

We give another example for which the relative progress of both $N$ and $N_+$ is tracked. Suppose only for the remainder of this section that $d$ is even. Consider

$$K := \left\{ \begin{pmatrix} x_0 \\ x \end{pmatrix} \in \mathbb{R}^{d+1} : x(S) \leq \frac{d}{2}x_0, \text{ for all } S \subset \{1, 2, \ldots, d\} \text{ such that } |S| = \frac{d}{2} + 1,\ 0 \leq x \leq x_0\bar{e} \right\}.$$

Then

$$K_I = \left\{ \begin{pmatrix} x_0 \\ x \end{pmatrix} \in \mathbb{R}^{d+1} : \sum_{i=j}^{d} x_j \leq \frac{d}{2}x_0,\ 0 \leq x \leq x_0\bar{e} \right\}.$$

Lemmas 7.12 and 7.8 imply that the $N$-rank of $\sum_{j=1}^d x_j \leq \frac{d}{2}x_0$ is at most $(d-2)$, while Theorem 8.3 implies that the $N_+$-rank is at most $\frac{d}{2}$. These bounds are tight.

**Theorem 8.8** *The $N$-rank of $\sum_{j=1}^{d} x_j \leq \frac{d}{2}$ relative to $K$ is $(d-2)$. The $N_+$-rank of the same inequality relative to $K$ is $\frac{d}{2}$. Moreover, for $r \leq \frac{d}{2} - \sqrt{d} + \frac{3}{2}$, the optimum values of*

$$\max\{\bar{e}^T x : x \in N^r(K)\} \text{ and } \max\{\bar{e}^T x : x \in N_+^r(K)\}$$

*are the same.*

There are much simpler examples with similar properties. For instance, take the unit hypercube $[0,1]^d$, chop off a single vertex by a halfspace, and call the resulting polytope $P$. For every such $P$,

$$N_0^k(P) = N^k(P) = N_+^k(P), \quad \forall k \geq 0,$$

and

$$N_+^{d-1}(P) \neq P_I.$$

## 8.2 Matching Polytope and the Related Polyhedra

Given a simple undirected graph $G = (V, E)$, the matching polytope defined by $G$ is

$$P_I = \left\{ x \in P : x(E(S)) \leq \frac{|S|-1}{2}, \text{ for all } S \subseteq V \text{ such that } |S| \text{ is odd} \right\},$$

where

$$P := \left\{ x \in \mathbb{R}^E : x(\delta(v)) \leq 1, \forall v \in V, x \geq 0 \right\}.$$

$P_I$ is one of the fundamental polytopes in combinatorial optimization. Among the combinatorial optimization problems that can be solved in strongly polynomial time, the maximum weight matching problem (maximizing a linear function over $P_I$) is one of the more sophisticated ones.

In this section, we deal with the problem of determining $N_+$-rank of $P_I$ relative to $P$. That is, we answer the question, "what is the smallest $k$ which guarantees $N_+^k(P) = P_I$ for all graphs $G$?"

Note that in terms of Gomory-Chvátal type of derivation, $P_I$ is very close to $P$.

**Proposition 8.9** *In the context of the matching polytope above, the Chvátal rank of $P_I$ relative to $P$ is at most one.*

**Proof** Let $S \subseteq V$ be such that $|S|$ is odd (and at least 3). Then summing up the inequalities $x(\delta(v)) \leq 1, \forall v \in S$, we arrive at

$$2 \sum_{e \in E(S)} x_e \leq |S|.$$

Now, for every $x \in \{0,1\}^E$ the LHS is an even integer and the RHS is odd. Therefore, the inequality

$$x(E(S)) \leq \frac{|S|-1}{2}$$

has Chvátal rank at most one as desired. □

Since $P$ is integral (that is, ranks with respect to all procedures is zero) iff the underlying graph, $G$ is bipartite, we conclude that $G$ is not bipartite iff the Chvátal rank of $P_I$ relative to $P$ is exactly one.

In our current setting with the matching problem, we have a distinct advantage (compared to a situation where we might be dealing with an $\mathcal{NP}$-hard problem) in that we have a complete understanding of all facets of $P_I$. It is easy to see that when $G$ is the complete graph we have the list of all facets that could possibly arise for any graph on the same set of nodes. Moreover, the complete graph has an extremely rich automorphism group (exploiting the symmetry should be fruitful). So, we focus on the case when $G$ is a complete graph.

First, we need to lay down some ground work. Note that

$$Q^* = \text{cone}\{e_1, \ldots, e_d, f_1, \ldots, f_d\},$$

where $f_i := e_0 - e_i$, for every $i$. Let us use $N_\sharp \in \{N_0, N, N_+\}$, and $M_\sharp \in \{M_0, M, M_+\}$, to state the results for all three operators $N_0, N, N_+$ (and $M_0, M, M_+$).

We will show how to construct points inside $N_\sharp^k(P)$ but outside $P_I$ for cliques on $(2k+3)$ nodes, using points inside $N_\sharp^{k-1}(P)$ but outside $P_I$ for cliques on $(2k+1)$ nodes.

Note that when we have a clique on $(2k+3)$ nodes and focus on an edge $e = \{u, v\}$, then $x_e$ is either 1 or 0. If $x_e = 1$ then the remaining entries of $x$ are completely determined by the underlying subproblem on the $(2k+1)$ clique (since we must set the values of all edges incident on $u$ or $v$ to zero, only exception being the edge $\{u, v\}$). See Figure 2.

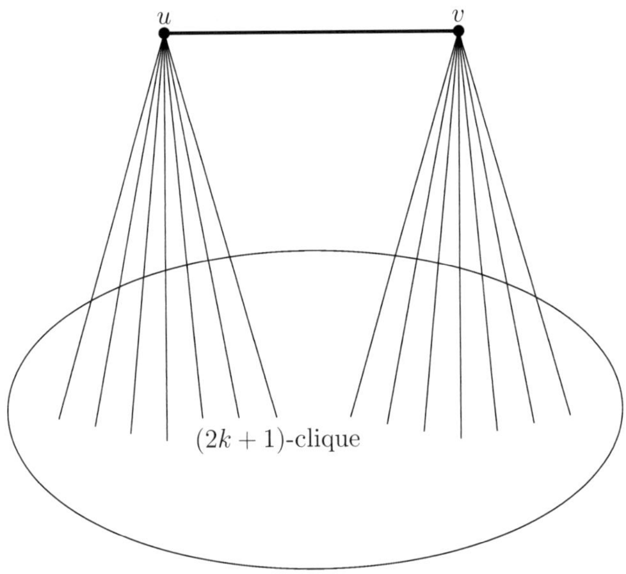


**Figure 2** Partition of the edges for assigning values on the $(2k+3)$ clique.

We can assign a uniform weight of $\frac{1}{2k}$ on all the edges of the $(2k+1)$-clique. For every choice of the pair $u \neq v$, we can consider two points (one sets $x_{\{u,v\}} := 1$ and the other sets $x_{\{u,v\}} := 0$ all other edges have value either 0 or $\frac{1}{2k}$ depending on whether they are incident on $u$ or $v$ as above). By taking suitable convex combinations of these points we can make the induction on $k$ work.

Even though we do not really need it for this section, we will work with a more general construction. To make some sense of the next definition, we note that $N_0, N, N_+$ procedures can be defined using two cones:

- cone $K_1 := K$ and

- cone $K_2$ such that $K \subseteq K_2 \subseteq Q$.

Then the condition

$$Ye_i, Y(e_0 - e_i) \in K, \quad \forall i \in \{1, 2, \ldots, d\}$$

becomes

$$YK_2^* \subseteq K.$$

**Definition 8.10** Suppose $L : \mathbb{R}^{d+1} \to \mathbb{R}^{d+1+t}$ is a linear map. Then, $L$ is said to be *$M_\sharp$- and $N_\sharp$-commutative*, respectively, if $LM_\sharp(K_1, K_2)L^T \subseteq M_\sharp(LK_1, LK_2)$ and $LN_\sharp(K_1, K_2) \subseteq N_\sharp(LK_1, LK_2)$ for every pair of closed convex cones $K_1, K_2 \subseteq Q$.

Let $\tilde{e}_i$'s and $\tilde{f}_i$'s be the extreme rays of the dual cone $\tilde{Q}^*$ of the $(d+1+t)$-dimensional cone $\tilde{Q}$ spanned by the $(d+t)$-dimensional hypercube. Related to the invariance of $N_\sharp$ under automorphisms of cone $Q$, we have the following analogous result.

**Proposition 8.11** *A linear map $L : x \in \mathbb{R}^{d+1} \mapsto \tilde{x} \in \mathbb{R}^{d+1+t}$ is $M_\sharp$-commutative if and only if, for every $j \in \{1, 2, \ldots, d+1+t\}$,*

$$L^T \tilde{f}_j \tilde{e}_j^T L \in \operatorname{span}\left\{f_i e_i^T : i \in \{1, 2, \ldots, d\}\right\}. \tag{8.1}$$

Here are some examples of $M_\sharp$-commutative and $N_\sharp$-commutative maps:

- *Embedding* (we want to take and $x \in \mathbb{R}^{d+1}$ and pad it with some zeros) $L : x \in \mathbb{R}^{d+1} \mapsto \tilde{x} \in \mathbb{R}^{d+1+t}$ such that, for some $l \in \{0, 1, \ldots, t\}$,

$$\tilde{x}_i := \begin{cases} x_i & \text{for } i \in \{0, 1, \ldots, d\}, \\ 0 & \text{for } i \in \{d+1, \ldots, d+l\}, \\ x_0 & \text{for } i \in \{d+l+1, \ldots, d+t\}. \end{cases} \tag{8.2}$$

- *Duplication* (we want to generate a new vector $\tilde{x}$ by repeating some entries of $x$) $L : x \in \mathbb{R}^{d+1} \mapsto \tilde{x} \in \mathbb{R}^{d+1+t}$ such that, for a subset $\{j_1, \ldots, j_t\} \subseteq \{1, 2, \ldots, d\}$,

$$\tilde{x}_i := \begin{cases} x_i & \text{for } i \in \{0, 1, \ldots, d\}, \\ x_{j_{i-d}} & \text{for } i \in \{d+1, \ldots, d+t\}. \end{cases} \tag{8.3}$$

- *Flipping* is an automorphism that maps $e_j \mapsto f_j$, $f_j \mapsto e_j$ for each $j \in J \subseteq \{1, 2, \ldots, d\}$.

In all of the above examples, it is easy to check that for every $j \in \{1, 2, \ldots, d+t\}$, there is $i \in \{1, 2, \ldots, d\}$ such that

$$\left\{L^T \tilde{e}_j, L^T \tilde{f}_j\right\} = \{e_0, 0\}, \text{ or } \{e_i, f_i\}. \tag{8.4}$$

Clearly, (8.4) is sufficient for (8.1). Moreover, (8.4) motivates the next fact.

**Corollary 8.12** *Suppose $L$ satisfies the following conditions:*
*1. The first row is $e_0$, and*
*2. the remaining rows are either, $0$, $e_0$, $e_i$, or $f_i$ for $i \in \{1, 2, \ldots, d\}$.*
*Then any positive multiple of $L$ is both $M_\sharp$- and $N_\sharp$-commutative.*

Next, we state a generalization of the result we saw in Chapter 7 about the invariance of $N_\sharp$ under the automorphisms of $Q$.

**Lemma 8.13** *Let $K \subseteq \mathbb{R}^{d+1}$ and $\tilde{K} \subseteq \mathbb{R}^{d+1+t}$, respectively, be the homogenizations of the convex sets $P \subseteq [0,1]^d$ and $\tilde{P} \subseteq [0,1]^{d+t}$. Assume $L : \mathbb{R}^{d+1} \to \mathbb{R}^{d+1+t}$ is an $N_\sharp$-commutative map. If $LK \subseteq \tilde{K}$, then for every $r \geq 0$, we have $LN_\sharp^r(K) \subseteq N_\sharp^r(LK) \subseteq N_\sharp^r(\tilde{K})$.*

We conclude our discussion of the basic properties of $N_\sharp$-commutative maps with a convexity theorem.

**Theorem 8.14** *Every convex combination of $N_\sharp$-commutative maps is also $N_\sharp$-commutative.*

Figure 3 explains the general strategy for a lower bound proof on $N_\sharp$-rank based on $N_\sharp$-commutative maps. We will proceed by induction. We will have a point

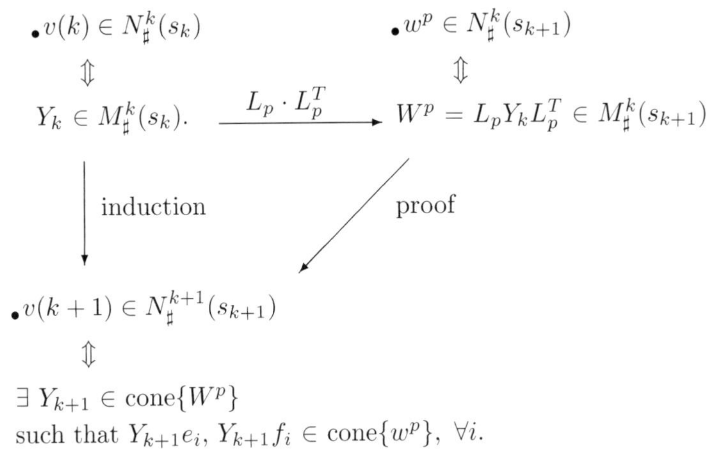


**Figure 3** A general strategy for using $N_\sharp$-commutative maps.

$$v^{(k)} \in N_\sharp^k(s_k) \setminus P_I$$

(corresponding to an instance of size $s_k$). We want to construct a point

$$v^{(k+1)} \in N_\sharp^{k+1}(s_{k+1}) \setminus P_I$$

(corresponding to an instance of size $s_{k+1} > s_k$). In the case of the matching polytope, and the $N_+$-operator, $s_k$ corresponds to the $(2k+1)$-clique and $s_{k+1}$ corresponds to the $(2k+3)$-clique.

Figure 4 shows conceptually how a matrix $Y_k \in M^{k-1}(s_k)$ is used to generate many matrices $L_p Y_k L_p^T \in M^{k-1}(s_{k+1})$ and then, how a strict convex combination of these can yield $Y_{k+1} \in M^k(s_{k+1})$.

Now, we discuss the construction of the points

$$v \in N_\sharp^k(s_k) \setminus P_I.$$

Suppose $v \in \mathbb{R}_+^d$ maximizes $\bar{e}^T x$ (we assume for this discussion that the underlying combinatorial optimization problem is a maximum cardinality problem) over $N_\sharp^k(P)$. Thus, if $N_\sharp^k(P)$ is invariant under all permutations $\mathcal{S}_d$ (represented as permutation matrices), i.e.,

$$\forall R \in \mathcal{S}_d : \ x \in N_\sharp^k(P) \iff Rx \in N_\sharp^k(P).$$

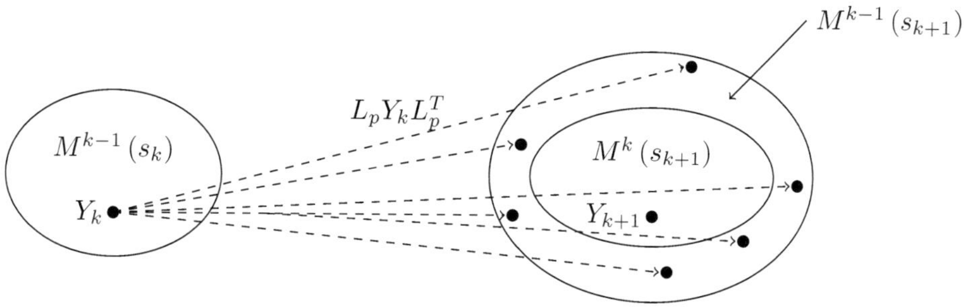


**Figure 4** A schematic illustration of the construction of $Y_{k+1} \in M^k(s_{k+1})$ from $Y_k \in M^{k-1}(s_k)$.

Then (analogous to our discussion about symmetry in Chapter 6),

$$\left(\frac{1}{|\mathcal{S}_d|} \sum_{R \in \mathcal{S}_d} Rv\right) \in N_\sharp^k(P),$$

by the convexity of $N_\sharp^k(P)$. Therefore, we can assume $v = \alpha\bar{e}$ for some $\alpha \geq 0$ (we used $\bar{e}^T Sv = \bar{e}^T v, \forall R \in \mathcal{S}_d$).

Using these commutative maps and averaging appropriately, we obtain by induction on $r$ the following theorem.

**Theorem 8.15** *Let $G = (V, E)$ be a complete graph with $|V|$ odd and at least 3. Further let $P$, $P_I$ be the corresponding polytopes. Then*

$$N_+^{r-1}(P) \supset P_I = N_+^r(P),$$

*where $r := \frac{|V|-1}{2}$.*

By our observation following Proposition 8.9 (or by Theorem 8.3), we deduce:

**Corollary 8.16** *Let $G = (V, E)$ be given. Then*

$$N_+^r(P) = P_I,$$

*for every $r \geq \lfloor \frac{|V|}{2} \rfloor$.*

Given a graph $G = (V, E)$, consider its *line graph* $H = (W, F)$. ($W := E$, and $\{e_1, e_2\} \in F$ if edges $e_1$ and $e_2$ share an end point in $G$.) Then the set of matchings in $G$ is precisely the set of stable sets in $H$.

Note that FRAC($H$) $\subseteq P$ ($P$ is the initial approximation of the matching polytope of $G$). In fact, equality holds iff $G$ is a cycle or a path, or a disjoint union of these. So, from our results about the matching polytope, we already learn that there are graphs $G = (V, E)$ for which the $N_+$-rank of STAB($G$) relative to FRAC($G$) is at least

$$\left\lfloor \frac{1}{4}\left(\sqrt{8|V(G)|+1} - 1\right)\right\rfloor.$$

Next, we study the $N_0$-, $N$-, $N_+$-ranks of STAB($G$) relative to FRAC($G$) in more detail.

## 8.3 Stable Set Problem and Graph Ranks

In this section we will use the notation $K_n$ to refer to a clique on $n$ nodes. Let us start with a very simple observation that follows from the basic properties of lift-and-project operators.

**Proposition 8.17** *If $G$ has $k$ nodes whose deletion leaves the graph bipartite, then*

$$N_0^k(G) = N^k(G) = \mathrm{STAB}(G).$$

For all $v \in V(G)$, let $G - v$ denote the graph defined by $V(G-v) := V(G) \setminus \{v\}$ and $E(G-v) := E(G) \setminus \{\{u,v\} \in E(G) : u \in V(G)\}$, and let $\Gamma_G(v) := \Gamma(v)$ denote the *neighbourhood* of $v$ in $G$:

$$\Gamma(v) := \{u \in V : \{u,v\} \in E\}.$$

Let $G \ominus v$ be defined by

$$V(G \ominus v) := V \setminus (\Gamma(v) \cup \{v\})$$

and

$$E(G \ominus v) := \{\{u,w\} \in E(G) : u, w \notin (\Gamma_G(v) \cup \{v\})\}.$$

This operation is the *destruction of* $v$. For any edge $e \in E$ let $G - e$ denote the graph obtained from $G$ by the deletion of the edge $e$. If the inequality

$$a^T x = \sum_{u \in V(G)} a_u x_u \le \beta$$

is valid for STAB($G$), so are

$$\sum_{u \in V(G-v)} a_u x_u \le \beta \text{ and } \sum_{u \in V(G \ominus v)} a_u x_u \le \beta - a_v,$$

obtained by the deletion and the destruction of the node $v$, respectively.

For a given graph $G = (V, E)$, its $N_0$*-rank* (and similarly its $N$*-rank* and $N_+$*-rank*) is defined as the smallest nonnegative integer $k$ for which the application of the $N_0$ ($N$ or $N_+$) operator $k$ times to FRAC($G$) gives STAB($G$).

Alternatively, this rank is the largest rank of a facet of STAB($G$) (the $N_0$-, $N$-, and $N_+$-rank of an inequality valid for STAB($G$) is defined as the minimum $k$ for which the inequality is valid for $N_0^k(\mathrm{FRAC}(G))$, $N^k(\mathrm{FRAC}(G))$, and $N_+^k(\mathrm{FRAC}(G))$ respectively).

We denote these ranks by $r_0(G)$, $r(G)$, and $r_+(G)$, respectively. To simplify the notation, from now on, we will write $N_0^k(G)$, $N^k(G)$, and $N_+^k(G)$ for $N_0^k(\mathrm{FRAC}(G))$, $N^k(\mathrm{FRAC}(G))$, and $N_+^k(\mathrm{FRAC}(G))$ respectively.

Recall Proposition 8.1. The *deletion* operation on the graph, corresponds to the polytope $P \cap \{x : x_j = 0\}$ and the *destruction* operation corresponds to the polytope $P \cap \{x : x_j = 1\}$ (this is why the destruction operation was originally called *contraction*). Proposition 8.1 leads to the next lemma.

**Lemma 8.18** *For all graphs $G$, we have*

$$r(G) \le r_0(G) \le \min_{v \in V}\{r_0(G-v)\} + 1 \quad \textit{and} \quad r(G) \le \min_{v \in V}\{r(G-v)\} + 1.$$

*Moreover, if the inequalities obtained from $a^T x \le \beta$ by the deletion and destruction of $v \in V$ are valid for $N_0^k(G)$, $N^k(G)$, respectively, then $a^T x \le \beta$ is valid for $N_0^{k+1}(G)$, $N^{k+1}(G)$, respectively.*

Destruction of a node with a very large neighbourhood leaves a much smaller graph; however, the tool analogous to Proposition 8.1 is Lemma 8.2 which has a condition for *every* node. Lemma 8.2 yields the following result.

**Lemma 8.19** *For all graphs $G$, we have*

$$r_+(G) \leq \max_{v \in V}\{r_+(G \ominus v)\} + 1.$$

Note that very powerful applications of this lemma usually requires relatively dense and symmetric graphs.

Let's turn to the node deletion operation. So, based on the above results, the deletion of a node can only decrease any of the ranks, and at most by one. Let us define the class of *odd-cycle inequalities*. Let $\mathcal{H}$ be the node set of an odd-cycle (a simple cycle with an odd number of nodes) in $G$ then the inequality

$$\sum_{i \in \mathcal{H}} x_i \leq \frac{|\mathcal{H}| - 1}{2}$$

is valid for STAB$(G)$. We define

$$\mathrm{OC}(G) := \{x \in \mathrm{FRAC}(G) : x \text{ satisfies all odd-cycle constraints for } G\}.$$

If $\mathcal{H}$ is an odd-anti-hole then the inequality

$$\sum_{i \in \mathcal{H}} x_i \leq 2$$

is valid for STAB$(G)$.

If we have an odd-wheel in $G$ with *hub* node represented by $x_{2k+2}$ and the *rim* nodes represented by $x_1, x_2, \ldots, x_{2k+1}$, then the *odd-wheel inequality*

$$kx_{2k+2} + \sum_{i=1}^{2k+1} x_i \leq k$$

is valid for STAB$(G)$.

Based on these classes of inequalities we define the polytopes

$$\mathrm{OC}(G), \mathrm{ANTI\text{-}HOLE}(G), \mathrm{WHEEL}(G).$$

Now we have all the tools necessary to prove the next two theorems.

**Theorem 8.20** *For every graph $G$,*

$$N_0(G) = N(G) = \mathrm{OC}(G).$$

Note that this theorem provides a compact lifted representations of the odd-cycle polytope of $G$ (in the spaces $\mathbb{R}^{(\{0\}\cup V)\times(\{0\}\cup V)}$ and $\Sigma^{\{0\}\cup V}$). This polytope can have exponentially many facets in the worst case. However, $M(G)$ is represented by

$$|V|(|V|+1)/2 \text{ variables and } O\left(|V|^3\right) \text{ linear inequalities.}$$

It turns out that there are graphs $G$ for which $N_0^2(G) \neq N^2(G)$.

**Open Problem 3:** Find characterizations of $N_0^2(G)$, $N^2(G)$ analogous to Theorem 8.20.

**Theorem 8.21** *For every graph $G$,*

$$N_+(G) \subseteq \mathrm{OC}(G) \cap \mathrm{ANTI\text{-}HOLE}(G) \cap \mathrm{WHEEL}(G) \cap \mathrm{TH}(G).$$

Notice that the above theorem implies that four classes of well-known, exponentially many inequalities (odd-cycle, odd-anti-hole, odd-wheel and clique inequalities) are valid for $N_+(G)$. Moreover, every orthonormal representation constraint is also valid for $N_+(G)$. Recall that optimizing any linear function over $N_+(G)$ can be achieved by linear optimization over $M_+(G)$ and the latter is an SDP problem!

**Open Problem 4:** Find characterizations of $N_+(G)$, $N_+^2(G)$ analogous to Theorem 8.20 (and of course, sharpening Theorem 8.21).

**8.3.1 A Simple Invariance.** A polyhedron $P$ is called *lower comprehensive* if $x \in P$ and $u \leq x$ imply $u \in P$. A convex set $C \subset \mathbb{R}^d$ is called a *convex corner* if

- $C \subset \mathbb{R}_+^d$ and
- for every $x \in C$, every $y$ satisfying $0 \leq y \leq x$ is also in $C$.

**Theorem 8.22** *For every graph $G$, $\mathrm{STAB}(G)$, $\mathrm{TH}(G)$, $\mathrm{CLQ}(G)$ and $\mathrm{FRAC}(G)$ are all convex corners. Moreover, in general if $P$ is a convex corner then so are $N_0^k(P)$, $N^k(P)$ and $N_+^k(P)$, for every $k \in \mathbb{Z}_+$.*

Note that all packing polyhedra ( $\{x \in \mathbb{R}_+^d : Ax \leq \bar{e}\}$, where $A \geq 0$) are lower comprehensive (or are convex corners) and this is trivial to see (hence so is the first part of the above theorem). The second part of the theorem establishes that the operators $N_\sharp$ maintain this convex corner property in all of the intermediate (and of course final) relaxations. To prove the second part, just let $P$ be a convex corner and $x \in N_\sharp(P)$. Then it suffices to show that $(x - x_j e_j) \in N_\sharp(P)$ for every $j$ such that $x_j > 0$. Given a $Y \in M_\sharp(P)$ which proves $x \in N_\sharp(P)$, construct $\tilde{Y}$ by making the $j$th row and column of $Y$ zero. Then $\tilde{Y}$ proves $(x - x_j e_j) \in N_\sharp(P)$.

**8.3.2 A Simple Decomposition.** It is well-known that if the graph $G$ can be decomposed into two parts, $G_1$ and $G_2$, so that their intersection is a complete graph, then the facets of $\mathrm{STAB}(G)$ are just the union of the facets of $\mathrm{STAB}(G_1)$ and $\mathrm{STAB}(G_2)$. We also get a similar property for the $N_0$-rank, $N$-rank and the $N_+$-rank:

**Lemma 8.23** *If $G = G_1 \cup G_2$ such that $G_1 \cap G_2$ is a complete graph, then*

$$\begin{aligned} r_0(G) &= \max\{r_0(G_1), r_0(G_2)\}, \\ r(G) &= \max\{r(G_1), r(G_2)\} \end{aligned}$$

*and*

$$r_+(G) = \max\{r_+(G_1), r_+(G_2)\}.$$

The most common application of Lemma 8.23 is when $V(G_1) \cap V(G_2) = \{v\}$ (i.e., when $G$ has a cut node). When $G$ has a cut node $v$, we have a significant strengthening of the about clique decomposition result.

**Theorem 8.24** *If $G = G_1 \cup G_2$ and $V(G_1) \cap V(G_2) = \{v\}$, then $N_0^k(G)$ ($N^k(G)$) is defined by the facets of the polytopes $N_0^k(G_1)$ and $N_0^k(G_2)$ ($N^k(G_1)$ and $N^k(G_2)$) for every $k \in \mathbb{Z}_+$.*

It is quite interesting that Theorem 8.24 does not generalize to the case when $G_1 \cap G_2$ is a larger clique. Consider the graph given by Figure 5. For a proper definition of $G_1$, $G_2$, this graph has $V(G_1) \cap V(G_2) = \{2, 6\}$.

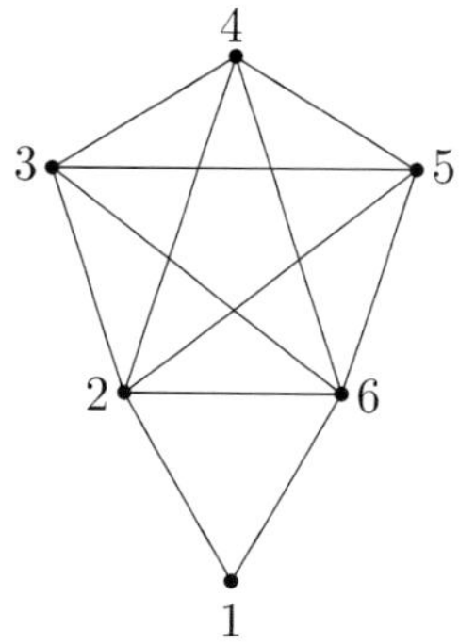


**Figure 5** $G = K_5 \cup K_3$ with $K_5 \cap K_3 = K_2$

**Proposition 8.25** *For the graph $G$ given in Figure 5, the inequality*

$$2x_1 + 3x_2 + 3x_6 + x_3 + x_4 + x_5 \leq 3 \tag{8.5}$$

*has $N_0$-rank 2. Moreover, it defines a facet of $N^2(G)$.*

**Proof** First, we prove that the inequality (8.5) has $N_0$-rank 2. Note that the deletion of node 1 gives an inequality which is the sum of three triangle inequalities. The destruction of node 1 gives a triangle inequality. By Proposition 8.1, the inequality (8.5) has $N_0$-rank at most two. It remains to prove that the inequality (8.5) has $N_0$-rank at least 2. To see this, note that the point $\frac{1}{3}\bar{e} \in [0,1]^6$ satisfies all triangle inequalities (that is, the point lies in $\mathrm{OC}(G)$), but it violates (8.5).

Secondly, we prove that the inequality (8.5) is a facet of $N^2(G)$. Since we just proved that the inequality has $N_0$-rank two, by definition, it is valid for for $N^2(G)$. For the graph in Figure 5, $\dim(\mathrm{STAB}(G)) = 6 = \dim(N^2(G))$, so we will exhibit 6 affinely independent vectors in $N^2(G)$ such that every one of the vectors satisfies the inequality (8.5) with equality. Consider the characteristic vectors of the stable sets $\{1,3\}, \{1,4\}, \{1,5\}, \{2\}$ and $\{6\}$. Since each of these vectors lie in $\mathrm{STAB}(G)$, they also lie in $N^2(G)$. For the sixth vector, consider $\left(\frac{3}{8}, \frac{1}{4}, \frac{1}{4}, \frac{1}{4}, \frac{1}{4}, \frac{1}{4}\right)^T$. To prove that the last vector lies in $N^2(G)$, we can exhibit $Y \in M^2(G)$ such that the projection of $Y$ onto the $x$-space is precisely our vector. We claim that the following matrix does the job:

$$Y := \frac{1}{8}\begin{bmatrix} 8 & 3 & 2 & 2 & 2 & 2 & 2 \\ 3 & 3 & 0 & 1 & 1 & 1 & 0 \\ 2 & 0 & 2 & 0 & 0 & 0 & 0 \\ 2 & 1 & 0 & 2 & 0 & 0 & 0 \\ 2 & 1 & 0 & 0 & 2 & 0 & 0 \\ 2 & 1 & 0 & 0 & 0 & 2 & 0 \\ 2 & 0 & 0 & 0 & 0 & 0 & 2 \end{bmatrix}.$$

By Theorem 8.20, $N_0(G) = N(G) = \mathrm{OC}(G)$, for every graph $G$. So, for the above matrix $Y$, we check that $Ye_i$ as well as $Y(e_0 - e_i)$ satisfy all the inequalities of $\mathrm{OC}(G)$ for every $i \in V$. It is easier to see that the six vectors mentioned above are affinely independent and they all satisfy (8.5) with equality. Thus, the inequality (8.5) is a facet of $N^2(G)$. □

In fact, for this example, $N_0^2(G) = N^2(G)$ and both are defined by all $K_4$ inequalities, the triangle inequality on $\{1,2,6\}$, and (8.5). However, $STAB(G)$ has just two nontrivial facets, the two maximal cliques (hence $r_0(G) = 3 = r_0(K_5)$), and we need both of them to get (8.5) as a linear combination of facets of $STAB(G)$.

While Theorem 8.24 does not generalize to the case of $G_1 \cap G_2$ being a clique of size at least 2, it does generalize to the $N_+$ operator.

**Theorem 8.26** *Suppose that $G = G_1 \cup G_2$ and $V(G_1) \cap V(G_2) = \{v\}$. Then for every $k \geq 0$, the convex set $N_+^k(G)$ is defined by the union of all valid inequalities for $N_+^k(G_1)$ and all valid inequalities for $N_+^k(G_2)$.*

The above decomposition theorem can be further generalized to polytopes (those that do not necessarily arise from a stable set problem) with a suitable notion of *cut coordinate* (playing the role of $v$ above, see [235]).

**8.3.3 Some Graph Operations and Graph Ranks.** Consider the simple graph operation of replacing an edge by a path. How does such operation affect the graph ranks? If we start with a triangle as our $G$, then $r(G) = r_0(G) = 1$. If we replace any edge in $G$ by a path of even length then the new graph $G'$ has $r(G') = r_0(G') = 0$. We could have started with an even cycle as our $G$ and replaced an edge by a path of even length. Then we would have $r(G') = r(G) + 1$.

For a more detailed study let's develop some rigorous definitions. A *subdivision* of a graph $G$ is obtained by replacing every edge by a path of length at least 1. The inserted new nodes, if any, on these paths should be all different. Note that a node of degree at least 3 in the subdivision must be also a node of the original graph. node of $G$.

An *odd subdivision of an edge* replaces the edge by a path of odd length (again, all new nodes on the path are different).

The *subdivision of a star* operation picks a node $v$ in $G$, and introduces a new node on every edge incident to $v$. The graph in Figure 6 is obtained from $K_4$ by applying the subdivision of a star operation at node 1.

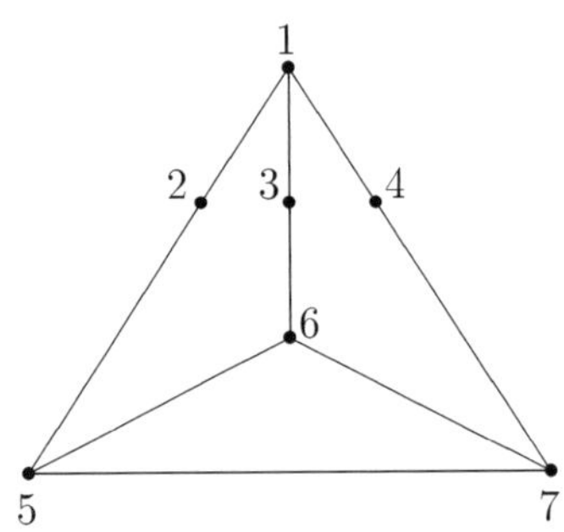


**Figure 6** Subdivision of a star applied to the 4-clique

**Theorem 8.27** *Suppose that graph $G'$ is obtained from graph $G$ using only the subdivision of a star or odd subdivision of an edge operations. Then, $r(G') \geq r(G)$ (similarly for the $N_0$- and the $N_+$-rank).*

When the underlying graph $G$ is a subdivision of a clique, we have a much sharper result.

**Theorem 8.28** *Let $G$ be a subdivision of $K_n$. Then TFAE*

(i) $r_0(G) = r_0(K_n) = n - 2$,
(ii) $r(G) = r(K_n) = n - 2$,

(iii) *$G$ is an odd-star subdivision of $K_n$.*

Let $v$ be a node with neighbourhood $\Gamma(v)$. Partition $\Gamma(v)$ into two nonempty, disjoint sets $A_1$ and $A_2$ (so $A_1 \cup A_2 = \Gamma(v)$, and $A_1 \cap A_2 = \emptyset$). A *stretching* of the node $v$ is obtained as follows:

Remove $v$, introduce two nodes instead, called $v_1$ and $v_2$, add an edge between $v_i$ and every node in $A_i$ for $i \in \{1, 2\}$, add the edge $\{v_1, v_2\}$, and then do one of the following:

(i) subdivide the edge $\{v_1, v_2\}$ with one node $w$; or
(ii) subdivide every edge between $v_2$ and $A_2$ with one node.

These operations are illustrated on Figure 7.

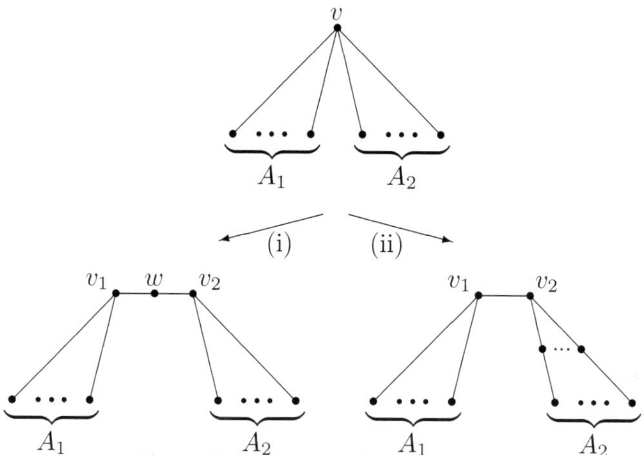


**Figure 7** Two types of stretching of $v$

Notice that when $A_1$ contains a single node, the stretching of a node operation reduces to the odd subdivision of an edge (in case of (i)) and the subdivision of a star (in case of (ii)) operations, so the stretching of a node operation is a common generalization of subdivision of a star and odd subdivision of an edge operations. Moreover, given any clique $K_n$ with $n \geq 4$, we can make sure that maximum degree of the final graph is at most three via only a sequence of stretching of a node operations. So, unlike exclusive applications of subdivision of a star and odd subdivision of an edge operations, a long sequence of stretching of a node operation can make a graph look very different than a clique. Nevertheless, we can still prove that $r(\cdot)$, and $r_0(\cdot)$ stay above $(n-2)$ if we start from $K_n$.

**Theorem 8.29** *If $G$ is a graph obtained from $K_n$ using the stretching of a node operation finitely many times, then $r_0(G) \geq r(G) \geq n-2$.*

We can further prove that all three graph ranks, $r(\cdot)$, $r_0(\cdot)$ as well as $r_+(\cdot)$ are monotone nondecreasing under the stretching of a node operation.

**Theorem 8.30** *If the graph $G'$ is obtained from $G$ using the stretching of a node operation, then $r_0(G') \geq r_0(G)$ (and similarly for the $N$- and the $N_+$-rank).*

**Open Problem 5:** Characterize the graphs for which

$$r_0(G) = r(G).$$

**8.3.4 Stable Set Problem and $r_+(G)$.** Motivated by the beautiful properties of TH$(G)$, theta body of $G$, and the strength of $N_+$ operator on the fractional stable set polytope FRAC$(G)$, we would like to understand "small" graphs for which $r_+(G)$ is "large." In particular, we should ask "what is the smallest graph whose $N_+$-rank is 1?" The answer is very easy: "the triangle."

Then, let's ask 'what is the smallest graph whose $N_+$-rank is 2?" An answer is in the next proposition.

**Proposition 8.31** *For the graph $G_{LT}$ on Figure 8 we have $r_+(G_{LT}) = 2$. Moreover, every graph $G$ with $V(G) \leq 5$ or $E(G) \leq 7$ satisfies $r_+(G) \leq 1$. Therefore, $G_{LT}$ is the smallest graph with $N_+$-rank equal to 2.*

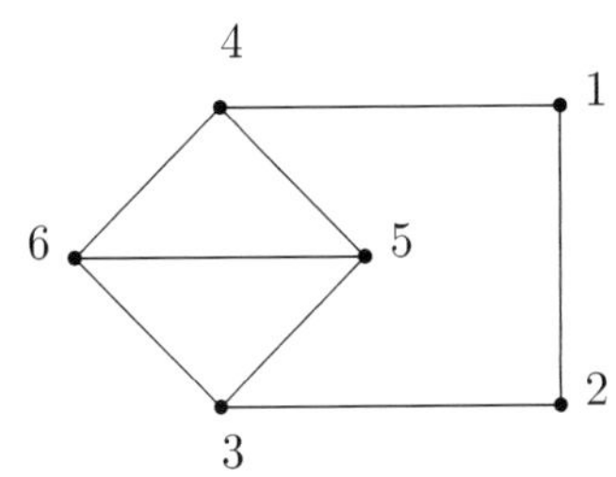


**Figure 8** Graph $G_{LT}$

**Proof** First, we will prove that for every graph $G$ with $|V(G)| \leq 5$ or $|E(G)| \leq 7$, we have $r_+(G) \leq 1$. Let $G$ be such that $|V(G)| \leq 5$ or $|E(G)| \leq 7$. Then by Lemma 8.23 we may assume that $G$ is 2-connected (otherwise we have a cut node and a smaller graph with the same $N_+$-rank). Whence, the degree of every node in $G$ is at least 2. If $|V(G)| \leq 5$, then the destruction of any node leaves at most an edge (whose rank is zero); thus, by Lemma 8.19, $r_+(G) \leq 1$. So, we may assume $|V(G)| \geq 6$ and $|E(G)| \leq 7$. Since the degree of every node in $G$ is at least two, we deduce $|V(G)| \in \{6, 7\}$. If $|V(G)| = 7$, then $G$ is an odd hole and $r_+(G) = 1$. If $|V(G)| = 6$, then $G$ is a 6-cycle with a chord. Now, we apply Lemma 8.18 at a degree 3 node of $G$ and conclude $r(G) \leq 1$. Since $r_+(G) \leq r(G)$, we are done.

Secondly, we prove that $r_+(G_{LT}) = 2$. To see that $r_+(G_{LT}) \leq 2$, we use Lemma 8.18. Deletion of the nodes 5 and 6 in Figure 8 leaves a path of length three whose $N$-rank is zero. Therefore, $r(G_{LT}) \leq 2$ which implies $r_+(G_{LT}) \leq 2$. To prove $r_+(G_{LT}) \geq 2$, we exhibit a matrix $Y \in M_+(G)$ such that its projection onto the $x$-space is

$$\bar{x} := \left(\frac{17}{40}, \frac{17}{40}, 1/3, 1/3, 1/4, 1/4\right)^T.$$

Note that $\bar{e}^T\bar{x} = 2 + \frac{1}{60}$, and $\alpha(G_{LT}) = 2$. Since $\bar{x}$ is not in $N_+^2(G) = \text{STAB}(G)$, this would prove $r_+(G_{LT}) \geq 2$. Let

$$Y(\xi) := \begin{bmatrix} 1 & \frac{17}{40} & \frac{17}{40} & 1/3 & 1/3 & 1/4 & 1/4 \\ \frac{17}{40} & \frac{17}{40} & 0 & \xi & 0 & 1/8 & 1/8 \\ \frac{17}{40} & 0 & \frac{17}{40} & 0 & \xi & 1/8 & 1/8 \\ 1/3 & \xi & 0 & 1/3 & 1/3-\xi & 0 & 0 \\ 1/3 & 0 & \xi & 1/3-\xi & 1/3 & 0 & 0 \\ 1/4 & 1/8 & 1/8 & 0 & 0 & 1/4 & 0 \\ 1/4 & 1/8 & 1/8 & 0 & 0 & 0 & 1/4 \end{bmatrix}.$$

It can be checked that

$$Y(\xi) \in M\left(G_{LT}\right) \text{ if and only if } \frac{11}{60} \leq \xi \leq \frac{3}{10}.$$

Setting $\xi := \frac{497+\sqrt{609}}{2400} \in \left(\frac{11}{60}, \frac{3}{10}\right)$ makes $Y$ positive semidefinite. □

Let's look at the graph $G_{LT}$ from some other viewpoints. First of all, we can obtain $G_{LT}$ from $K_4$ by an odd subdivision of an edge. Since $r_+(K_4) = 1$, this observation and Proposition 8.31 imply that the odd subdivision of an edge can increase the $N_+$-rank of a graph.

Now, consider for a node $v$ in $G$, replacing it by a clique of size at least 2 and connecting every new node to every original neighbour of $v$. This operation is called *cloning a node*. Cloning a node can also increase the $N_+$-rank of a graph: Start with a 5-cycle. The $N_+$-rank is 1. Cloning any node gives $G_{LT}$ which has $N_+$-rank two. We summarize such facts in the next theorem.

**Theorem 8.32** *Each of the following operations can increase $r_+(G)$: odd subdivision of an edge, subdivision of an edge, cloning of a node, adding an edge, deleting an edge, contracting an edge.*

**Proof** We already proved before the statement of the theorem that odd subdivision of an edge, subdivision of an edge and cloning of a node can increase the $N_+$-rank. To see that adding an edge can increase the $N_+$-rank, consider a path of length 2 ($N_+$-rank is 0) and add the edge that joins the endpoints of the path, resulting in a triangle which has $N_+$-rank 1. To see that deleting an edge can increase the $N_+$-rank, start with $K_6$ (which has $N_+$-rank 1) and notice that $G_{LT}$ can be obtained from $K_6$ by deleting some edges. Since $r_+(G_{LT}) = 2$, at some point throughout these deletions, the $N_+$-rank must increase. Finally, contraction of an edge can also increase the $N_+$-rank, since contracting an edge from a 4-cycle ($N_+$-rank is zero), leaves a triangle. □

Now, we continue with the line of thought that started the current subsection. Let $n_+(k)$ denote the minimum number of nodes needed in a graph $G$ to have $r_+(G) = k$, for $k \in \mathbb{Z}_+$. We would like to understand this function $n_+(k)$. For instance, is it true that "for every $n$ large enough, there exists a graph $G$ with $|V(G)| = r_+(G) = n$?" Next theorem provides an answer to this question and a lower bound on $n_+(k)$.

**Theorem 8.33** *For every graph $G = (V, E)$, $r_+(G) \leq \left\lfloor \frac{|V|}{3} \right\rfloor$.*

**Proof** We will prove $n_+(k) \geq 3k$, for every $k \in \mathbb{Z}_{++}$ by induction on $k$. As we noted before, clearly, $n_+(1) = 3$. (In fact, Proposition 8.31 already established $n_+(2) = 6$.) Suppose $n_+(k) \geq 3k$. Let $G$ satisfy $|V(G)| = n_+(k+1)$ and $r_+(G) = k+1$. Then $G$ does not contain a leaf node or an isolated node (since removing the isolated nodes and the leaf nodes does not decrease the $N_+$-rank, this would contradict the choice of $G$). Thus, $\deg(v) \geq 2$ for every $v \in V(G)$. Since $r_+(G) = k+1$, by Lemma 8.19, there exists a node $v$ in $G$ such that $r_+(G \ominus v) \geq k$. Let $G' := G \ominus v$. Note that $r_+(G') \leq k$ (since $|V(G)| = n_+(k+1)$). Thus, $r_+(G') = k$ and we deduce $|V(G')| \geq n_+(k)$. Whence,

$$n_+(k+1) = |V(G)| \geq |V(G')| + 3 \geq n_+(k) + 3 \geq 3(k+1),$$

where the first inequality uses the facts $G' = G \ominus v$, $\deg(v) \geq 2$ for every $v \in V(G)$, the last inequality uses the induction hypothesis. □

It is an interesting open question whether the relation $r_+(G) \leq \left\lfloor \frac{|V|}{3} \right\rfloor$ is tight for an infinite family of graphs, or whether $n_+(k) = 3k$, for all $k \geq 1$.

**Open Problem 6:** What are the values of $n_+(k)$ for every $k \in \mathbb{Z}_+$? In particular, is it true that $n_+(k) = 3k$, for all $k \in \mathbb{Z}_+$?

## 8.4 Graph Ranks for Max Cut

Let us revisit Max Cut in the context of lift-and-project methods. We will denote $P_I$ by $\mathrm{CUT}(G)$ (the cut polytope), and take $P$ the initial approximation of $\mathrm{CUT}(G)$ to be the odd-cycle polytope or equivalently the polytope defined by the triangle inequalities on the complete graph on $V(G)$. The latter is called the *metric polytope* which we denote by $\mathrm{MET}(G)$.

**Proposition 8.34** *If there are $k$ nodes in $G$ such that the contraction of these $k$ nodes leaves a graph with no $K_5$-minor, then*

$$N_0^k(\mathrm{MET}(G)) = N^k(\mathrm{MET}(G)) = \mathrm{CUT}(G).$$

After applying a lift-and-project operator to $\mathrm{MET}(G)$, we can further project the resulting convex set onto the space of variables defined only by the original graph $G$. In some cases, this seems convenient. Let us denote the resulting ranks by $r(G)$, $r_+(G)$. It turns out that the graph ranks are monotone with respect to graph minor operations. For the current context, we say that graph $H$ is a *minor* of graph $G$ if $H$ can be obtained from $G$ by a sequence of edge deletions and/or edge contractions.

**Theorem 8.35** *Let $H$ be a minor of $G$. Then*

$$r(H) \leq r(G) \text{ and } r_+(H) \leq r_+(G).$$

For the stable set problem we showed that the operators $N$, $N_+$ behave very well under the simple graph decomposition (when the intersection of the two subgraphs is a clique). This property extends to the Max Cut setting as well: Suppose that the graph $G$ can be decomposed into two parts, $G_1$ and $G_2$, so that their intersection is a complete graph, then

$$r(G) \leq \max\{r(G_1), r(G_2)\},$$

and

$$r(G) \leq \max\{r(G_1), r(G_2)\}.$$

## 8.5 TSP Polytope

TSP has been one of the most popular and most widely-studied problems in combinatorial optimization. There are various reasons we can cite here. However, whenever there is a theorem that will make one of these points, it is best to leave it with that:

**Theorem 8.36** *Every 0,1 polytope is isomorphic to a face of a TSP polytope.*

How would $N_+$ procedure do on the TSP polytope when we start with the subtour elimiation polytope SEP? Since we can separate over SEP in polynomial time, we can optimize over $N_+^k(\text{SEP})$ in polynomial time for every $k = O(1)$. Therefore, unless $\mathcal{P} = \mathcal{NP}$, $N_+$-rank of the TSP polytope is not bounded by a constant. It turns out the situation is worse for the lift-and-project methods.

**Theorem 8.37** *Let $G = (V, E)$ a simple, undirected graph be given. Consider the symmetric TSP polytope of $G$ and a combined lift-and-project operator for which in every iteration, we generate all valid inequalities for $N_+$ as well as all valid inequalities for the Gomory-Chvátal procedure. Then the combined operator requires at least $\left\lfloor \frac{|V|}{8} \right\rfloor$ iterations to compute the TSP polytope starting from the subtour elimination polytope.*

Currently, one of the most popular open problems on TSP is the $\frac{4}{3}$-*conjecture* (that is, *the integrality gap of* SEP *for metric TSPs is* $\frac{4}{3}$). We can ask a related question. Does there exist tractable LP or SDP relaxations of the TSP polytope with integrality gap less than $\frac{4}{3}$?

To answer such a question in the affirmative, all one needs to do is to come up with such a formulation of LP or SDP, prove that it is tractable and prove that the integrality gap is smaller than $\frac{4}{3}$. (We have seen such processes throughout the course.) However, what if such formulations do not exist? Proving the nonexistence of such formulations seem considerably harder. One way of attack would be to lower our standards of proof for "non-existence" and try to prove that the existence of such formulations would imply solving some $\mathcal{NP}$-hard problem in polynomial time and hence the unlikely outcome of $\mathcal{P} = \mathcal{NP}$. While this kind of result can be very useful, in some cases it is hard to attain. So, then we might want to lower our standards of proof for "non-existence" one more notch and show that $O(1)$ applications of a lift-and-project procedure does not improve the integrality ratio. Indeed, it turns out that the latter is provable.

**Theorem 8.38** *For the metric TSP, the integrality gap of $N_+^k(\text{SEP})$ is $\frac{4}{3}$ for every $k = O(1)$.*

Therefore, obtaining tractable LP or SDP relaxations starting with SEP, with better integrality gaps is not possible via Lovász-Schrijver procedures.

## 8.6 Bibliographical Notes

The study of ranks of lift-and-project methods were initiated by Lovász and Schrijver in early 1990s. However, the upper bound $d$ on the number of iterations was already proved by Balas in early 1970s for a lift-and-project method even weaker than $N_0$ (therefore, Balas' convergence result is stronger). Some general ground work can be found in Cook and Dash [92], Goemans and Tunçel [143] and Hong and Tunçel [178]. Characterization of the facets of the matching polytope is due to Edmonds [111]. $N_+$-rank of the matching polytope was determined by Stephen and Tunçel [337]; later further related results were obtained by Aguilera et al. [1] who also make a connection to the work of Doob [108]. Mathieu and Sinclair determined the Sherali-Adams rank of the matching poltope as $(2k-1)$ when applied to the clique on $(2k+1)$ nodes [252]. The relative strength of $N$ vs. $N_+$ was studied by Cook and Dash [92] as well as Goemans and Tunçel [143]. Some results on the stable set problem are due to Lovász and Schrijver [245]; others are due to Lipták and Tunçel [235]. Lemma 8.23 is motivated by a result of Chvátal [86] which showed that if $G_1 \cap G_2$ is a complete graph, then the facets of $\text{STAB}(G_1 \cup G_2)$ is the union of the facets of $\text{STAB}(G_1)$ and the facets of $\text{STAB}(G_2)$. For other facet generating operations, see Wolsey [368]. For a general look at polyhedral as well as lift-and-project cuts, see the survey paper [94] by Cornuéjols.

For results on the critical facets of $\text{STAB}(G)$, see Lipták [232] and Lipták and Lovász [233].

For results on the stable set problem based on copositive programming relaxations, see de Klerk and Pasechnik [101], Peña, Vera and Zuluaga [289].

Our example in Figure 5 and Proposition 8.25 (due to [235]) solve Problem 5 of Lovász [241]. Lovász asked (paraphrased here) whether $N^k(G)$ is defined only by the clique inequalities (up to $K_{k+2}$) and nonnegativity constraints when $G$ is perfect. Our example and claim prove that the answer to this question is "no." As we mentioned, there are graphs $G$ for which $N_0^2(G) \neq N^2(G)$ (see Au [15] or [16]). Related to Open Problem 5 of this chapter, it is conjectured in [235] "$r_0(G) = r(G)$, for all graphs $G$."

A question of Knuth [203] (Problem 5 in [203]) is answered by Proposition 8.31 which is originally from [235]. Graph $G_{LT}$ from this Proposition has been generalized by Escalante, Montelar and Nasini [117]; they proved $n_+(3) = 9$ and that odd subdivisions of cliques will not give the equality $n_+(k) = 3k$ in general, hence making some progress towards the solution of *Open Problem 6.*

For a connection between the tree-width of graphs and RLT operator rank, see Bienstock and Ozbay [44]. For an application of RLT to set partitioning problems see Sherali and Lee [327].

BCC operator on the stable set problem, behaves very nicely under polarity as shown by Aguilera, Escalante and Nasini [2]; for shorter proofs, see [139] and [234]; for related results, see Escalante, Nasini and Varaldo[118].

Graph ranks for Max Cut were studied by Laurent [225, 223]. Theorem 8.36 is from Billera and Sarangarajan [46]. The rank result on the TSP polytope (Theorem 8.37) is due to Cook and Dash [92] and Theorem 8.38 is from Cheung [77].

While it is impressive that $N_+$ operator does so well (on some respects) on $\text{FRAC}(G)$, in general one should expect not so great results from such generally applicable algorithms. E.g., Alekhnovich, Aurora and Tourlakis [3] showed that for problems such as Max-3Sat and Min. Set Cover, we cannot obtain bounds within an absolute constant of the true optimum objective value with polynomial time computations only utilizing the $N_+$ operator. Note that Cheung's result mentioned above is of similar character. Along this direction, also see Arora, Bollobás and Lovász [11] and Karlin, Mathieu and Nguyen [69].

For a probabilistic analysis of the performance of Lovász-Schrijver operators on the stable set problem, see [124]. For a connection between lift-and-project operators and multiparty communication complexity see Beame, Pitassi and Segerlind [28]. Grigoriev, Hirsch and Pasechnik [150] consider the lift-and-project procedures in the context of proof systems for systems of polynomial inequalities which is a subject of the next chapter.

One of the interesting directions to continue along here would be to compare the strength of the various relaxations $N_{\sharp}^{k}(P)$ using the measures utilized by Goemans [141], also see [16].

CHAPTER 9

# Successive Convex Relaxation Methods

The lift-and-project methods seem very general in that they apply to any combinatorial optimization problem. In fact, they have been further generalized to solve essentially *any* optimization problem in finite dimensional Euclidean spaces.

Let $F \subset \mathbb{R}^d$ be a compact set. Let $f : \mathbb{R}^d \to \mathbb{R}$ be a continuous function. Then the optimization problem

$$\max \{f(x) : x \in F\}$$

can be reduced to

$$\max \{t : f(x) \geq t, x \in F\}.$$

So, without loss of generality, we can assume that the objective function is linear. So, our problem is

$$\max \left\{c^T x : x \in F\right\},$$

for a given vector $c$. Now, as in the special case of combinatorial optimization, we have

$$\max \left\{c^T x : x \in F\right\} = \max \left\{c^T x : x \in \operatorname{conv}(F)\right\}.$$

How do we "compute" the convex hull?

We take an approach similar to that of *lift-and-project methods* for combinatorial optimization. We start with a convex compact set $C_0$ which contains $F$ and utilizing liftings and stronger valid inequalities in the lifted space and positive semidefiniteness, we generate nested convex compact sets $C_i$ such that $C_i$ converge to $\operatorname{conv}(F)$:

$$C_0 \supseteq C_1 \supseteq C_2 \supseteq \cdots \supseteq \operatorname{conv}(F)$$

and

$$\bigcap_{i=0}^{\infty} C_i = \operatorname{conv}(F).$$

Note that the topology in the space of convex compact sets in finite-dimensional Euclidean spaces is just like the point topology in such spaces.

Let us reformulate a fact from convex geometry. We denote by $D_1$ the $d$-dimensional unit hypersphere (in the context of the next theorem this is the set of all nontrivial objective functions).

**Theorem 9.1** *The set of all $c \in D_1$ for which the optimal solution of*

$$\max \left\{c^T x : x \in \operatorname{conv}(F)\right\}$$

*is not unique has zero, $(d-1)$-dimensional Hausdorff measure.*

If we are able to get our hands on a convex compact set $C$ which is tractable (that is, we can optimize a linear function over it in polynomial time) and $C = \operatorname{conv}(F)$, then for almost all objective function vectors $c$, the problem

$$\max\left\{c^T x : x \in C\right\}$$

will have a unique minimizer $\hat{x}$. Thus, $\hat{x} \in \text{ext}(C) \subseteq F$. Therefore, $\hat{x}$ will also be an optimal solution of our original (non-convex) problem

$$\max\left\{c^T x : x \in F\right\}.$$

Whence, for the rest of this chapter, we will focus on solving (or approximately solving) the optimization problem

$$\max\left\{c^T x : x \in \text{conv}(F)\right\}.$$

## 9.1 Fundamental Framework

We first choose a representation of $F$. There are many possibilities. However, systems of quadratic inequalities (possibly infinitely many) form a well-behaved class and they have the power to express any compact set in an Euclidean space. Moreover, as we generate new, stronger, valid convex inequalities, staying in the same class (of quadratic inequalities) seems very easy to do. We represent quadratic inequalities with the following algebraic structure: $Q \in \Sigma^d$, $q \in \mathbb{R}^d$, $\gamma \in \mathbb{R}$ are given; the inequality is

$$x^T Q x + 2q^T x + \gamma \leq 0.$$

We outline some reasons for choosing the quadratic inequality representation in the following series of lemmas.

**Lemma 9.2** *Every compact set $F \subset \mathbb{R}^d$ admits a representation as the feasible region of a system of quadratic inequalities.*

**Proof** Let $F$ be a compact set in $\mathbb{R}^d$. Then the complement of $F$ is open, hence admits a representation as the union of open balls. The complement of this representation is the intersection of closed balls which proves the existence of a quadratic inequality system representation of $F$. □

The next lemma is a trivial generalization of the usual way we have been deriving our SDP relaxations.

**Lemma 9.3** *For every $(Q, q, \gamma)$ as above,*

$$\begin{aligned}&\left\{x \in \mathbb{R}^d : x^T Q x + 2q^T x + \gamma \leq 0\right\}\\ &\subseteq \left\{x \in \mathbb{R}^d : \left\langle \begin{bmatrix} \gamma & q^T \\ q & Q \end{bmatrix}, \begin{bmatrix} 1 & x^T \\ x & X \end{bmatrix} \right\rangle \leq 0, \begin{bmatrix} 1 & x^T \\ x & X \end{bmatrix} \in \Sigma_+^{d+1} \right\}.\end{aligned}$$

*Equality holds above if* $\begin{bmatrix} 1 & x^T \\ x & X \end{bmatrix}$ *is rank one.*

We assume that we are given a quadratic inequality description of $F$. We will denote the data set, that is the set of $(Q, q, \gamma)$ by $\mathcal{P}_F$. As the last lemma makes it clear, we can write a quadratic inequality in at least two ways:

$$x^T Q x + 2q^T x + \gamma \leq 0$$

or

$$\left\langle \begin{bmatrix} \gamma & q^T \\ q & Q \end{bmatrix}, \begin{bmatrix} 1 & x^T \\ x & xx^T \end{bmatrix} \right\rangle \leq 0.$$

So, we will use these two *formats* interchangeably. E.g., we may say $(Q, q, \gamma) \in \mathcal{P}_F$ or $\begin{bmatrix} \gamma & q^T \\ q & Q \end{bmatrix} \in \mathcal{P}_F$ both meaning the same thing.

Since $F$ is compact, there exists a ball which contains $F$. We assume that the quadratic inequality description $\mathcal{P}_F$ contains an inequality such as

$$x^T x \leq M,$$

where $x^T x < M$, for every $x \in F$.

Let $\mathcal{P} \subset \Sigma^{d+1}$ be a set representing the coefficients of a quadratic inequality system as above. Then,

- on the one hand, we may assume that $\mathcal{P}$ is a closed convex cone, without changing the actual set that it represents as the feasible solutions to the quadratic system;
- on the other hand, we may replace $\mathcal{P}$ by any subset of it that generates the same cone as above (e.g., a basis for the lineality space of the cone and its negation and the normalized extreme rays of the cone).

Let $\mathcal{Q}_+ \subset \Sigma^{d+1}$ denote

$$\left\{ \begin{bmatrix} \gamma & q^T \\ q & Q \end{bmatrix} : Q \in \Sigma_+^d \right\}.$$

Then, notice that the set

$$\{x \in \mathbb{R}^d : x^T Q x + 2q^T x + \gamma \leq 0\}$$

is convex iff $Q \succeq 0$. Therefore,

$$\{x \in \mathbb{R}^d : x^T Q x + 2q^T x + \gamma \leq 0, \forall (Q, q, \gamma) \in (\mathcal{P} \cap \mathcal{Q}_+)\}$$

is a convex set containing the convex hull of whatever set was represented by $\mathcal{P}$. In this setting, replacing $\mathcal{P}$ by the cone it generates can improve the resulting convex set (making it a tighter relaxation).

We have the following representation theorem establishing an equivalence between such convex quadratic relaxations and SDP relaxations of the solution sets of quadratic inequality systems.

**Theorem 9.4** *Let $\mathcal{P}$ be a convex cone as above (containing $(I, 0, M)$). Then the convex sets*

$$\{x \in \mathbb{R}^d : x^T Q x + 2q^T x + \gamma \leq 0, \forall (Q, q, \gamma) \in (\mathcal{P} \cap \mathcal{Q}_+)\}$$

*and*

$$\left\{ x \in \mathbb{R}^d : \left\langle \begin{bmatrix} \gamma & q^T \\ q & Q \end{bmatrix}, \begin{bmatrix} 1 & x^T \\ x & X \end{bmatrix} \right\rangle \leq 0, \forall (Q, q, \gamma) \in \mathcal{P}; \begin{bmatrix} 1 & x^T \\ x & X \end{bmatrix} \in \Sigma_+^{d+1} \right\}$$

*are identical. Moreover, in the second description, we can replace $\mathcal{P}$ by its generators.*

This theorem already tells us how to start. Given $\mathcal{P}_F$, we write the SDP relaxation given in the last theorem. If $\mathcal{P}_F$ is finitely generated, then we can have the feasible region of an SDP (otherwise, we have a semi-infinite SDP).

So, we get a compact convex set $C_0$ containing $F$. In general, at our $k$th iteration, we have a compact convex set $C_k$ containing $F$, where $C_k$ is defined by the "implicit" quadratic system and the SDP relaxation given in the last theorem.

Recall that $D_1$ denotes the unit hypersphere in $\mathbb{R}^d$. Let $D_2$ denote any basis for $\mathbb{R}^d$ and its negation (e.g.,

$$D_2 = \{e_i, -e_i : i \in \{1, 2, \ldots, d\}\}).$$

Compute

$$\alpha := \max\left\{a^T x : x \in C_k\right\}$$

and

$$\beta := \max\left\{b^T x : x \in C_k\right\}$$

for every $a \in D_1$ and for every $b \in D_2$. Then the quadratic inequality

$$\left(\alpha - a^T x\right)\left(\beta - b^T x\right) \geq 0$$

is a valid quadratic inequality for $C_k$ and hence a valid quadratic inequality for $F$. Denote by $\mathcal{P}_k$ the set of all such valid quadratic inequalities. We define $C_{k+1}$ by the convex relaxation where $\mathcal{P}$ is replaced by $\mathcal{P}_F \cup \mathcal{P}_k$ (or the generators of the cone$(\mathcal{P}_F \cup \mathcal{P}_k)$). This gives a very general method. Let us call it Successive SDP Relaxation (SSDPR) method. Moreover, the $N_+$ operator that we discussed in the last two chapters is a very special case.

Let us take one more look at Theorem 9.4. It says that the SDP relaxation describes the same region as all nonnegative linear combinations of those quadratic inequalities that turn out to be convex quadratic inequalities. Between these two equivalent definitions, each has its own advantages; however, in terms of tractability, the SDP representation is indeed striking.

We have a convergence theorem.

**Theorem 9.5** *With the above definitions, the sequence of convex compact sets $C_i$ generated by SSDPR method satisfies*

(a) *for every $k \geq 0$, $\text{conv}(F) \subseteq C_{k+1} \subseteq C_k$ (monotonicity), in fact $C_{k+1} = C_k$ iff $C_k = \text{conv}(F)$;*

(b) $\bigcap_{k=1}^{k^*} C_k = \emptyset$ *for some finite number $k^*$ if $F = \emptyset$ (detecting infeasibility in finitely many steps);*

(c) $\bigcap_{k=1}^{\infty} C_k = \text{conv}(F)$ *(asymptotic convergence).*

The above theorem was extended to the case when $D_1$ is replaced by a finite mesh of the unit hypersphere (as we will summarize in the next section). There are many other related results. But first, we will sketch a proof of the convergence theorem.

For the simplicity of presentation, we will analyze the successive convex relaxation algorithm that constructs all valid Euclidean ball constraints on $C_k$ at iteration $k$ instead of the quadratic constraints described above in SSDPR method. (It is relatively easy to show that the SSDPR constraints are strong enough that in the following proof they could be used to draw the desired contradiction.) Given a compact convex set $C_k$, let us denote by $\mathcal{P}_k^S$, the set of all Euclidean ball constraints that are valid for $C_k$.

The main tool is a separation theorem. Using a standard separation theorem from Chapter 2, we can easily prove the following fact (Separating Hypersphere Theorem).

**Proposition 9.6** *Let $G \subset \mathbb{R}^d$ be a nonempty, convex compact set and $\hat{x} \in \mathbb{R}^d \setminus G$. Then there exists a hypersphere (with center $\bar{x}$, radius $R$) which separates $\hat{x}$ from $G$ as follows:*

$$G \subseteq \left\{ x \in \mathbb{R}^d : \langle x - \bar{x}, x - \bar{x} \rangle \leq R^2 \right\} \text{ and } \langle \hat{x} - \bar{x}, \hat{x} - \bar{x} \rangle > R^2.$$

Getting back to Theorem 9.5, establishing part (a) is easy. Part (b) follows from the compactness of $C_k$ and part (c). So, we will only justify part (c). By part (a) and the compactness of $C_k$ we will have a limit $C$ a compact convex set contained in every $C_k$. Clearly, $\text{conv}(F) \subseteq C$. So, for a contradiction, suppose there exists $u \in C \setminus \text{conv}(F)$. See Figure 9.

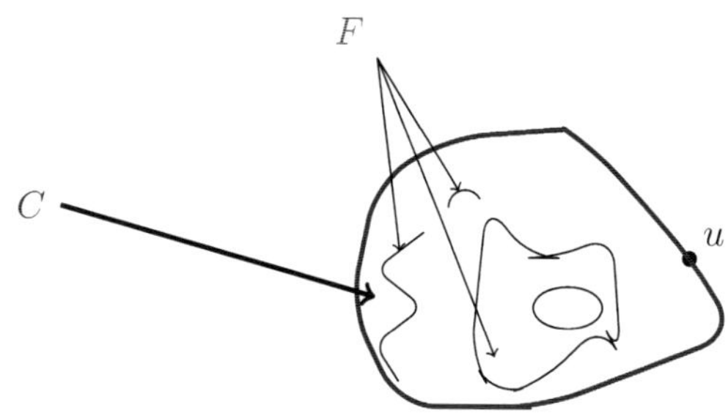


**Figure 9** Suppose (for a contradiction) that $\bigcap_{k=1}^{\infty} C_k =: C \neq \text{conv}(F)$. Then, $\exists u \in C \setminus \text{conv}(F)$.

Then by Proposition 9.6, there exist $\bar{x}$ and $R$ such that

$$\|x - \bar{x}\|_2 < R < \|u - \bar{x}\|_2, \text{ for every } x \in \text{conv}(F).$$

See Figure 10.

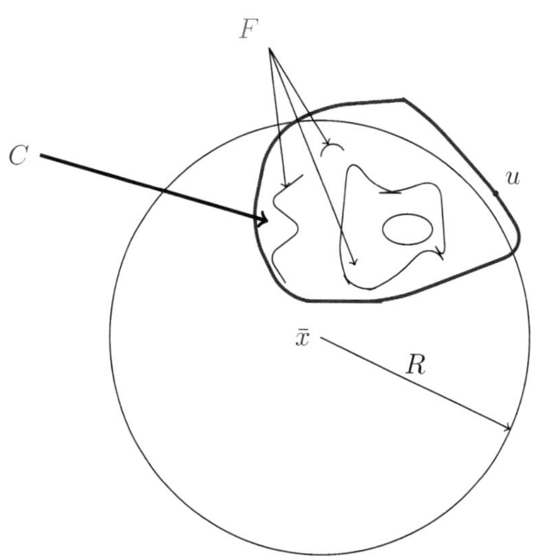


**Figure 10** There exists a separating hypersphere for $u$, $\text{conv}(F)$.

Let $\bar{R} > R$ be the radius of the smallest ball centered at $\bar{x}$ containing $C$. See Figure 11.

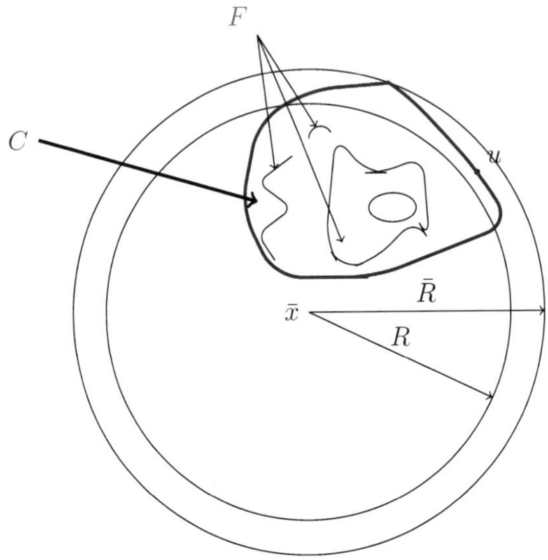


**Figure 11** Blow-up the hypersphere to just enclose $C$.

Then pick any $v \in C$ that is on the boundary of this ball ($\|v - \bar{x}\|_2 = \bar{R}$). Thus, $v \notin F$. See, Figure 12.

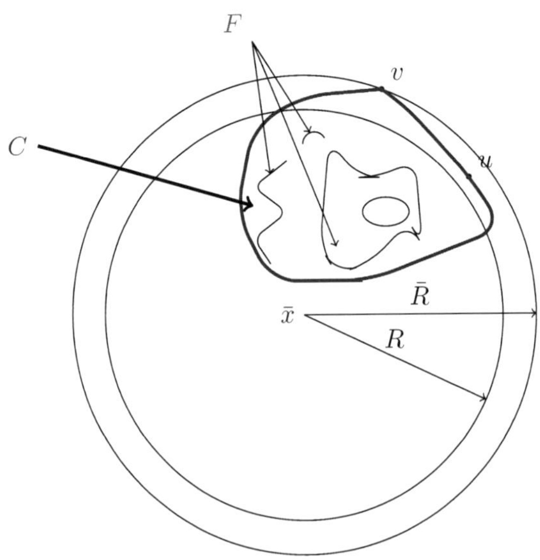


**Figure 12** $\exists v \in C \setminus \text{conv}(F)$ lying on the blown-up hypersphere.

So, there exists $p$ in the original formulation of $\mathcal{P}_F$ which is violated by $v$ (see Figure 13). Let $(Q, q, \gamma)$ be the data for the quadratic function $p$. Since our formulation is homogeneous, we can scale $p$ and assume without loss of generality that $\lambda_n(Q) = -1$ (note that $p$ has to be nonconvex, since all points in all $C_k$s satisfy every convex inequality in $\mathcal{P}_F$ by the definition of $C_k$). We denote the scaled inequality by $p(x) \leq 0$ and the amount of violation of this scaled inequality by $\epsilon$. I.e., $p(v) = \epsilon > 0$. Now, we can construct a ball (and the corresponding quadratic constraint) whose interior contains $C$:

$$\tilde{p}(x) := \|x - \bar{x}\|_2^2 - \bar{R}^2 - \frac{\epsilon}{2} \leq -\frac{\epsilon}{2}, \text{ for every } x \in C.$$

Since $C_k \to C$, there exists a finite $\bar{k}$ such that $\tilde{p}(x) \leq 0$ becomes a valid ball constraint for $C_{\bar{k}}$. This means, $(p + \tilde{p}) \in \operatorname{cone}\left(\mathcal{P}_F \cup \mathcal{P}_{\bar{k}+1}\right) \cap \mathcal{Q}_+$. But, we have

$$p(v) + \tilde{p}(v) = \frac{\epsilon}{2} > 0.$$

Therefore, by definition of the method, $v \notin C_{\bar{k}+1}$ which implies $v \notin C$, a contradiction. Whence, part (c) is established.

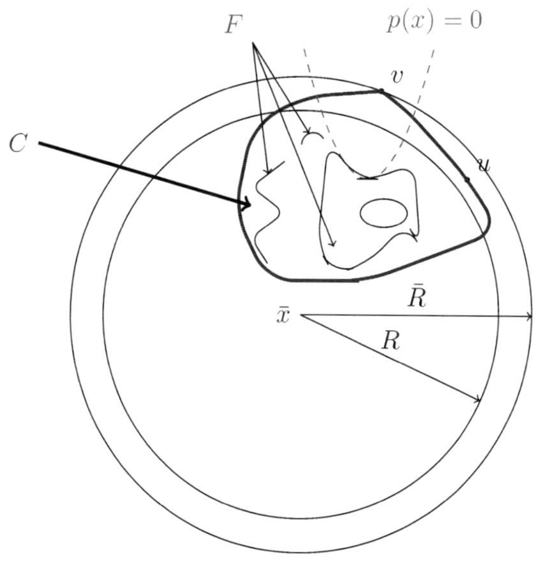


**Figure 13** $\exists$ original constraint $p(x) \geq 0$ violated by $v$.

Note that a more constructive proof is also possible. We essentially proved, for every point $u \in C_j \setminus \operatorname{conv}(F)$, there exists an iteration number $k$ such that $u \notin C_k$. (One difference is that from a given $u$ we moved to a potentially different $v$ with the same kind of property.)

SSDPR operator is a direct generalization of $N_+$ operator. Let $C_0 \subseteq [0,1]^d$ be a compact convex set which is the initial relaxation of $F \subseteq \{0,1\}^d$. I.e.,

$$F = C_0 \cap \{0,1\}^d.$$

Then we know that $N_+^d(C_0) = \operatorname{conv}(F)$.

To apply the SSDPR method in this very special case, we include the quadratic inequalities

$$x_j^2 - x_j \leq 0, \quad -x_j^2 + x_j \leq 0, \quad \forall j \in \{1, 2, \ldots, d\}$$

in $\mathcal{P}_F$.

**Theorem 9.7** *Let $C_1, C_2, \ldots$ be the sequence generated by the SSDPR method. Then*

$$C_k = N_+^k(C_0), \quad \forall k \in \{0, 1, \ldots\}.$$

## 9.2 Discretized/Localized Method

We gave a very general method to solve essentially any optimization problem in $\mathbb{R}^d$. If the initial system $\mathcal{P}_f$ is finite, then the first SDP relaxation of the SSDPR method is an SDP problem. However, the descriptions of the convex relaxations $C_k$ obtained for the following iterations can have infinitely many linear constraints on the finite dimensional matrix variable. This leaves us with a *semi-infinite* SDP problem. (More on this issue in Chapter 10, when we revisit the subset lattice interpretation of lift-and-project methods.) An obvious idea is to *discretize* the set $D_1$. E.g., replace $D_1$ by a finite mesh.

A related issue is that we are primarily interested in solving the optimization problem

$$\max\left\{c^T x : x \in F\right\}.$$

So, why do we have to compute $\text{conv}(F)$? Our approximation of $\text{conv}(F)$ need not be very accurate, it suffices to be accurate in a neighborhood of a maximizer $x^*$ of $\max\left\{c^T x : x \in F\right\}$ (we are assuming $c \neq 0$). An unsophisticated idea would be to make the mesh (the one replacing $D_1$) very fine near $\frac{c}{\|c\|_2}$, but very rough elsewhere. This approach (which we call *localization*) does not help in the worst-case; however, it may be worthwhile for some problems.

Discretized SSDPR Algorithm : Given a mesh parameter $\delta > 0$, in the SSDPR algorithm, replace $D_1$ by $D_1(\delta)$, where $D_1(\delta)$ is a finite subset of $D_1$ such that for every point in $u \in D_1$, there exists a point in $v \in D_1(\delta)$ such that $\|u - v\|_2 \leq \delta$. We call the resulting algorithm: $D_1(\delta)$-SSDPR algorithm.

**Theorem 9.8** *Let $\mathcal{P}_F$ be given (including a constraint $x^T x \leq M$). Then for every $\epsilon > 0$, there exists a $\delta > 0$ such that the sequence $C_0, C_1, \ldots$ of compact convex sets generated by the $D_1(\delta)$-SSDPR algorithm satisfies*

(a) $\text{conv}(F) \subseteq C_{k+1} \subseteq C_k, \quad \forall k \in \{0, 1, \ldots\}$;
(b) *there exists $\bar{k}$ such that*

$$\text{conv}(F) \subseteq C_{\bar{k}} \subseteq \text{relax}\left(\text{conv}(F), \epsilon\right).$$

Localized SSDPR Algorithm : Given an objective function vector $c \neq 0$, a localization parameter $\kappa > 0$ and a mesh parameter $\delta > 0$, in the SSDPR algorithm, replace $D_1$ by $D_1(c; \kappa, \delta)$, where $D_1(c; \kappa, \delta)$ is a finite subset of $D_1$ such that for every point in $u \in D_1$, with $\|u - c\|_2 \leq \kappa$, there exists a point in $v \in D_1(c; \kappa, \delta)$ such that $\|u - v\|_2 \leq \delta$. We call the resulting algorithm: $D_1(c; \kappa, \delta)$-SSDPR algorithm.

**Theorem 9.9** *Let $\mathcal{P}_F$ be given (including a constraint $x^T x \leq M$), suppose $c \neq 0$. Then for every $\epsilon > 0$, and for every $\kappa > 0$, there exists a $\delta > 0$ such that the sequence $C_0, C_1, \ldots$ of compact convex sets generated by the $D_1(c; \kappa, \delta)$-SSDPR algorithm satisfies*

(a) $\text{conv}(F) \subseteq C_{k+1} \subseteq C_k, \quad \forall k \in \{0, 1, \ldots\}$;
(b) *there exists $\bar{k}$ such that*

$$\max\left\{c^T x : x \in C_{\bar{k}}\right\} \leq \max\left\{c^T x : x \in F\right\} + \epsilon.$$

## 9.3 Finite Convergence

We just proved that a version of SSDPR method can work with finite SDP problems and guarantee asymptotic convergence (or convergence to any desired accuracy, in principle). We also proved that SSDPR method coincides with $N_+$ in the case of combinatorial optimization problems. On combinatorial optimization problems, $N_+$ and hence SSDPR method terminates in at most $d$ iterations. We can then ask "how general can this type of finite convergence result be?" Can we attain $C_k = \text{conv}(F)$ exactly, for a finite $k$?"

To provide a partial answer, we consider a necessary condition.

**Proposition 9.10** *Suppose that the sequence $C_0, C_1, \ldots$ generated by the SS-DPR method converges to* $\text{conv}(F)$ *in a finite number of iterations. Then*

$$(F \cap \partial\left[\text{conv}(F)\right]) \subseteq \partial C_1.$$

The necessary condition given by the above proposition is trivially satisfied for combinatorial optimization problems. Note that in this case, we always have $C_0 \subseteq [0,1]^d$ and since $F \subseteq \{0,1\}^d$, every $x \in F \cap \partial[\text{conv}(F)]$ is a 0,1 vector and must be an extreme point of every convex set $C$ such that $F \subset C \subset [0,1]^d$.

We conclude this section with the following result. To state it properly, we need the following notation first: $\text{face}(x, C)$ denotes the minimal face of $C$ containing $x$, $\text{aff}(x, C)$ denotes the smallest affine space spanned by $\text{face}(x, C)$. $\mathcal{P}_F^+(U)$ denotes the set of all quadratic forms in $\mathcal{P}_F$ which are positive at some point in $U$. $\mathcal{Q}_+(U)$ denotes the set of quadratic functions in $\mathbb{R}^d$ that are convex on $U$.

**Theorem 9.11** *Suppose the following conditions hold for some $x^* \in F$:*

(i) $d > \dim(\text{aff}(x^*, C_k)) > \dim(\text{aff}(x^*, C_{k'}))$ *for some $k$, $k'$ such that* $0 \leq k < k'$;

(ii) *if* $p(\cdot) \in \text{cl}\left[\text{cone}\left(\mathcal{P}_F^+(\text{aff}(x^*, C_k))\right)\right] \cap \mathcal{Q}_+(\text{aff}(x^*, C_k))$ *is nonzero on* $\mathbb{R}^d$ *and* $p(x^*) = 0$, *then it is nonzero on* $\text{aff}(x^*, C_k)$;

(iii) $\text{face}(x^*, C_k)$ *is an exposed face of* $C_k$.

*Then* $\dim(\text{aff}(x^*, C_k)) > \dim(\text{aff}(x^*, C_{k+1}))$.

## 9.4 Complexity Analysis

We have been working with an extremely general method which is guaranteed to converge. However, any method as general as SSDPR must perform very poorly any many problem instances unless $\mathcal{P} = \mathcal{NP}$. This is still so, even if we restrict our attention to combinatorial optimization problems. One bright spot though is SSDPR generalizes $N_+$ which is very impressive on certain stable set problems (recall Theorem 8.21). In the worst-case, the convergence rate for SSDPR method can be very painstakingly slow. There are even examples in $\mathbb{R}^2$ with 3 quadratic constraints (one of them is convex) on which SSDPR converges slower than linearly.

In this section, we give an overview of what is known in terms of complexity measures and bounds for the SSDPR method.

We assume that some of the given constraints are nonlinear and some are nonconvex (otherwise, we have $C_1 = \text{conv}(F)$). Since we have been writing our constraints in homogeneous form, we can scale each $p_i$ such that

$$\|\text{vec}(Q_i)\|_1 = 1.$$

(Note that if $Q_i = 0$ for some $i$, then the corresponding constraint $p_i(x) \leq 0$ is a linear inequality and hence is a convex constraints, so it does not affect the complexity bound of the method).

Let us define the *maximum Lipschitz constant* of the given formulation:

$$\nu_\ell := \max\left\{\frac{|p_i(u) - p_i(v)|}{\|u - v\|_2} : u, v \in C_0, u \neq v, i \in \{1, 2, \ldots, m\}\right\}.$$

We measure the *nonconvexity* of a constraint $p_i(x) \leq 0$ by $\lambda_d(Q_i)$ and nonconvexity of the whole formulation $\mathcal{P}_F$ by

$$\nu_n := \max\{|\lambda_d(Q_i)| : \lambda_d(Q_i) < 0, i \in \{1, 2, \ldots, m\}\}.$$

Note that $\nu_n \in (0, 1]$.

Let us define

$$F(\epsilon) := \{x \in C_0 : p_i(x) \leq \epsilon, \quad \forall i \in \{1, 2, \ldots, m\}\}.$$

**Theorem 9.12** *For every $\epsilon > 0$, there exists a $\delta > 0$ such that $D_1(\delta)$-SSDPR algorithm generates a compact convex set $C$ such that*

$$\operatorname{conv}(F) \subseteq C \subseteq \operatorname{conv}(F(\epsilon))$$

*in*

$$O\left(\frac{d^2 \nu_\ell^2 \nu_n^2 \left[\operatorname{diam}(C_0)\right]^6}{\epsilon^4} \ln\left(\frac{\nu_\ell \left[\operatorname{diam}(C_0)\right]^2}{\epsilon \operatorname{diam}(F)}\right)\right)$$

*iterations.*

Let us go back to combinatorial optimization problems. We will assume $\operatorname{diam}(F) \geq 1$ (there are at least two feasible solutions). Clearly, $\operatorname{diam}(C_0) \leq \sqrt{d}$. Our *difficult* constraints are

$$-x_j^2 + x_j \leq 0.$$

Whence, $\nu_\ell \leq 1$, and $\nu_n = 1$. We obtain the iteration bound

$$O\left(\frac{d^5}{\epsilon^4} \ln\left(\frac{1}{\epsilon}\right)\right).$$

Of course, we know that if we use $D_1$ then $d$ iterations suffice. However, for the above bound, we are allowed to use $D_1(\delta)$ which might yield easier to solve relaxations.

What does $\epsilon$-approximation mean here? Well, $\forall x \in F(\epsilon), \forall j$,

$$-x_j^2 + x_j \leq \epsilon \text{ iff } x_j \in \left[0, \frac{1}{2} - \sqrt{\frac{1}{4} - \epsilon}\right] \cup \left[\frac{1}{2} + \sqrt{\frac{1}{4} - \epsilon}, 1\right].$$

Note that for small $\epsilon > 0$, $\operatorname{conv}(F(\epsilon))$ is a very good approximation of $\operatorname{conv}(F)$.

Suppose we want to compute the optimum value

$$v^* := \max\left\{\bar{e}^T x : x \in F\right\}.$$

If $\epsilon < \frac{d-1}{d^2}$ then $v^* = \lfloor \bar{v} \rfloor$, where

$$\bar{v} := \max\left\{\bar{e}^T x : x \in \operatorname{conv}(F(\epsilon))\right\}.$$

The bound then becomes

$$O\left(d^9 \ln(d)\right),$$

since we can pick $\frac{1}{\epsilon} = O(d)$.

## 9.5 Applications to Systems of Polynomial Inequalities

Since any system of polynomial inequalities can be reformulated as a system of quadratic inequalities, all results of this chapter can be translated to the setting of polynomial optimization problems (POP) :

$$\begin{array}{llll} \min & p_0(x) & & \\ \text{subject to:} & p_i(x) & \leq & 0, \quad i \in \{1, 2, \ldots, m\}, \end{array}$$

where $p_0, p_1, \ldots, p_m : \mathbb{R}^d \to \mathbb{R}$ are polynomials.

Iteration complexity bounds of the previous section for POP depend on the maximum degree of the polynomials $p_0, p_1, \ldots, p_m$ (the number of variables $d$ in $\mathcal{P}_F$ grows with the logarithm of this maximum degree), $\nu_\ell$ the Lipschitz constant of the resulting quadratic functions, and the way the monomials making up these polynomials use the original variables (common monomials in $p_0, p_1, \ldots, p_m$ and/or common factors in these monomials help decrease the bound).

## 9.6 Bibliographical Notes

The material in the first section is based on Fujie and Kojima [133] and Kojima and Tunçel [210]. The second section is based on Kojima and Tunçel [211]. For an application to linear complementarity problems, see [213]. The material in Sections 9.1 and 9.2 were the first answers to a question of Lovász [241] (which essentially asked for generalizations of the lift-and-project methods to handle systems of polynomial inequalities). Many other answers followed in the area of optimization and in fact most of the fundamental tools necessary for these answers pre-existed in some other areas of mathematics (see Chapter 10).

Section 9.3 is based on [212] and Theorem 9.12 is based on Kojima and Takeda [209] and Tunçel and Xu [351].

For POP, Lasserre [221, 220] as well as Parrilo [282] developed SDP relaxation schemes that lead to stronger relaxations than the ones we described here. Also see Laurent [222]. For some combinatorial optimization problems on which Lasserre's scheme requires many iterations, see [76], [316]. Many other extensions for POP exist, including those which exploit sparsity, see Kojima et al. [204], Lasserre [219] and Waki et al. [362].

There are also SSILPR methods, where in each iteration we have an LP relaxation instead of SDP. The theoretical results above generally extend to these LP, semi-infinite LP relaxation methods.

Results of the last three chapters and the results that we quoted here imply that SDP can be utilized in a *universal* way for solving all combinatorial optimization problems. In fact LP is also universal in the same sense (using the polyhedral lift-and-project operators). In terms of the computational effort per major iteration, SDP based methods generally require more effort than those based on polyhedral relaxations; however, in terms of number of major iterations SDP based operator can perform better at least in some instances.

CHAPTER 10

# Connections to Other Areas of Mathematics

Suppose we are given two expressions $f_1(x), f_2(x)$ and we want to relate them. How do we prove an inequality between them?

Without loss of generality, we can assume that we want to prove the nonnegativity or positivity of some given expression (namely $f(x) := [f_1(x) - f_2(x)]$ and/or its negation). How do we go about doing this?

For instance, suppose $f$ is a polynomial of finitely many real variables. Our first attempt would be to try to express $f$ as a square

$$f(x) = [g(x)]^2,$$

where $g$ is a real-valued function with the same domain as $f$, or a sum of squares (SoS):

$$f(x) = \sum_{i=1}^{\ell} [g_i(x)]^2,$$

where $g_i$ are real-valued functions with the same domain as $f$.

When it works, this approach is indeed wonderful, it provides a good certificate of nonnegativity and usually a great insight into why the inequality holds (assuming the functions $f_1$, $f_2$ had been used to formulate some phenomenon). Of course it is not surprising that such fundamental approaches are prevalent in virtually every subject in mathematics. In 1894, Hermite [171] asked whether every scalar (one-variable) polynomial $p$ of degree at most $n$, with property

$$p(t) > 0, \forall t \in (-1, 1),$$

can be expressed as

$$p(t) = \sum_{i,j \geq 0: i+j \leq n} a_{ij}(1-t)^i(1+t)^j,$$

where $a_{ij} \geq 0$, for every $i, j$. It was quickly answered "no." However, in 1921, Hausdorff [165] proved that if the restriction $i + j \leq n$ on the maximum degree of the representing polynomial is relaxed, then the answer becomes "yes." Whence, there exist $a_{ij} \geq 0$, for all $i, j$ such that

$$p(t) = \sum_{i,j \geq 0} a_{ij}(1-t)^i(1+t)^j.$$

Hilbert's 17th problem which was answered by Artin [14] was about representing nonnegative polynomials (of many variables) as a sum of squares (SoS). Hilbert asked whether every nonnegative polynomial of $d$ real variables can be expressed as a sum of squares of finitely many rational functions. Artin answered the question in the affirmative and hence the theorem:

**Theorem 10.1** *Let $f : \mathbb{R}^d \to \mathbb{R}$ be a polynomial. Then $f(x) \geq 0$ for every $x \in \mathbb{R}^d$ iff there exist $g_i : \mathbb{R}^d \to \mathbb{R}$ rational functions such that*

$$f(x) = \sum_{i=1}^{m} [g_i(x)]^2, \quad \forall x \in \mathbb{R}^d.$$

The next question is whether there always exists a nicer certificate of nonnegativity. For instance, can we take $g_i$ to be polynomials?

Motzkin's example:

$$f(x_1, x_2, x_3) := x_1^4 x_3^2 + x_1^2 x_2^4 - 3x_1^2 x_2^2 x_3^2 + x_3^6$$

is a nonnegative polynomial that cannot be expressed as a sum of squares of polynomials. So, the answer is "no." However, this area is extremely rich and contains many powerful and elegant results. To get into this area more, see for instance, Delzell [105], Prestel and Delzell's book [297], Kuhlmann and Marshall [216], Marshall's monograph [251], Marshall [250], Reznick [304], Powers and Reznick [296, 295], Ruiz [310], Schweighofer [321, 320].

Given $x \in \mathbb{R}^d$ and polynomial $f : \mathbb{R}^d \to \mathbb{R}$ of degree $2n$, let

$$h(x) := [1, x_1, x_2, \ldots, x_d, x_1^2, x_1x_2, x_1x_3, \ldots, x_2^2, \ldots, x_d^n]^T \in \mathbb{R}^N,$$

where $N := \binom{n+d}{n}$. We are interested in

$$\mathcal{F}(f) := \left\{ X \in \Sigma^N : [h(x)]^T X h(x) = f(x) \right\}.$$

The following well-known fact connects SoS and semidefinite optimization.

**Theorem 10.2** *Let $\bar{z} \in \mathbb{R}$. Then $[f(x) - \bar{z}]$ is SoS iff*

$$\left\{ X \in \mathcal{F}(f) : X \succeq \bar{z} e_1 e_1^T \right\} \neq \emptyset.$$

Note that for each given $f$, $\mathcal{F}(f)$ is an affine subspace of $\Sigma^N$. (Simply equate the coefficients of matching monomials in the equation $[h(x)]^T X h(x) = f(x)$.) Therefore, for given $f$ and $\bar{z}$, $\left\{ X \in \mathcal{F}(f) : X \succeq \bar{z} e_1 e_1^T \right\}$ is the feasible region of an SDP. We will see a related result in Chapter 12 for univariate polynomials. Of course, the multi-variate case above is much more complicated than the univariate case. Also, in terms of computational complexity, there seem to be stark differences. (Think of $f$ and $\bar{z}$ as the input, where $f$ is sparse—say number of nonzero coefficients are $O(d^2)$—but $n = \Omega(d)$...)

Consider the Cauchy-Binet formula for the $\det(XX^*)$, where $X \in \mathbb{C}^{m \times n}$. Let $E_k(X)$ denote the $k$th elementary symmetric function of the eigenvalues of $X$. Then

$$E_k(XX^*) = \text{SoS}\,(\text{moduli of all } k \times k \text{ subdeterminants of } X)\,.$$

This provides another connection to one of our first propositions (Proposition 1.10).

Let us start from optimization. At the end of the previous chapter, we were discussing polynomial optimization problems:

$$\begin{array}{rrcl} \inf & f(x) & & \\ \text{subject to:} & g_i(x) & \leq & 0, \quad i \in \{1, 2, \ldots, m\}. \end{array}$$

Suppose we know that the optimal objective value of the above POP exists and is in the interval $[a, b]$. Then, if we had an oracle which would correctly answer questions "does there exist $x \in \mathbb{R}^d$ such that $g_i(x) \leq 0$, for every $i \in \{1, 2, \ldots, m\}$ and $f(x) \leq z$ ($z$ is the input that we provide for the oracle) ?" We could use such an oracle to determine the optimum objective function value of the POP, approximately. What if we also wanted a proof from the oracle together with its answers?

If the oracle said "Yes," it could return $\bar{x} \in \mathbb{R}^d$ satisfying $g_i(\bar{x}) \leq 0$, for every $i \in \{1, 2, \ldots, m\}$ and $f(\bar{x}) \leq z$. However, this is of course the easy direction, what if the answer is "no," then what kind of proof would we accept? Based on our little discussion above, one choice would be to have functions $h_i : \mathbb{R}^d \to \mathbb{R}$ and some constant $\epsilon > 0$ such that

$$f(x) \geq z + \epsilon - \sum_{i=1}^{m} [h_i(x)]^2 \, g_i(x),$$

for every $x$ in the domain of the POP.

In terms of solving the POP, it would be great, if whenever we claim to have an optimal solution $\bar{x}$ with objective value $\bar{z}$ we could also produce functions $h_i$ such that

$$f(x) = \bar{z} - \sum_{i=1}^{m} [h_i(x)]^2 \, g_i(x).$$

There are many theorems of this type in algebraic geometry. The following is a theorem of Putinar [298]:

**Theorem 10.3** *Suppose $F := \{x \in \mathbb{R}^d : p_i(x) \leq 0\}$ is compact, the polynomials $p_i$ have even degree, and their highest degree homogeneous parts do not have common zeroes in $\mathbb{R}^d$ except 0. Then every polynomial that is positive on $F$ can be written as a nonnegative combination of polynomials of the form*

$$[h_0(x)]^2 - \sum_{i=1}^{m} [h_i(x)]^2 \, p_i(x).$$

The above characterization and similar "witnesses" of positivity (or infeasibility of an SDP problem, or certificate of optimality of a solution of an SDP problem) are closely connected to Positivstellensatz which can be viewed as a common generalization of Hilbert's Nullstellensatz and Farkas' Lemma. Below, we state the version of Positivstellensatz due to Stengle [336]:

**Theorem 10.4** *Let the polynomials $g_i : \mathbb{R}^d \to \mathbb{R}, i \in J_1$ and $p_i : \mathbb{R}^d \to \mathbb{R}, i \in J_2$ be given. Then exactly one of the following two systems has a solution:*

(I)

$$\begin{array}{rcl} g_i(x) & = & 0, \quad \forall i \in J_1 \\ p_i(x) & \leq & 0, \quad \forall i \in J_2; \end{array}$$

(II) *There exist polynomials $y_i : \mathbb{R}^d \to \mathbb{R}, i \in J_1$ and $s_J : \mathbb{R}^d \to \mathbb{R}, J \subseteq J_2$, $s_J$ is SoS for every $J \subseteq J_2$ such that*

$$\sum_{i \in J_1} y_i(x) g_i(x) + \sum_{J \subseteq J_2} s_J(x) \prod_{i \in J} p_i(x) = 1.$$

In the above statement when $J = \emptyset$, we interpret $\prod_{i \in \emptyset} (\cdot) := -1$.

Moment problems, (recall our moment matrix from Chapter 7) form a dual problem to the representation of polynomials by SoS. Foundations of these results go back to Stieltjes (1894), A. A. Markov (1895), Perron (1913) and Hamburger (1916-1919). Perhaps one of the most fundamental problems here is the *$K$-moment problem* which is, given $K \subset \mathbb{R}^d$ to decide when a real valued function $f$ of set of monomials in $d$ variables is a moment function $\int_K x^m d\mu$ for some nonnegative Borel measure $\mu$ on $K$. Schmüdgen [314] characterized the solutions to the $K$-moment problem (called *$K$-moment sequences*) for all compact semi-algebraic sets $K$ in terms of the positive definiteness of matrices arising from the moment functions. Schmüdgen's proof utilizes Positivstellensatz in proving the above-mentioned algebraic fact. The result also generalizes many other preexisting beautiful results such as Handelman's theorem [162]; some of these connections are old and they generalize results some of which go all the way back to Minkowski in late 1800's. For some related results, see Curto and Fialkow [99].

We said that Schmüdgen's theorem gives a necessary and sufficient condition for the $K$-moment problem for the case that the underlying set is a compact, basic semialgebraic set. Our setting above is precisely this setting!

Applying the above-mentioned theorems, Lasserre [221, 220] and Parrilo [282] gave SDP based approaches generating relaxations similar to those generated by SSDPR (but in many cases much tighter than those given by the SSDPR). Also see Henrion and Lasserre [170], Jibetean and de Klerk [183], Jibetean and Laurent [184], Kojima and Muramatsu [206] as well as Parrilo and Sturmfels [284].

Now, let us turn to a very elegant result of Bröcker [68] and Scheiderer [313]:

**Theorem 10.5** (i) *Every basic, closed semi-algebraic set in $\mathbb{R}^d$ can be represented by the solution set of $\frac{d(d+1)}{2}$ polynomial inequalities.*

(ii) *Every basic, open semi-algebraic set in $\mathbb{R}^d$ can be represented by the solution set of $d$ polynomial inequalities.*

Even when one considers an extremely special case of the basic semi-algebraic set being a polytope, the result is still nontrivial and very interesting. Grötschel and Henk [152] gave explicit constructions for simple polytopes. The representations they worked with (based on the results of Handelman [162] and Powers and Reznick [295]) have the form

$$\sum_{w \in \mathbb{Z}_+^m} c_w \prod_{i=1}^{m} \left(b_i - \langle a^{(i)}, x\rangle\right)^{w_i}$$

where the polytope is given by

$$\left\{x \in \mathbb{R}^d : \langle a^{(i)}, x\rangle \leq b_i, \quad \forall i \in \{1, 2, \ldots, m\}\right\}.$$

Also see, Firey [125], Bosse, Grötschel and Henk [60]. For an exposure to related subjects, see [10] and [49]. In a recent work, Nie, Ranestad and Sturmfels [278] study the entries of optimal solutions of SDPs by focusing on the degree of the minimal polynomials of these entries (when the entries are considered as algebraic numbers). They go on to compute this degree using projective duality and determinantal identities.

Next, we revisit graph representations and metric embeddings to mention another important result that is connected to graph representations. Let $V$ be a finite set and $d : V \times V :\longrightarrow \mathbb{R}_+$ be such that $(V, d)$ is a metric space. $(V, d)$ is called *negative type* if $(V, d^{1/2})$ is isometric to a subset of a Euclidean space. For instance, $L_1$ is of negative type. *The Euclidean distortion of $V$* is

$$\operatorname{distort}_2(V) := \min_{f:V \to L_2} \left\{\nu_\ell(f), \nu_\ell(f^{-1})\right\},$$

where the minimum runs over all bijections of $f$ (and the Lipschitz constants $\nu_\ell(f)$, $\nu_\ell(f^{-1})$ are computed using the Euclidean 2-norm in $L_2$ and the distance function $d(\cdot,\cdot)$ in $V$). Bourgain [61] proved:

**Theorem 10.6** *For every $n$ point metric space $(V, d)$,*

$$\mathrm{distort}_2(V) = O(\ln(n)).$$

For some related work, also see Bourgain, Milman and Wolfson [62], Talagrand [340], Johnson and Lindenstrauss [185], Matoušek [253], Linial and Magen [230] and Linial, Magen and Naor [231]. Bourgain's theorem above is essentially optimal. However, there are many, commonly used, special classes of metrics with the property that $\mathrm{distort}_2(V) = O\left(\sqrt{\ln(n)}\right)$. A very nice result along this direction is due to Arora, Lee and Naor [12]:

**Theorem 10.7** *For every $n$ point, negative type metric space $(V, d)$,*

$$\mathrm{distort}_2(V) = O\left(\sqrt{\ln(n)}\ln\left(\ln(n)\right)\right).$$

In Chapter 6, we talked about the connections among the sparsest cut problem, the embeddability of finite metric spaces, spectral gap and conductance in Markov Chains. There is also a connection along these lines to extremal graph theory. See for instance, Chung and Graham [84].

Stable sets in distance graphs obtained from hamming cubes give rise to error correcting codes. This connection may lead to seme interesting applications of lift-and-project methods based on SDP. As we are on the subject of stable sets and distance graphs, we should also mention that $\vartheta(G)$, when $G$ is a distance graph defined on the unit hypersphere, can give useful bounds on the kissing numbers (these bounds are known to be tight in dimensions 1,2,8 and 24) see Bachoc and Vallentin [17].

Quantum computing is another area where there are many applications of SDP theory. See for instance, Kitaev [195], Barnum, Saks and Szegedy [23], Childs, Landahl and Parrilo [78], Gutoski and Watrous [159]; for an introduction using a mixture of quantum computing and optimization notation, see Sikora [329] and Gutoski [158].

Next, we will see an application in proving a theorem in number theory.

CHAPTER 11

# An Application to Discrepancy Theory

Some subfields in number theory involve measuring *regularities* (uniform distribution of structures) and *irregularities.* Discrepancy theory is a field of mathematics which is useful in this regard in many other fields such as number theory, graph theory and combinatorics in general.

## 11.1 Introduction to Discrepancy Theory via Optimization

Let $\mathcal{F}$ be a family of subsets of $\{0, 1, 2, \ldots, (n-1)\}$. We want

$$\begin{pmatrix} x_0 \\ x_1 \\ \vdots \\ x_{n-1} \end{pmatrix} \in \{-1, 1\}^n$$

such that the *discrepancy of* $x$, that is,

$$\max_{J \in \mathcal{F}} \left| \sum_{i \in J} x_i \right|$$

is minimized. Let's denote the *discrepancy of* $x$ by

$$\Delta(x) := \max_{J \in \mathcal{F}} \left| \sum_{i \in J} x_i \right|.$$

The corresponding discrepancy for the family of subsets given by $\mathcal{F}$ is defined as:

$$\Delta(\mathcal{F}) := \min_{x \in \{-1,1\}^n} \Delta(x).$$

To bound $\Delta(\mathcal{F})$, we will focus on a squared Euclidean norm analogue of discrepancy; namely,

$$\Delta_2(x) := \sum_{J \in \mathcal{F}} \left( \sum_{i \in J} x_i \right)^2,$$

and

$$\Delta_2(\mathcal{F}) := \min_{x \in \{-1,1\}^n} \left\{ \frac{1}{|\mathcal{F}|} \Delta_2(x) \right\}.$$

Note that we immediately have

$$\Delta_2(\mathcal{F}) \leq (\Delta(\mathcal{F}))^2,$$

since the maximum of $\left|\sum_{i \in J} x_i\right|^2$ over $J \in \mathcal{F}$ is at least their average over $J \in \mathcal{F}$.

We can easily write an integer programming formulation for $\Delta(\mathcal{F})$. We will soon see how $\Delta_2(\mathcal{F})$ arises naturally via our SDP relaxation. The optimal objective value of the following combinatorial optimization problem is precisely $\Delta(\mathcal{F})$:

$$\begin{array}{lrcl} \min & & t & \\ \text{subject to:} & \sum_{i \in J} x_i & \leq & t, \forall J \in \mathcal{F}; \\ & -\sum_{i \in J} x_i & \leq & t, \forall J \in \mathcal{F}; \\ & x & \in & \{-1, 1\}^n. \end{array}$$

We can immediately write down an SDP relaxation whose optimal value will provide a lower-bound on $\Delta_2(\mathcal{F})$:

$$\begin{array}{lrcl} \inf & \frac{1}{|\mathcal{F}|} \sum_{J \in \mathcal{F}} \left( \sum_{i \in J} \sum_{j \in J} X_{ij} \right) & & \\ \text{subject to:} & \operatorname{diag}(X) & = & \bar{e}, \\ & X \succeq & 0. & \end{array}$$

Note that the objective function averages $\bar{e}^T Y \bar{e}$ over all symmetric minors $Y$ of $X$ that come from the family $\mathcal{F}$. Clearly, the above SDP has the form

$$(P) \ \min \left\{ \frac{1}{|\mathcal{F}|} \langle C, X \rangle : \operatorname{diag}(X) = \bar{e}, X \succeq 0 \right\}$$

with the dual problem

$$(D) \ \max \left\{ \bar{e}^T y : \operatorname{Diag}(y) \preceq \frac{1}{|\mathcal{F}|} C \right\}.$$

Just as before, both $(P)$ and $(D)$ have Slater points. We have $\bar{X} := I \succ 0$ feasible in $(P)$ and $\bar{y} := -\left(\|C\|_2 + 1\right) \bar{e}$ feasible in $(D)$. Thus, by the corresponding corollary of the Strong Duality Theorem, both problems attain their optimum values, and the optimum values are the same.

This chapter involves a study of a theorem of Roth [308] about arithmetic progressions:

**Theorem 11.1** *For every partition $[N_1, N_2]$ of the integers $\{1, 2, \ldots, n\}$, there exists an arithmetic progression $\mathcal{S} := \{\ell, \ell + \alpha, \ell + 2\alpha, \ldots, \ell + k\alpha\} \subseteq \{1, 2, \ldots, n\}$ such that*

$$\big| |\mathcal{S} \cap N_1| - |\mathcal{S} \cap N_2| \big| > \frac{n^{1/4}}{20}.$$

It is indeed interesting that Semidefinite Optimization techniques are useful here. Many of the approaches we represent have components that are familiar to us from the previous chapters. For instance, in Chapters 5 and 6 the SDP relaxations we considered were providing $L_2$-embeddings which approximated $L_1$-embeddings. Here, in the current chapter, we are utilizing an $L_2$-norm like function to approximate an $L_\infty$-norm type function.

## 11.2 Lovász-Sós' Approach to Roth's Theorem

Consider arithmetic progressions of size $\lfloor \sqrt{\frac{n}{6}} \rfloor$. Let us characterize such progressions by a *starting point* $\ell$, and a *step size* $\alpha$:

$$\{\ell, \ell + \alpha, \ell + 2\alpha, \ldots, \ell + k\alpha\},$$

where $k := \lfloor\sqrt{\frac{n}{6}}\rfloor - 1$ and we allow a wrap-around (going passed $n$ is interpreted as continuing from the beginning). Now, let $A_j \in \{0,1\}^{n\times n}$ denote the matrix whose $i$th row is the characteristic vector of the arithmetic progression with starting point $i$ and step size $j$. Note that the matrix $A_j$ is a circulant matrix. Form

$$M := \sum_{j=1}^{6\lfloor\sqrt{\frac{n}{6}}\rfloor} A_j A_j^T.$$

Note that $A_j$ is circulant for all $j$ implies ($A_j A_j^T$ is a circulant matrix for all $j$, and whence) $M$ is a circulant matrix.

Now, it is easy to work with the eigenvectors and eigenvalues of $M$ over the complex numbers (we skip the details as the next section will deal with a similar situation). We can show that

$$\lambda_n(M) \geq \frac{\lfloor\sqrt{\frac{n}{6}}\rfloor^2}{4}$$

which yields the desired result.

## 11.3 Lovász' SDP Approach to Roth's Theorem

We will focus on the case when $\mathcal{F}$ is the family of all arithmetic progressions in $\{0,1,\ldots,n-1\}$.

Using this SDP formulation, we will prove something slightly stronger than the original statement of Roth. We will prove that there exists an arithmetic progression $\mathcal{S}$ of length $k := \lfloor\sqrt{\frac{n}{8}}\rfloor$ such that $\alpha \in \{1,2,\ldots,8k\}$ and

$$\big|\,|\mathcal{S}\cap N_1| - |\mathcal{S}\cap N_2|\,\big| > \frac{n^{1/4}}{20}.$$

We consider progressions *modulo* $n$; that is, as in the previous section, we allow them to wrap-around. However,

$$(k-1)8k \leq \left\lfloor\sqrt{\frac{n}{8}} - 1\right\rfloor \sqrt{8n} \leq n-1 < n.$$

Thus, a progression, as the one described above, cannot wrap around more than once.

Let us denote by $\mathcal{H}$ the above described family of progressions (which are allowed to wrap around). Suppose that we prove

$$\Delta(\bar{x}) = \max_{J\in\mathcal{H}} \left|\sum_{i\in J} \bar{x}_i\right| \geq K,$$

where $K$ is a constant. Then we could conclude that there exists a true progression (no wrap-around allowed) $\bar{J}$ such that

$$\left|\sum_{i\in\bar{J}} \bar{x}_i\right| \geq \frac{K}{2},$$

where we pick $J \in \mathcal{H}$ which achieved the maximum above and if it wraps around once we take one of the pieces (whichever piece gives a larger absolute value of the sum) as $\bar{J}$.

Let us focus on $\mathcal{H}$. Since a progression can start from any of the numbers $0, 1, \ldots, n-1$ and can have any of the step sizes $1, 2, \ldots, 8k$, we have that

$$|\mathcal{H}| = 8kn.$$

Again, we are in a situation where we can take advantage of the symmetries of the feasible region and the objective function of the SDP.

**Lemma 11.2** *The feasible region and the objective function of $(P)$ are both invariant under the cyclic shifts:*

$$X_{ij} \mapsto X_{i+1,j+1},$$

*where the indices are interpreted modulo $n$.*

Let $\Gamma$ represent the set of these cyclic shifts. Let $\bar{X}$ be an optimal solution of $(P)$. Then by the last lemma and the convexity of the feasible region,

$$\hat{X} := \frac{1}{|\Gamma|} \sum_{f \in \Gamma} f(\bar{X})$$

is also an optimal solution of $(P)$. The latter matrix has a lot of special structure. First of all,

$$\hat{X} = \begin{bmatrix} 1 & \beta_1 & \beta_2 & \cdots & \beta_{n-1} \\ \beta_{n-1} & 1 & \beta_1 & \cdots & \beta_{n-2} \\ \beta_{n-2} & \beta_{n-1} & 1 & \cdots & \beta_{n-3} \\ \vdots & \vdots & \vdots & \ddots & \vdots \\ \beta_1 & \beta_2 & \beta_3 & \cdots & 1 \end{bmatrix};$$

that is, a circulant matrix. Secondly, $\hat{X}$ is symmetric. So,

$$\beta_i = \beta_{n-i}, \forall i \in \{1, 2, \ldots, n-1\}.$$

Of course, in addition, $\hat{X}$ is positive semidefinite.

Let $\hat{X}$ be an optimal solution of $(P)$. By the above observation we can choose $\hat{X}$ to satisfy the above property. Let us write for some $B \in \mathbb{R}^{n \times d}$,

$$\hat{X} = BB^T,$$

We took $d$ different then $n$ $(d < n)$ so that we can now assume $B^T$ has full row rank. Since $\hat{X}$ is circulant, we have for every $i, j \in \{0, 1, 2, \ldots, n-1\}$,

$$\langle v^{(i)}, v^{(j)} \rangle = \langle v^{(i+1)}, v^{(j+1)} \rangle,$$

where the indices are interpreted mod $n$. Therefore, there exists a linear transformation, represented by $L : \mathbb{R}^d \to \mathbb{R}^d$, preserving the lengths of $v^{(0)}, v^{(1)}, \ldots, v^{(n-1)}$ and the angles between any pair of these vectors such that

$$Lv^{(i)} = v^{(i+1)}, \ \forall i \in \{0, 1, 2, \ldots, n-1\},$$

again the indices are interpreted modulo $n$. Therefore, $L^n = I$. Let us diagonalize $L$ over $\mathbb{C}$. Let $\lambda_i$ denote the not necessarily distinct eigenvalues of $L$. From now on we express $L$ and $v^{(i)}$ with respect to these new coordinates (in which $L$ is diagonal). By the above properties of $L$, every eigenvalue of $L$ is an $n$th root of unity.

Before we finish the analysis, we need to recall a classical result of Dirichlet from 1842:

**Lemma 11.3** *For every $\alpha \in \mathbb{R}$, there exist $p \in \mathbb{Z}$, $Q \in \mathbb{N}$ and $q \in \{1, 2, \ldots, Q\}$ such that*

$$\left|\alpha - \frac{p}{q}\right| \leq \frac{1}{qQ}.$$

That is, given a real number $\alpha$, we want to approximate it by a rational number. However, we would like to keep the integer in the denominator $q$ in the rational representation small as well. Dirichlet's lemma tells us an upper bound on the quantity $\left|\alpha - \frac{p}{q}\right|$, while keeping $q$ under control. The result can be proved by using a continued fraction expansion of $\alpha$. Further related classical results are due to Hurwitz [180] and Khintchine [194].

Now, we are ready to begin analyzing the optimal objective value of $(P)$:

$$\begin{aligned}
\frac{1}{8kn}\sum_{J\in\mathcal{H}}\left(\sum_{i\in J}\sum_{j\in J}\hat{X}_{ij}\right) &= \frac{1}{8kn}\sum_{J\in\mathcal{H}}\left\|\sum_{i\in J} v^{(i)}\right\|_2^2 \\
&= \frac{1}{8kn}\sum_{J\in\mathcal{H}}\left\|\sum_{i\in J} L^i v^{(0)}\right\|_2^2 \\
&= \frac{1}{8kn}\sum_{J\in\mathcal{H}}\sum_{j=1}^{d}\left|\sum_{i\in J}\lambda_j^i\right|^2 \left|v_j^{(0)}\right|^2.
\end{aligned}$$

Note that $\lambda_j = \exp\left(-\frac{2\pi a}{n}\iota\right)$ defines a rotation of the plane with angle $\frac{2\pi a}{n}$, where $\iota := \sqrt{-1}$, and $a \in \{1, 2, \ldots, n\}$.

By Dirichlet's Lemma above, there exist $p \in \mathbb{Z}$, $q \in \{1, 2, \ldots, 8k\}$ such that

$$\left|\frac{a}{n} - \frac{p}{q}\right| < \frac{1}{8kq}.$$

Whence,

$$\left| q\frac{2\pi a}{n} - \underbrace{2\pi p}_{\equiv 0, \text{ since } p\in\mathbb{Z}} \right| < \frac{2\pi}{8k}.$$

Therefore, the maximum angle between any pair of $\lambda_j^i$s for $i \in J$, is strictly less than $\frac{\pi}{4}$. Thus, we obtain for each fixed $j$ and fixed $J$ with step size $q$,

$$\left|\sum_{i\in J}\lambda_j^i\right|^2 > \frac{k^2}{2}.$$

Now, using the fact that there are $n$ arithmetic progressions in $\mathcal{H}$ with step size $q$, we arrive at the following:

$$\begin{aligned}
\frac{1}{8kn}\sum_{J\in\mathcal{H}}\left(\sum_{i\in J}\sum_{j\in J}\hat{X}_{ij}\right) &> \frac{1}{8kn}\frac{k^2 n}{2}\sum_{j=1}^{d}\left|v_j^{(0)}\right|^2 \\
&= \frac{k}{16} \\
&\geq \frac{\sqrt{\frac{n}{8}} - 1}{16}.
\end{aligned}$$

Hence, utilizing the previous arguments and accounting for the wrap-around, we conclude, for large $n$,

$$\Delta(\mathcal{F}) > \frac{n^{1/4}}{14}.$$

## 11.4 A Primal-Dual SDP Approach to Roth's Theorem

We saw in the previous section that to prove Roth's Theorem, it suffices to give a good lower-bound on the optimal objective value of $(P)$.

As we already noted, both $(P)$ and $(D)$ have Slater points. Therefore, by a corollary of the Strong Duality Theorem, both attain their optimum values and there is no duality gap. Thus, it should be possible to prove Roth's theorem by simply giving a good feasible solution $\hat{y}$ of $(D)$ and concluding by the Weak Duality Relation that $\left(\bar{e}^T\hat{y}\right)$ is a valid lower-bound on the optimal objective value of $(P)$.

In fact, it suffices to focus on very specially structured dual solutions $y$. Namely, it suffices to take $\hat{y} := \eta\bar{e}$, where $\eta \in \mathbb{R}$. Then, the problem $(D)$ becomes

$$\max\left\{\eta : \quad \eta I \preceq \frac{1}{|\mathcal{F}|}C\right\}.$$

The optimum objective value of this SDP is very easy to compute:

$$\eta^* := \frac{\lambda_n(C)}{|\mathcal{F}|}.$$

Thus, it suffices to give a lower-bound on $\lambda_n(C)$. In fact, this $C$ is very similar to the matrix $M$ from Section 11.2. Here we have

$$C = \sum_{j=1}^{8\lfloor\sqrt{\frac{n}{8}}\rfloor} A_j A_j^T.$$

Moreover, there should not be a need to go over to the complex space $\mathbb{C}^n$, since $C$ has only real eigenvalues. in any case, we can prove that

$$\lambda_n(C) \geq \frac{\lfloor\sqrt{\frac{n}{8}}\rfloor}{16}.$$

The latter also yields the desired result.

## 11.5 Bibliographical Notes

There is a very nice chapter on Discrepancy Theory in the Handbook of Combinatorics. Section 11.2 is based on Beck and Sós [29] (also see, Chazelle [74]). Section 11.3 is based on Lovász [240]. For a trace bound on discrepancy with an application to range searching, see Chazelle and Lvov [75]. Another application of Semidefinite Optimization to the analysis of arithmetic sequences can be found in proving a lower bound for the minimum number of monochromatic 3-term arithmetic progressions in any 2-colouring of $\{1, 2, \ldots, n\}$ (see [283]).

CHAPTER 12

# SDP Representability

A fundamental question about the power and reach of Semidefinite Optimization is:

*what kind of convex sets can be represented as the feasible region of an SDP?*

Equivalently,

*what kind of functions can be modeled in the SDP framework?*

Indeed, as we mentioned before in Chapter 2, a function $f : \mathbb{R}^d \to \mathbb{R} \cup \{+\infty\}$ is convex iff its epigraph

$$\operatorname{epi}(f) := \left\{ \begin{bmatrix} t \\ x \end{bmatrix} \in \mathbb{R} \oplus \mathbb{R}^d : f(x) \leq t \right\}$$

is a convex set.

Suppose $f$ above is a convex function. Then the problem of minimizing $f(x)$ over the intersection of the positive semidefinite cone with any affine subspace can be reformulated as a SDP problem if the epigraph of $f$ can be represented as the feasible region of some SDP problem. Our objective function would be the simple linear function $t$.

## 12.1 Functions whose Epigraphs are SDP Representable

(i) Linear Functions:

$f(x) := a^T x + \beta$. Then the epigraph of $f$ is

$$\operatorname{epi}(f) = \{(x,t) \in \mathbb{R}^d \oplus \mathbb{R} : t \geq a^T x + \beta\}.$$

The representation is

$$\begin{aligned} a^T x + \beta + s &= t \\ s &\in \Sigma_+. \end{aligned}$$

I.e., a finite collection of such functions can be represented as an LP problem.

(ii) Maximum of finitely many variables:

$f(x) := \max\{x_1, x_2, \ldots, x_d\}$. Then the epigraph of $f$ is

$$\operatorname{epi}(f) = \left\{ (x,t) \in \mathbb{R}^d \oplus \mathbb{R} : t \geq \max_j \{x_j\} \right\}.$$

The representation is

$$\begin{bmatrix} t - x_1 & 0 & \ldots & 0 \\ 0 & t - x_2 & \ldots & 0 \\ \vdots & \vdots & \ddots & \vdots \\ 0 & 0 & \ldots & t - x_d \end{bmatrix} \succeq 0.$$

Note that we can replace each variable $x_i$ by an affine function $\left(\langle a^{(i)}, x\rangle - b_i\right)$, where $a^{(1)}, a^{(2)}, \ldots, a^{(m)} \in \mathbb{R}^d$ and $b \in \mathbb{R}^m$ are given as data. Indeed, this also can be represented as an LP problem, but we use the SDP format, not only to show that LP is a special case of SDP, but also to warm up to the notation and the basic properties. For instance, the next one **cannot** be represented as the feasible region of an LP problem.

(iii) Euclidean Norm:

$f(x) := \|x\|_2 = \sqrt{x_1^2 + x_2^2 + \ldots + x_d^2}$. Then the epigraph of $f$ is

$$\operatorname{epi}(f) = \left\{(x,t) \in \mathbb{R}^d \oplus \mathbb{R} : t \geq \|x\|_2\right\}.$$

The representation is

$$M := \begin{bmatrix} t & x^T \\ x & tI \end{bmatrix} \succeq 0.$$

Note that if $t = 0$ then

$$M \succeq 0 \iff x = 0.$$

Also note that if $t < 0$, then $M$ is not positive semidefinite ($e_1^T M e_1 = t < 0$). If $t > 0$, then we have

$$\begin{aligned} M \succeq 0 &\iff t - \frac{1}{t}x^T x \succeq 0. \\ &\iff \|x\|_2^2 \leq t^2. \end{aligned}$$

We used the Schur complement of $tI$ in $M$. If we instead use the Schur complement of $t$ in $M$ then the arguments need a few more lines. We give the arguments here since they are essentially the same for the next formulation.

$$\begin{aligned} M \succeq 0 &\iff tI - \frac{1}{t}xx^T \succeq 0. \\ &\iff t^2 I - xx^T \succeq 0 \\ &\iff \lambda_1(xx^T) \leq t^2 \\ &\iff \|x\|_2^2 \leq t^2. \end{aligned}$$

We used the fact that if $A \in \mathbb{R}^{m \times n}$ then

$$\begin{aligned} \|A\|_2 &:= \max_{z \in \mathbb{R}^n \setminus \{0\}} \frac{\|Az\|_2}{\|z\|_2} \\ &= \sqrt{\lambda_1(AA^T)} \\ &= \sqrt{\lambda_1(A^T A)}. \end{aligned}$$

(iv) Matrix 2-norm:

$f : \mathbb{R}^{m \times n} \to \mathbb{R}$, $f(X) := \|X\|_2$. Then the epigraph of $f$ is

$$\operatorname{epi}(f) = \left\{(X,t) \in \mathbb{R}^{m \times n} \oplus \mathbb{R} : t \geq \|X\|_2\right\}.$$

The representation is

$$M := \begin{bmatrix} tI & X^T \\ X & tI \end{bmatrix} \succeq 0.$$

Same arguments as in the previous example prove that the representation above is correct.

(v) Canonical Example of a Homogeneous Cone:

The last example generalizes further to a very interesting cone:

$$\operatorname{cl}\left\{\begin{bmatrix} t \\ U \\ X \end{bmatrix} \in \mathbb{R} \oplus \mathbb{R}^{n\times m} \oplus \Sigma^n : tX - UU^T \succeq 0,\ t > 0\right\}.$$

This cone is a homogeneous cone (it is not self-dual, unless $m = 1$ or $n = 1$) and it admits the SDP representation

$$\left\{\begin{bmatrix} t \\ U \\ X \end{bmatrix} \in \mathbb{R} \oplus \mathbb{R}^{n\times m} \oplus \Sigma^n : \begin{bmatrix} tI & U^T \\ U & X \end{bmatrix} \succeq 0\right\}.$$

(vi) Maximum Eigenvalue of a Symmetric Matrix:

$f : \Sigma^n \to \mathbb{R}$, $f(X) := \lambda_1(X)$. Then the epigraph of $f$ is

$$\operatorname{epi}(f) = \{(X, t) \in \Sigma^n \oplus \mathbb{R} : t \geq \lambda_1(X)\}.$$

The representation is

$$M \ := \ tI - X \ \succeq \ 0.$$

(vii) Sum of $k$-largest Eigenvalues of a Symmetric Positive Semidefinite Matrix:

$f : \Sigma^n_+ \to \mathbb{R}$, $f(X) := \sigma_k(X) := \sum_{j=1}^{k} \lambda_j(X)$. Then the epigraph of $f$ is

$$\operatorname{epi}(f) = \left\{(X, t) \in \Sigma^n_+ \oplus \mathbb{R} : t \geq \sigma_k(X)\right\}.$$

The representation is

$$M \ := \begin{bmatrix} (t - k\eta) - \operatorname{Tr}(Y) & 0 & 0 \\ 0 & Y & 0 \\ 0 & 0 & Y - X + \eta I \end{bmatrix} \ \succeq \ 0.$$

**Proposition 12.1** *Let $f$ and $M$ be as above. Then $(X, t) \in \operatorname{epi}(f)$ iff there exist $Y \in \Sigma^n$ and $\eta \in \mathbb{R}$ such that $M \succeq 0$.*

**Proof** Suppose that there exist $Y \in \Sigma^n$ and $\eta \in \mathbb{R}$ such that $M \succeq 0$. Then

$$\begin{aligned} \operatorname{Tr}(Y) \ &\leq \ t - k\eta \qquad (12.1) \\ Y \ &\succeq \ 0 \\ Y - X \ &\succeq \ -\eta I. \end{aligned}$$

Now, Theorem 1.20 has the following consequence: $X, Y \in \Sigma^n$, $Y - X \succeq 0$ implies

$$\lambda_j(Y) \geq \lambda_j(X), \text{ for all } j \in \{1, 2, \ldots, n\}.$$

This consequence, combined with $Y - X \succeq -\eta I$ implies

$$\lambda_j(Y) - \lambda_j(X) \geq -\eta$$

for all $j \in \{1, 2, \ldots, n\}$. Thus, we obtain,

$$\sigma_k(X) \leq \sigma_k(Y) + k\eta.$$

Since $Y \succeq 0$, $\sigma_k(Y) \leq \operatorname{Tr}(Y)$, and by relation (12.1), we arrive at

$$\sigma_k(X) \leq \operatorname{Tr}(Y) + k\eta \leq t.$$

We proved $(X, t) \in \operatorname{epi}(f)$.

For the converse, let $(X,t) \in \text{epi}(f)$. That is $\sigma_k(X) \le t$. Let

$$\eta := \lambda_k(X).$$

Thus, we have $\lambda_j(X-\eta I) \ge 0$ for all $j \in \{1,2,\ldots,k\}$ and $\lambda_j(X-\eta I) < 0$ for all $j \in \{k+1, k+2, \ldots, n\}$. Find symmetric positive semidefinite matrices $Y$ and $Z$ such that $Y - Z = X - \eta I$ and

$$\lambda_j(Y) = \lambda_j(X - \eta I) \text{ for all } \ j \in \{1,2,\ldots,k\}$$

and

$$\lambda_{k+1}(Y) = \lambda_{k+2}(Y) = \cdots = \lambda_n(Y) = 0.$$

Note that we can always find such $Y, Z$ simply by doing the following. Let

$$\sum_{j=1}^{n} \lambda_j(X - \eta I) u^{(j)} \left(u^{(j)}\right)^T$$

be the eigenvalue decomposition of $(X - \eta I)$. Define

$$Y \ := \ \sum_{j=1}^{k} \lambda_j(X - \eta I) \left(u^{(j)}\right)^T$$

and

$$Z := Y - (X - \eta I).$$

Now, we have $Y \succeq 0$, and

$$Y - (X - \eta I) = Z \succeq 0,$$

and

$$t - k\eta - \text{Tr}(Y) = t - k\lambda_k(X) - \sigma_k(X) + k\lambda_k(X) \ = \ t - \sigma_k(X) \ \ge \ 0$$

as desired. Now, we see that $M \succeq 0$ for some choice of $Y, \eta$. $\square$

The above SDP representation for the epigraph of the sum of $k$-largest eigenvalues of a symmetric matrix can be generalized to give an SDP representation for the epigraph of the sum of $k$-largest singular values of a matrix $X \in \mathbb{R}^{m\times n}$ for arbitrary $m$ and $n$ via the fact that the eigenvalues of the $(m+n)$-by-$(m+n)$ symmetric matrix

$$\begin{bmatrix} 0 & X^T \\ X & 0 \end{bmatrix}$$

are precisely the singular values of $X$, their negation and (unless $X$ is square and nonsingular) some additional zeros.

(viii) More Eigenvalue Problems:

Let $A_0, A_1, \ldots, A_m \in \Sigma^n$ be given. We want to find $y \in \mathbb{R}^m$ such that

$$\lambda_1\left(A_0 + \sum_{i=1}^{m} y_i A_i\right)$$

is as small as possible. Equivalently, we can solve the SDP problem

$$\inf\left\{t : \sum_{i=1}^{m} y_i A_i - tI \preceq -A_0\right\}$$

For another variation of the problem, suppose we want the matrix $(A_0 + \sum_{i=1}^{m} y_i A_i)$ to be positive definite (say the minimum eigenvalue is at least 1) and want to minimize its condition number. Solving this problem is equivalent to solving the SDP problem

$$\begin{array}{llcll} \inf & & t & & \\ \text{subject to:} & I & \preceq & A_0 + \sum_{i=1}^m y_i A_i & \preceq \; tI. \end{array}$$

(ix) Geometric Mean:

$f : \mathbb{R}^n_+ \to \mathbb{R}$, $f(x) := -\left(x_1 x_2 \ldots x_n\right)^{1/n}$. Then the epigraph of $f$ is

$$\operatorname{epi}(f) = \left\{(x,t) \in \mathbb{R}^n \oplus \mathbb{R} : t \geq -\left(x_1 x_2 \ldots x_n\right)^{1/n}\right\}.$$

We will represent this convex set as the feasible region of a Second Order Cone Programming (SOCP) problem. This is a special case of the problem $(CP)$ (or $(CD)$) from Chapter 1, where the convex cone $K$ is taken to be

$$K := K_1 \oplus K_2 \oplus \cdots \oplus K_r$$

and each $K_i$ is linearly isomorphic to

$$\left\{ \begin{pmatrix} t \\ x \end{pmatrix} \in \mathbb{R} \oplus \mathbb{R}^{n_i} : t \geq \|x\|_2 \right\},$$

i.e., is a Second Order Cone. Our arguments in this chapter already established that SOCP problem is a special case of SDP.

Let us see a slightly different form in which a second order cone can arise.

$$K = \operatorname{cl}\left\{(x,t) \in \mathbb{R}^n \oplus \mathbb{R} : t > 0,\ \ (t+x_n)(t-x_n) > x_1^2 + \ldots + x_{n-1}^2\right\}$$

$$= \operatorname{cl}\left\{(x,t) \in \mathbb{R}^n \oplus \mathbb{R} : t + x_n > 0,\ \ (t+x_n)(t-x_n) > x_1^2 + \ldots + x_{n-1}^2\right\}.$$

Defining $\tilde{x}_1 := x_1, \ldots, \tilde{x}_{n-1} := x_{n-1}$, $\tilde{x}_n := t - x_n$, and $\tilde{t} := t + x_n$, we see that

$$K = \operatorname{cl}\left\{(x,t) \in \mathbb{R}^n \oplus \mathbb{R} : (\tilde{x},\tilde{t}) \in \tilde{K}\right\},$$

where

$$\tilde{K} := \left\{(\tilde{x},\tilde{t}) \in \mathbb{R}^n \oplus \mathbb{R} : \tilde{t} > 0,\ \ \tilde{t}\tilde{x}_n > \tilde{x}_1^2 + \ldots + \tilde{x}_{n-1}^2\right\}.$$

Thus, we have $K = A^{-1}(\operatorname{cl}(\tilde{K}))$ (or $\operatorname{cl}(\tilde{K}) = A(K)$), where

$$A = \begin{bmatrix} I & 0 & 0 \\ 0 & -1 & 1 \\ 0 & 1 & 1 \end{bmatrix}.$$

So, we conclude that

$$K_{1,2} \;:=\; \operatorname{cl}\left\{(x, s_{1,2}) \in \mathbb{R}^3 : x_1 > 0,\ x_1 x_2 > s_{1,2}^2\right\}$$

is a second order cone. Consider

$$K_{3,4} \;:=\; \operatorname{cl}\left\{(x, s_{3,4}) \in \mathbb{R}^3 : x_3 > 0,\ x_3 x_4 > s_{3,4}^2\right\}$$

and

$$K_{1,2,3,4} \;:=\; \operatorname{cl}\left\{(s_{1,2}, s_{3,4}, s_{1,2,3,4}) \in \mathbb{R}^3 : s_{1,2} > 0,\ s_{1,2}s_{3,4} > s_{1,2,3,4}^2\right\}.$$

Now, enforce the conditions

$$(x_1, x_2, \ldots, x_9) \in K_{1,2} \oplus K_{3,4} \oplus K_{1,2,3,4}$$

and

$$x_3 = x_7,\ \ x_6 = x_8.$$

It follows that $(x_1, x_2, \dots, x_9)$ satisfies the above conditions iff

$$x_9 \geq -(x_1 x_2 x_4 x_5)^{1/4}.$$

So, if we assume that $n = 2^q$ for some positive integer $q$ then we are done. The general case is an easy extension.

Let us note here that the above 3-dimensional second order cones can be expressed as $2 \times 2$ semidefinite cones:

$$\begin{bmatrix} t - x_1 & x_2 \\ x_2 & t + x_1 \end{bmatrix} \succeq 0 \iff t - x_1 \geq 0 \text{ and } (t - x_1)(t + x_1) \geq x_2^2.$$

Indeed, the representation of the epigraph of the Euclidean norm above can be used to establish the fact that the second order cone admits a semidefinite representation, in general.

Current representation of the geometric mean motivates the next three SDP representations: epigraph of the $p$th power function, $p$-norm and the determinant of a symmetric positive definite matrix.

(x) $p$th Power Function (for $p \geq 1$, $p \in \mathbb{Q}$):

We saw that the hypograph of the geometric mean

$$\left\{(x, t) \in \mathbb{R}^n \oplus \mathbb{R} : t \leq (x_1 x_2 \dots x_n)^{1/n}\right\} \tag{12.2}$$

admits an SDP representation. In fact, we went over a second order cone problem (SOCP) representation. Using this result and affine restriction, it is easy to show that for any rational number $p \geq 1$,

$$\{(x, t) \in \mathbb{R}_+ \oplus \mathbb{R} : t \geq x^p\}$$

can be expressed as the feasible region of an SOCP (and whence SDP). Suppose $p = \frac{p_1}{p_2}$, where $p_1 > p_2$, $p_1, p_2 \in \mathbb{Z}_+$ (and they are relatively prime). We simply take the SOCP representation of (12.2) for $n := p_1$, add the linear equations $x_j = \tau$ for $p_2$ of $x_j$s, and the linear equations $x_j = 1$ for the remaining $(p_1 - p_2)$ of $x_j$s. Finally, add the equation $t = x$ and the inequality $x \geq 0$. Then we have an SOCP representation of the set

$$\left\{(x, \tau) \in \mathbb{R} \oplus \mathbb{R} : 0 \leq x \leq \tau^{\frac{p_1}{p_2}}\right\} = \left\{(x, t) \in \mathbb{R}_+ \oplus \mathbb{R} : t \geq x^{\frac{p_1}{p_2}}\right\}.$$

(xi) Epigraph of $p$-norm (for $p \geq 1$, $p \in \mathbb{Q}$):

SOCP representation can easily be constructed by using the techniques of the last two constructions.

(xii) Determinant of a Symmetric Positive Definite Matrix:

$f : \Sigma^n_{++} \to \mathbb{R}$, $f(X) := -(\det(X))^{1/n}$. Then the epigraph of $f$ is

$$\text{epi}(f) = \left\{(X, t) \in \Sigma^n_{++} \oplus \mathbb{R} : t \geq -(\det(X))^{1/n}\right\}.$$

Again, we will assume that $n = 2^q$ for some positive integer $q$. As before, the general case is an easy extension.

Let $\xi$ denote an $\frac{n(n+1)}{2}$ vector with entries $\xi_{ij}$ indexed with $i, j \in \{1, 2, \ldots, n\}$, $i \geq j$. That is, $\xi$ represents the entries of a lower triangular matrix $Y(\xi)$.

$$Y(\xi) := \begin{bmatrix} \xi_{11} & 0 & \cdots & 0 \\ \xi_{21} & \xi_{22} & \cdots & 0 \\ \vdots & \vdots & \ddots & \vdots \\ \xi_{n1} & \xi_{n2} & \cdots & \xi_{nn} \end{bmatrix}.$$

Let $Z(\xi, t)$ be the SDP representation of the set

$$\left\{(\xi, t) : \xi_{ii} \geq 0, \text{ for all } i \in \{1, 2, \ldots, n\} \quad t \geq -\left(\xi_{11}\xi_{22}\ldots\xi_{nn}\right)^{1/n}\right\}.$$

Now, we can define the SDP representation of the determinant of $X$:

$$M := \begin{bmatrix} Z(\xi, t) & 0 & 0 \\ 0 & I & [Y(\xi)]^T \\ 0 & Y(\xi) & X \end{bmatrix} \succeq 0.$$

**Proposition 12.2** *Let $f$ and $M$ be as above. Then $(X, t) \in \operatorname{epi}(f)$ iff there exist $\xi \in \mathbb{R}^{n(n+1)/2}$, $\xi_{ii} \geq 0$, for all $i \in \{1, 2, \ldots, n\}$ such that $M \succeq 0$.*

**Proof** Note that $M \succeq 0$ if and only if $Z(\xi, t) \succeq 0$ and $X - Y(\xi)[Y(\xi)]^T \succeq 0$. Also note that $\det\left(Y(\xi)[Y(\xi)]^T\right) = \xi_{11}^2\xi_{22}^2\ldots\xi_{nn}^2$. We have that

$$\det(X) \geq \det\left(Y(\xi)[Y(\xi)]^T\right) = \prod_{i=1}^{n} \xi_{ii}^2.$$

For the converse, suppose $(X, t) \in \operatorname{epi}(f)$. Then letting $\xi_{ii} := \lambda_i(X)$, for $i \in \{1, 2, \ldots, n\}$ and $\xi_{ij} := 0$ for $i \neq j$, yields $M \succeq 0$. $\square$

(xiii) Max. Volume Ellipsoid Contained in a Given Polytope:

The above construction can be used to "find" a maximum volume ellipsoid contained in a convex polytope.

$$Q := \{x \in \mathbb{R}^n : Ax \leq b\},$$

where

$$A^T := \left[a^{(1)}, a^{(2)}, \ldots, a^{(m)}\right].$$

Let $u \in \mathbb{R}^n$ and $U \in \Sigma_{++}^n$. Then $u$ and $U$ define an ellipsoid in $\mathbb{R}^n$ (ellipsoid centered at $u$):

$$E(U, u) := \left\{u + U^{1/2}y : \|y\|_2 \leq 1\right\}.$$

The ellipsoid $E(U, u)$ is contained in the polytope $Q$ if and only if

$$\langle a^{(i)}, u\rangle + \langle a^{(i)}, U^{1/2}y\rangle \leq b_i, \ \forall \ \|y\|_2 \leq 1, \ \forall \ i \in \{1, 2, \ldots, m\}.$$

Noting that for each fixed $i$, the left-hand-side above is maximized by

$$y := \frac{U^{1/2}a^{(i)}}{\|U^{1/2}a^{(i)}\|_2},$$

we have

$$E(U, u) \subseteq Q \iff \|U^{1/2}a^{(i)}\|_2 + \langle a^{(i)}, u\rangle - b_i \leq 0 \ \forall i \in \{1, 2, \ldots, m\}.$$

Therefore, the maximum volume ellipsoid contained in a polytope can be computed by solving the following optimization problem:

$$\begin{array}{llllllll} \min & -\det(X) & & & & & & \\ \text{subject to:} & \|Xa^{(i)}\|_2 & + & \langle a^{(i)}, u\rangle & - & b_i \leq & 0, & \text{for all } i \in \{1,2,\ldots,m\} \\ & X & \succeq & 0, & & u \in & \mathbb{R}^n. & \end{array}$$

In current practice, it is more efficient to solve

$$\begin{array}{lll} \inf & -\ln\det(X) & \\ \text{subject to:} & \mathcal{A}(X) & = b \\ & X & \succeq 0, \end{array}$$

directly, instead of representing $t \geq -\det(X)$ in the constraints and minimizing $t$. However, the above constructions indicate that the representations

$$\left\{ \begin{array}{c} \mathcal{A}(X) = b \\ X \succeq 0 \end{array} \right\}$$

are far reaching.

(xiv) Min. Volume Ellipsoids Containing a Given Set of Points:

What we described above is sufficient to solve the problem of finding the **minimum** volume ellipsoid containing a given set of points $\{v^{(1)}, v^{(2)}, \ldots, v^{(m)}\} \subset \mathbb{R}^n$. The idea is to use the inverse of the matrix $U$ in the definition of the related ellipsoid. Then maximizing the determinant of $U^{-1}$ is equivalent to minimizing the volume of the ellipsoid defined by $U$.

Just this example has many important applications in other fields. For example, many *matrix completion* problems (given a symmetric matrix , certain entries fixed, fill in the unspecified entries so that the resulting *completion* is positive definite), some experimental design problems from statistics and design theory, information and communication theory can be formulated in this way.

(xv) A Simple Control Theory Example:

Let $x_j(t)$, $x_j : \mathbb{R} \to \mathbb{R}$ be differentiable. Suppose that the vector function

$$x(t) := \begin{bmatrix} x_1(t) \\ x_2(t) \\ \vdots \\ x_n(t) \end{bmatrix}$$

satisfies the differential inclusion

$$\dot{x} \in [\operatorname{conv}(A_1, A_2, \ldots, A_n)]\, x,$$

with the initial condition

$$x(0) = \bar{x},$$

where $\bar{x} \in \mathbb{R}^n$ and $A_1, A_2, \ldots, A_m \in \mathbb{R}^{n\times n}$ are all given.

A main problem is to determine whether $x(t)$ can be contained in a bounded set for every $t \geq 0$. The answer is "yes" iff there exists $\hat{X} \succ 0$ such that

$$v(x(t)) := [x(t)]^T \hat{X}\, [x(t)]$$

remains uniformly bounded.

Recall that as we proved, for every $\hat{X} \succ 0$ and every $\alpha$, the set

$$\left\{X \in \Sigma_+^n : \langle \hat{X}, X\rangle \leq \alpha\right\}$$

is compact.

We will go after a simple sufficient condition. If $v(x(t))$ is nonincreasing, then clearly $v(x(t))$ is uniformly bounded. Thus, a sufficient condition for $x(t)$ to be contained in a bounded set is (called *stability*):

$$\frac{d}{dt}v(x(t)) = \dot{x}^T \hat{X} x + x^T \hat{X} \dot{x} \leq 0, \quad \forall t \geq 0.$$

To make the approach more widely applicable, let us not put any restrictions on $x(0)$ and $\dot{x}(0)$ except that

$$\dot{x}(0) \in \left[\operatorname{conv}(A_1, A_2, \ldots, A_m)\right] x(0).$$

So, $\dot{x}$ can be equal to $A_i x$ for any $i$. Thus, we would expect $\hat{X}$ to satisfy

$$x^T A_i \hat{X} x + x^T \hat{X} A_i x \leq 0, \text{ for all } i \in \{1, 2, \ldots, m\}.$$

Hence, it is sufficient to ensure

$$A_i \hat{X} + \hat{X} A_i \preceq 0, \quad \forall i \in \{1, 2, \ldots, m\}.$$

The set of all $\hat{X} \succ 0$ which satisfy the condition above is a convex cone. Let us use $X$ for $\hat{X}$. We can actually minimize the condition number of such a solution (by normalizing $\lambda_n(X) = 1$ and minimizing $\lambda_1(X)$) utilizing an SDP problem:

$$\begin{array}{lrcll} \inf & & t & & \\ \text{subject to:} & & A_i X + X A_i & \preceq & 0, \quad \forall i \in \{1, 2, \ldots, m\} \\ & I \preceq & X & \preceq & tI. \end{array}$$

(xvi) Univariate, Non-negative Polynomials:

Here, we show that the set of coefficients of a degree $n$ univariate polynomial, which generate polynomials with non-negative values for every choice of the argument can be represented as the intersection of the positive semi-definite cone with an affine space.

$\mathcal{P}^n$ denotes the $(n+1)-$dimensional vector space. $p \in \mathcal{P}^n$ is written

$$p(t) \; = \; \sum_{k=0}^{n} p_k t^k.$$

Defining

$$\tau_n \; := \; (1, t, t^2, \ldots, t^n)^T \in \mathcal{P}^n,$$

we have $p(t) = \langle p, \tau_n \rangle$.

The cone of non-negative polynomials is the cone of all coefficient vectors $p$ for which the underlying polynomial is non-negative for all values of $t$. That is

$$K_{2n} \; := \; \left\{p \in \mathcal{P}^{2n} : \; p(t) \geq 0, \text{ for all } t \in \mathbb{R}\right\}.$$

Clearly, we are only interested in even values of the degree. As it will become clearer, it seems more natural to treat such cones as squares of some other object. There are many versions of this. For instance, we may be interested only in non-negative values of $t$ or only in $t \in [0, 1]$, etc.

Let $\tilde{E}_k$ denote the $k$th $(n+1)\times(n+1)$ cross diagonal matrix:

$$\tilde{E}_0 := \begin{bmatrix} 1 & 0 & \dots & 0 \\ 0 & 0 & \dots & 0 \\ \vdots & \vdots & \ddots & \vdots \\ 0 & 0 & \dots & 0 \end{bmatrix}, \quad \tilde{E}_1 := \begin{bmatrix} 0 & 1 & 0 & \dots & 0 \\ 1 & 0 & 0 & \dots & 0 \\ 0 & 0 & 0 & \dots & 0 \\ \vdots & \vdots & \vdots & \ddots & \vdots \\ 0 & 0 & 0 & \dots & 0 \end{bmatrix},$$

$$\tilde{E}_2 := \begin{bmatrix} 0 & 0 & 1 & 0 & \dots & 0 \\ 0 & 1 & 0 & 0 & \dots & 0 \\ 1 & 0 & 0 & 0 & \dots & 0 \\ 0 & 0 & 0 & 0 & \dots & 0 \\ \vdots & \vdots & \vdots & \vdots & \ddots & \vdots \\ 0 & 0 & 0 & 0 & \dots & 0 \end{bmatrix}, \dots \tilde{E}_{2n} := \begin{bmatrix} 0 & 0 & \dots & 0 \\ 0 & 0 & \dots & 0 \\ \vdots & \vdots & \ddots & \vdots \\ 0 & 0 & \dots & 1 \end{bmatrix}.$$

**Theorem 12.3** *Let $K_{2n}$, $\mathcal{P}^{2n}$ and $\tilde{E}_k$ be as above. Then*
(i) $K_{2n} = \left\{p \in \mathcal{P}^{2n} : \ p_k = \langle \tilde{E}_k, X\rangle, \ \forall k \in \{0,1,\dots,2n\}, \ X \succeq 0\right\}$;
(ii) *if $p \in \operatorname{relint}(K_{2n})$ then the set*

$$\left\{X \succeq 0 : \ \langle X, \tilde{E}_k\rangle = p_k, \ \forall k \in \{0,1,\dots,2n\}\right\}$$

*is bounded and there exists $X \succ 0$ such that $\langle X, \tilde{E}_k\rangle = p_k$ for all $k \in \{0,1,\dots,2n\}$.*

Polynomials that are non-negative on the half-line, $\mathbb{R}_+$, or on an interval $[0,1]$ can be treated similarly.

Suppose that $C \in \Sigma^{n+1}$, $c \in \mathbb{R}^{2n+1}$, $b \in \mathbb{R}^{m_1}$, $q \in \mathbb{R}^{m_2}$, linear maps $\mathcal{A} : \Sigma^{n+1} \to \mathbb{R}^{m_1}$, $\mathcal{B} : \mathbb{R}^{2n+1} \to \mathbb{R}^{m_2}$ are given and we have an optimization problem of the form:

$$\begin{array}{rrcl} \inf & \langle C, X\rangle + \langle c, p\rangle & & \\ \text{subject to:} & \langle \tilde{E}_k, X\rangle - p_k & = & 0, \ \ \forall k \in \{0,1,\dots,2n\}, \\ & \mathcal{A}(X) & = & b, \\ & \mathcal{B}(p) & = & q, \\ & X & \in & \Sigma_+^{n+1}. \end{array}$$

Its dual will include the constraints

$$\sum_{k=0}^{2n} y_k \tilde{E}_k + \mathcal{A}^*(u) \preceq C.$$

Note that $\sum_{k=0}^{2n} y_k \tilde{E}_k$ describes a *Hankel matrix*. This is indeed connected to our discussion in Chapter 10. Moreover, our little discussion about the dual SDP above is no coincidence. There is a much clearer connection to Hankel matrices:

**Proposition 12.4** *Let $K_{2n}$ be as above. Then*

$$K_{2n}^* = \left\{s \in \mathcal{P}^{2n} : \sum_{k=0}^{2n} s_k \tilde{E}_k \succeq 0\right\}.$$

## 12.2 Generalized Epigraphs with respect to a Cone

Consider the following matrix-valued generalization of the epigraph:

$$\mathrm{EPI}(f) := \{(X,T) \in \mathbb{R}^{m\times n} \oplus \Sigma^d : f(X) \preceq T\}.$$

Let $\mathcal{S}$ be a convex subset of $\mathbb{R}^d$. Let $f : \mathcal{S} \to \Sigma^n$. We say that $f$ *is PSD-convex on* $\mathcal{S}$ if for every $\lambda \in [0,1]$ and every $x^{(1)}, x^{(2)} \in \mathcal{S}$, we have

$$f(\lambda x^{(1)} + (1-\lambda)x^{(2)}) \preceq \lambda f(x^{(1)}) + (1-\lambda) f(x^{(2)}).$$

**Proposition 12.5** *Let $\mathcal{S}$ be a convex subset of $\mathbb{R}^d$. Let $f : \mathcal{S} \to \Sigma^n$. Then $f$ is PSD-convex iff EPI($f$) is a convex set.*

For example, let $X \in \mathbb{R}^{m\times n}$, define $f : \mathbb{R}^{m\times n} \to \Sigma^d$,

$$f(X) := (AXB)(AXB)^T + CXD + (CXD)^T + E,$$

where $A, B, C, D, E$ are given matrices (arbitrary, except, their dimensions should be compatible with the above definitions; e.g., $E \in \Sigma^d$ is implied, $A \in \mathbb{R}^{d\times m}$ is implied). Then the matrix-valued epigraph of $f$ admits an SDP representation as follows:

$$\left\{(X,T) \in \mathbb{R}^{m\times n} \oplus \Sigma^d : f(X) \preceq T\right\} = \left\{(X,T) \in \mathbb{R}^{m\times n} \oplus \Sigma^d : \mathcal{L}(X,T) \succeq 0\right\},$$

where $\mathcal{L} : \mathbb{R}^{m\times n} \oplus \Sigma^d \to \Sigma^{2d}$ is the following linear map:

$$\mathcal{L}(X,T) := \begin{bmatrix} I & (AXB)^T \\ (AXB) & T - CXD - (CXD)^T - E \end{bmatrix}.$$

A related subject is that of PSD-monotone functions. Given $\mathcal{S} \subseteq \Sigma^n$ and $f : \mathcal{S} \to \Sigma^m$, we say that $f$ is *PSD-monotone on* $\mathcal{S}$ if

$$\forall U, V \in \mathcal{S}, \quad U \succeq V \text{ implies } f(U) \succeq f(V).$$

To make up some examples of PSD-monotone functions, we might want to start with operator generalizations of simple scalar functions.

Given $g : (a,b) \to \mathbb{R}$, we can define an operator $G : \Sigma^n \to \Sigma^n$ by

$$G(X) := \sum_{i=1}^n g(\lambda_i(X)) u^{(i)} \left(u^{(i)}\right)^T,$$

where $u^{(i)}$ is the eigenvector of $X$ corresponding to the eigenvalue $\lambda_i(X)$.

Let us consider $g$ differentiable on $(a,b)$. Given $v \in (a,b)^n$ such that all entries are distinct, we define a matrix $G(v)$ by

$$[G(v)]_{ij} := \begin{cases} \frac{g(v_i)-g(v_j)}{v_i - v_j}, & i \neq j; \\ g'(v_i), & i = j. \end{cases}$$

Löwner [246] proved that $G$ is PSD-monotone iff

$$G(v) \succeq 0 \text{ and } G(v) \geq 0, \quad \forall v \in (a,b)^n \text{ such that all entries are distinct.}$$

For example, let us take a look at

$$g : \mathbb{R}_{++} \to \mathbb{R}, \quad g(x) := -\frac{1}{x}.$$

Then, $G : \Sigma^n_{++} \to \Sigma^n$ is

$$G(X) = -X^{-1}.$$

Since, for every $U, V \in \Sigma^n_{++}$ with $U \succeq V$, we have $V^{-1} \succeq U^{-1}$, $G$ above is PSD-monotone on $\Sigma^n_{++}$.

Another example is $f : \Sigma^n_{++} \to \Sigma^m$ defined as

$$f(X) := \left(MX^{-1}M^T\right)^{-1},$$

where $M \in \mathbb{R}^{m \times n}$ such that $\text{rank}(M) = m$ is given. This $f$ is also PSD-monotone on $\Sigma^n_{++}$.

## 12.3 Representing Convex Cones as Feasible Regions of SDP Problems

Under mild assumptions, any convex optimization problem can be put into the following conic form:

$$\inf \left\{ \langle c, x \rangle : \mathcal{A}(x) = b, x \in K \right\},$$

where $c \in \mathbb{R}^d$, $b \in \mathbb{R}^m$, linear map $\mathcal{A} : \mathbb{R}^d \to \mathbb{R}^m$ are given and $K \subset \mathbb{R}^d$ is a pointed, closed convex cone with nonempty interior. To understand the modeling power of SDP, we restate the main questions that started this chapter and we ask

*which conic optimization problems can be expressed as SDP problems?*

We need to be a bit more rigorous. We will use two different notions of SDP representation.

- the cone $K$ is represented as $\Sigma^n_+$ intersected with a linear subspace;
- the cone $K$ is represented as the projection onto a linear subspace of $\Sigma^n_+$ intersected with a linear subspace.

We call the first one *SDP representation* and the second one *lifted-SDP representation*. As far as a problem being formulated as an SDP, these two notions are not that different. However, the latter notion, the lifted-SDP representability, is strictly more general.

**Theorem 12.6** *Every homogeneous cone is SDP representable.*

The next family in the hierarchy is *hyperbolic cones*. A homogeneous polynomial $p : \mathbb{R}^d \to \mathbb{R}$ is *hyperbolic in the direction* $h \in \mathbb{R}^d$, if $p(h) > 0$ and the univariate polynomial (in $t \in \mathbb{R}$)

$$p(x + th)$$

has only real roots for every $x \in \mathbb{R}^d$. (More general notions exist for nonhomogeneous polynomials over $\mathbb{C}$; see Güler [160].) A convex cone $K$ is a *hyperbolic cone* if it is

$$\left\{x \in \mathbb{R}^d : p(x + th) \neq 0, \ \forall t \in \mathbb{R}_+\right\}$$

for a polynomial $p$ which is hyperbolic in the direction $h \in \mathbb{R}^d$. Homogeneous cones make up a proper subset of hyperbolic cones. For a very nice exposure to hyperbolic cones, see Renegar [301] as well as Bauschke, Güler, Lewis and Sendov [27]. The latter reference also contains a generalization of Proposition 12.1 where the eigenvalues of the given symmetric matrix $X$ is substituted by the suitable generalization of the real roots of the hyperbolic polynomial $p(x+th)$ for the given $x \in \mathbb{R}^d$. For some applications of the theory of hyperbolic polynomials to combinatorics, see Choe, Oxley, Sokal and Wagner [79], Gurvits [157] and the references therein.

**Open Problem 7:** (Generalized Lax conjecture) "*All hyperbolic cones can be expressed as a linear subspace intersected with a positive semidefinite cone*" is perhaps one of the most interesting open problems here.

Equivalently to our two notions of SDP representation and lifted-SDP representation above, for the dual form

$$\left\{ y \in \mathbb{R}^d : \sum_{i=1}^{d} y_i A_i \preceq C \right\}$$

we will use the terminology *LMI representation.*

Let us define the more general notion in this dual form of representations. Namely, those which also allow projecting away some of the variables (for some fundamental results on these representations, pertaining to optimization, see [34], [81]):

**Definition 12.7** $G \subset \mathbb{R}^d$ is said to *admit a lifted-LMI representation* if there exists $\mathcal{L} : \mathbb{R}^d \oplus \mathbb{R}^m \to \mathbb{R}^n$ a linear map such that

$$x \in \text{int}(G) \iff \mathcal{L}(x, u) \succ 0 \text{ for some } u \in \mathbb{R}^m.$$

Note that a convex cone $G$ admits a lifted-LMI representation iff its dual cone $G^*$ does. In addition to this symmetry property, lifted-LMI representations cover a larger class of convex sets than the LMI representations and in terms of optimization algorithms which are used to solve the underlying optimization problems, lifted-LMI representations provide no additional difficulty (at least from a theoretical viewpoint).

$G$ admits a *poly-time, lifted-LMI representation* if

$$\max\{m, n\} = O(\text{poly}(d)),$$

where poly$(\cdot)$ is a polynomial.

**Open Problem 8:** Characterize all $d$-dimensional, pointed, closed, convex cones in $\mathbb{R}^d$ which admit a lifted-LMI representation. I believe that this set of cones strictly contain the hyperbolic cones.

**Open Problem 9:** Characterize all $d$-dimensional, pointed, closed, convex cones in $\mathbb{R}^d$ which admit a poly-time, lifted-LMI representation.

Most specifically:

**Open Problem 10:** Are all Hyperbolic Feasibility Problems polynomial-time equivalent to LMI problems?

This last question needs some definitions and clarifications.

**Definition 12.8** Let $p_1, p_2, \dots, p_m : \mathbb{R}^d \to \mathbb{R}$ be given polynomials. Then the problem "does there exist $x \in \mathbb{R}^d$ such that $p_i(x) \geq 0, \forall i \in \{1, 2, \dots, m\}$ is a *Hyperbolic Feasibility Problem (HFP)* if every $p_i$ is a hyperbolic polynomial.

Next, we define the size$(HFP)$. The "size" should involve the basic complexity measures needed to bound the amount of computational effort required (in the Blum-Shub-Smale real computation model [48]) to "solve" HFP to $\epsilon \in (0,1)$ accuracy using some general class of well-established algorithms. For instance, let $\eta$ denote the maximum degree of polynomials $p_1, p_2, \dots, p_m$, we can define

$$\text{size}(HFP) := \max\{m, d, \eta, \ln(1/\epsilon), \ln(R)\},$$

where $R > 1$ denotes the volume of a given ellipsoid $E_0$ which determines the region in which we will decide the solvability of HFP. I.e., our problem is to find $\bar{x} \in E_0$ satisfying all the inequalities. We require that after

$$\text{poly}(\text{size}(HLP)) \text{ operations}$$

the algorithm either outputs $\bar{x} \in \mathbb{R}^d$ such that $p_i(\bar{x}) \geq 0, \forall i \in \{1, 2, \ldots, m\}$ or it outputs

"there does not exist a ball of volume at least $\epsilon$ which is contained in

$$E_0 \cap \{x \in \mathbb{R}^d : p_i(x) \geq 0, \forall i \in \{1, 2, \ldots, m\}\}."$$

In this context, when we say *HFP is polynomial-time equivalent to LMI* we mean that for every HFP (with $m$, $d$, $\eta$, $R$ and a given $\epsilon \in (0, 1)$), we can explicitly describe an LMI such that

- the formulated LMI can be solved to $\epsilon$ accuracy in time poly (size($HFP$)),
- solving the LMI within accuracy $\epsilon$, solves the original HFP.

This notion of poly-time equivalence is quite important in optimization theory. A problem analogous to Open Problem 10, was solved in [35] by showing that Second Order Cone Programming is poly.-time equivalent to Linear Programming.

## 12.4 Bibliographical Notes

Many of the fundamental SDP representations can be found in Nesterov and Nemirovskii's book [275]. The representation of the epigraph of sum of $k$ largest eigenvalues was also independently found by Alizadeh [5]. A good reference on system and control theory applications is the book by Boyd, El Ghaoui, Feron and Balakrishnan [63]. This book on Linear Matrix Inequalities in System and Control Theory also has many nonlinear SDP problems. For more applications, also see a more recent book by Boyd and Vandenberghe [64] and [18], [311]. For eigenvalue optimization problems, see the survey by Lewis and Overton [227].

It has been long known that every symmetric cone is SDP representable. Theorem 12.6 (which is about every homogeneous cone being SDP-representable) is due to Chua [80] and Faybusovich [121] independently. Lax conjecture was also proved recently (Helton and Vinnikov [169], Vinnikov [361], Lewis, Parrilo and Ramana [228]), hence putting more pressure on the generalized Lax conjecture we just discussed. For some related results, see Helton and Nie [168, 167], Lasserre [218, 217].

Some experimental design problems from statistics and design theory, information and communication theory can be formulated based on the techniques we discussed. For some such applications utilizing determinant maximization, see Vandenberghe, Boyd and Wu [355].

The representation of nonnegative polynomials of a single variable as SDPs follows Nesterov's presentation [273]; but many of the related results go back at least hundred years.

There are also many applications in financial mathematics. To begin, see the chapter by Bertsimas and Sethuraman in the SDP Handbook [366] and the book by Cornuéjols and Tütüncü [97].

# Bibliography


[1] N. E. Aguilera, S. M. Bianchi and G. L. Nasini, Lift and project relaxations for the matching and related polytopes, *Discrete Appl. Math.* 134 (2004) 193–212.

[2] N. E. Aguilera, M. S. Escalante and G. L. Nasini, A generalization of the perfect graph theorem under the disjunctive index, *Math. Oper. Res.* 27 (2002) 460–469.

[3] M. Alekhnovich, S, Arora and I. Tourlakis, Towards strong nonapproximability results in the Lovász-Schrijver hierarcy, *Proc. of the 37th Annual ACM Symp. on Theory of Computing* (2005) 294–303.

[4] S. Al-Homidan and H. Wolkowicz, Approximate and exact completion problems for Euclidean distance matrices using semidefinite programming, *Linear Algebra Appl.* 406 (2005) 109–141.

[5] F. Alizadeh, Interior point methods in semidefinite programming with applications to combinatorial optimization, *SIAM J. Optim.* 5 (1995) 13–51.

[6] F. Alizadeh, J-P. A. Haeberly, and M. L. Overton, Primal-dual interior-point methods for semidefinite programming: Convergence rates, stability and numerical results, *SIAM J. Optim.* 8 (1998) 746–768.

[7] F. Alizadeh, J-P. A. Haeberly and M. L. Overton, Complementarity and nondegeneracy in semidefinite programming, *Math. Program.* 77 (1997) 111–128.

[8] N. Alon and A. Naor, Approximating the cut-norm via Grothendieck's inequality, *SIAM J. Comput.* 35 (2006) 787–803.

[9] E. J. Anderson and P. Nash, *Linear Programming in Infinite-Dimensional Spaces. Theory and Applications*, Wiley-Interscience Series in Discrete Mathematics and Optimization, John Wiley & Sons, Ltd., Chichester, 1987.

[10] C. Andradas, L. Bröcker and J. M. Ruiz, *Constructible Sets in Real Geometry*, Springer-Verlag, Berlin, 1996.

[11] S. Arora, B. Bollobás and L. Lovász, Proving integrality gaps without knowing the linear program, *Proc. 43rd IEEE Symp. FOCS* 2002, 313–322.

[12] S. Arora, J. R. Lee and A. Naor, Euclidean distortion and the sparsest cut, *J. Amer. Math. Soc.* 21 (2008) 1–21.

[13] S. Arora, S. Rao and U. Vazirani, Expander flows, geometric embeddings and graph partitioning, *Proc. 36th Annual ACM Symposium on Theory of Computing*, pp. 222–231 (electronic), ACM, New York, 2004.

[14] E. Artin, Über die zerlegung definiter functionen in quadrate, *Abh. Math. Sem. Univ. Hamburg* 5 (1927) 100–115.

[15] Y. Au, *On the Polyhedral Lift-and-Project Rank Conjecture for the Fractional Stable Set Polytope,* M.Math. Thesis, Dept. of Combinatorics and Optimization, University of Waterloo, Canada, 2008.

[16] Y. Au and L. Tunçel, On the polyhedral lift-and-project methods and the fractional stable set polytope, *Discrete Optim.* 6 (2009) 206–213.

[17] C. Bachoc and F. Vallentin, New upper bounds for kissing numbers from semidefinite programming, *J. Amer. Math. Soc.* 21 (2008) 909–924.

[18] V. Balakrishnan and S. Boyd, Existence and uniqueness of optimal matrix scalings, *SIAM J. Matrix Anal. Appl.* 16 (1995) 29–39.

[19] E. Balas, Disjunctive programming: Properties of the convex hull of feasible points, *Management Science Research Report* 348 GSIA, Carnegie Mellon University, Pittsburgh, PA, USA, 1974. (Appeared as: E. Balas, Disjunctive programming: properties of the convex hull of feasible points, *Discrete Appl. Math.* 89 (1998) 3–44.)

[20] E. Balas, S. Ceria and G. Cornuéjols, A lift-and-project cutting plane algorithm for mixed 0-1 programs, *Math. Program.* 58 (1993) 295–323.

[21] G. P. Barker, The lattice of faces of a finite dimensional cone, *Linear Algebra Appl.* 7 (1973) 71–82.

[22] G. P. Barker and D. Carlson, Cones of diagonally dominant matrices, *Pacific J. of Math.* 57 (1975) 15–32.

[23] H. Barnum, M. Saks and M. Szegedy, Quantum query complexity and semidefinite programming, in *Proc. IEEE Conf. on Computational Complexity*, 2003.

[24] A. Barvinok, *A Course in Convexity*, Graduate Studies in Mathematics, 54. American Mathematical Society, Providence, RI, 2002.

[25] A. Barvinok, A remark on the rank of positive semidefinite matrices subject to affine constraints, *Discrete Comput. Geom.* 25 (2001) 23–31.

[26] A. Barvinok, Problems of distance geometry and convex properties of quadratic maps, *Discrete Comput. Geom.* 13 (1995) 189–202.

[27] H. H. Bauschke, O. Güler, A. S. Lewis and H. S. Sendov, Hyperbolic polynomials and convex analysis, *Canad. J. Math.* 53 (2001) 470–488.

[28] P. Beame, T. Pitassi and N. Segerlind, Lower bounds for Lovász-Schrijver systems and beyond, using multiparty communication complexity, *Automata, languages and programming* pp. 1176–1188, Lecture Notes in Comput. Sci., 3580, Springer, Berlin, 2005.

[29] J. Beck and V. T. Sós, Discrepancy theory, In *Handbook of Combinatorics* Vol. 1, 2, pp. 1405–1446, Elsevier, Amsterdam, 1995.

[30] M. Belk, Realizability of graphs in three dimensions, *Discrete Comput. Geom.* 37 (2007) 139–162.

[31] M. Belk and R. Connelly, Realizability of graphs, *Discrete Comput. Geom.* 37 (2007) 125–137.

[32] M. Bellare and P. Rogaway, The complexity of approximating a nonlinear program, *Math. Program.* 69 (1995) 429–441.

[33] S. J. Benson and Y. Ye, Approximating maximum stable set and minimum graph coloring problems with the positive semidefinite relaxation, In *Complementarity: Applications, Algorithms and Extensions* (Madison, WI, 1999) Kluwer Academic Publ., Dordrecht, 2001, pp. 1–17.

[34] A. Ben-Tal and A. Nemirovski, *Lectures on Modern Convex Optimization Analysis, Algorithms, and Engineering Applications*, MPS-SIAM Series in Optimization, SIAM, Philadelphia, PA, USA, 2001.

[35] A. Ben-Tal and A. Nemirovski, On polyhedral approximations of the second-order cone, *Math. Oper. Res.* 26 (2001) 93–205.

[36] A. Ben-Tal, A. Nemirovski and C. Roos, Extended matrix cube theorems with applications to $\mu$-theory in control, *Math. Oper. Res.* 28 (2003) 497–523.

[37] A. Ben-Israel, A. Ben-Tal and S. Zlobec, *Optimality in Nonlinear Programming: A Feasible Directions Approach*, John Wiley & Sons Inc., New York, USA, 1981.

[38] A. Berman, *Cones, Matrices and Mathematical Programming*, Lecture Notes in Economics and Mathematical Systems, Vol. 79. Springer-Verlag, Berlin-New York, 1973.

[39] A. Berman and R. J. Plemmons, *Nonnegative Matrices in the Mathematical Sciences*, Revised reprint of the 1979 original, Classics in Applied Mathematics, SIAM, Philadelphia, PA, USA, 1994.

[40] D. P. Bertsekas and P. Tseng, Set intersection theorems and existence of optimal solutions, *Math. Program.* 110 (2007) 287–314.

[41] D. P. Bertsekas and A. E. Özdağlar, Pseudonormality and a Lagrange multiplier theory for constrained optimization, *J. Optim. Theory Appl.* 114 (2002) 287–343.

[42] D. Bertsimas and Y. Ye, Semidefinite relaxations, multivariate normal distributions, and order statistics, *Handbook of combinatorial optimization*, vol. 3, pp. 1–19, Kluwer Acad. Publ., Boston, MA, 1998.

[43] R. Bhatia, *Matrix Analysis*, Springer. New York, NY, USA, 1997.

[44] D. Bienstock and N. Ozbay, Tree-width and the Sherali-Adams operator, *Discrete Optim.* 1 (2004) 13–21.

[45] D. Bienstock and M. Zuckerberg, Subset algebra lift operators for 0-1 integer programming, *SIAM J. Optim.* 15 (2004) 63–95.

[46] L. J. Billera and A. Sarangarajan, All 0-1 polytopes are traveling salesman polytopes, *Combinatorica* 16 (1996) 175–188.
[47] R. G. Bland, D. Goldfarb and M. J. Todd, The ellipsoid method: a survey, *Operations Research* 29 (1981) 1039–1091.
[48] L. Blum, F. Cucker, M. Shub and S. Smale, *Complexity and Real Computation*, Springer-Verlag, New York, NY, U.S.A., 1998.
[49] J. Bochnak, M. Coste and M.-F. Roy, *Real Algebraic Geometry*, Springer-Verlag, 1998.
[50] S. Bochner, Hilbert distances and positive definite functions, *Ann. of Math.* 42 (1941) 647–656.
[51] A. Bockmayr, F. Eisenbrand, M. Hartmann, and A. S. Schulz, On the Chvátal rank of polytopes in the 0/1 cube, *Discrete Appl. Math.* 98 (1999) 21–27.
[52] F. Bohnenblust, Joint positiveness of matrices, Technical report, 1948. Unpublished manuscript.
[53] I. M. Bomze and E. de Klerk, Solving standard quadratic optimization problems via linear, semidefinite and copositive programming, *J. Global Optim.* 24 (2002) 163–185.
[54] J. M. Borwein and W. B. Moors, Stability of closedness of convex cones under linear mappings, *Tech. Report*, Dalhousie Univ., N.S., Canada, 2008.
[55] J. M. Borwein and A. S. Lewis, *Convex Analysis and Nonlinear Optimization. Theory and Examples*, Second edition. CMS Books in Mathematics, Springer, New York, 2006.
[56] J. M. Borwein and H.Wolkowicz, Characterizations of optimality without constraint qualification for the abstract convex program, *Math. Program. Stud.* 19 (1982) 77–100.
[57] J. M. Borwein and H.Wolkowicz, Regularizing the abstract convex program, *J. Math. Anal. Appl.* 83 (1981) 495–530.
[58] J. M. Borwein and H.Wolkowicz, Characterization of optimality for the abstract convex program with finite dimensional range, *J. Austral. Math. Soc. Ser. A* 30 (1980/1981) 390–411.
[59] J. M. Borwein and H.Wolkowicz, Facial reduction for a cone-convex programming problem, *J. Austral. Math. Soc. Ser. A* 30 (1980/1981) 369–380.
[60] H. Bosse, M. Grötschel and M. Henk, Polynomial inequalities representing polyhedra, *Math. Program.* 103 (2005) 35–44.
[61] J. Bourgain, On Lipschitz embedding of finite metric spaces in Hilbert space, *Israel J. Math.* 52 (1985) 46–52.
[62] J. Bourgain, V. Milman and H. Wolfson, On type of metric spaces, *Trans. Amer. Math. Soc.* 294 (1986) 295–317.
[63] S. Boyd, L. E. Ghaoui, E. Feron and V. Balakrishnan, *Linear Matrix Inequalities in System and Control Theory,* SIAM, Philadelphia, 1994.
[64] S. Boyd and L. Vandenberghe, *Convex Optimization*, Cambridge University Press, Cambridge, UK, 2004.
[65] J. W. Brewer, Kronecker products and matrix calculus in system theory, *IEEE Trans. Circuits and Systems* 25 (1978) 772–781.
[66] J. W. Brewer, Correction to: "Kronecker products and matrix calculus in system theory" (IEEE Trans. Circuits and Systems 25 (1978) 772–781). *IEEE Trans. Circuits and Systems* 26 (1979) 360.
[67] A. Brøndsted, *An Introduction to Convex Polytopes*, Graduate Texts in Mathematics, Springer-Verlag, New York-Berlin, 1983.
[68] L. Bröcker, On basic semialgebraic sets, *Expo. Math.* 9 (1991) 289–334.
[69] A. R. Karlin, C. Mathieu, C. T. Nguyen, Integrality gaps of linear and semidefinite programming relaxations for knapsack, *manuscript*, 2009.
[70] S. Ceria, Lift-and-project cuts and perfect graphs, *Math. Prog.* 98 (2003) 309–317.
[71] S. Ceria, *Lift-and-Project Methods for Mixed 0-1 Programs*, Ph.D. Dissertation, Carnegie Mellon University, 1993.
[72] M. Charikar, On semidefinite programming relaxations for graph coloring and vertex cover, In *SODA 2002: Proc. 13th annual ACM-SIAM Symp.on Disc. Algorithms*, SIAM, Philadelphia, PA, USA, pp. 616–620.
[73] S. Chawla, R. Krauthgamer, R. Kumar and D. Sivakumar, On the hardness of approximating multicut and sparsest-cut, *Comput. Complexity* 15 (2006) 94–114.
[74] B. Chazelle, *The Discrepancy Method. Randomness and Complexity*, Cambridge University Press, Cambridge, 2000.

[75] B. Chazelle and A. Lvov, A trace bound for the hereditary discrepancy, *Discrete Comput. Geom.* 26 (2001) 221–231.

[76] K. K. H. Cheung, Computation of the Lasserre ranks of some polytopes, *Math. Oper. Res.* 32 (2007) 88–94.

[77] K. K. H. Cheung, On Lovász-Schrijver lift-and-project procedures on the Dantzig-Fulkerson-Johnson relaxation of the TSP, *SIAM J. Optim.* 16 (2005) 380–399.

[78] A. M. Childs, A. J. Landahl and P. A. Parrilo, Improved quantum algorithms for the ordered search problem via semidefinite programming, *Physical Review A* 75 (2007), article number:032335.

[79] Y.-B. Choe, J. G. Oxley, A. D. Sokal and D. G. Wagner, Homogeneous multivariate polynomials with the half-plane property, *Adv. in Appl. Math.* 32 (2004) 88–187.

[80] C. B. Chua, Relating homogeneous cones and positive definite cones via $T$-algebras, *SIAM J. Optim.* 14 (2003) 500–506.

[81] C. B. Chua and L. Tunçel, Invariance and efficiency of convex representations, *Math. Program.* 111 (2008) 113–140.

[82] M. Chudnovsky, N. Robertson, P. Seymour, R. Thomas, The strong perfect graph theorem, *Ann. of Math.* 164 (2006) 51–229.

[83] M. Chudnovsky, N. Robertson, P. Seymour, R. Thomas, Progress on perfect graphs, *Math. Program.* 97 (2003) 405–422.

[84] F. Chung and R. Graham, Sparse quasi-random graphs, *Combinatorica* 22 (2002) 217–244.

[85] V. Chvátal, *Linear Programming*, W. H. Freeman and Co., USA, 1983.

[86] V. Chvátal, On certain polytopes associated with graphs, *J. Combin. Theory B* 18 (1975) 138–154.

[87] V. Chvátal, Edmonds polytopes and a hierarchy of combinatorial problems, *Discrete Math.* 4 (1973) 33–41.

[88] V. Chvátal, W. Cook and M. Hartmann, On cutting plane proofs in combinatorial optimization, *Linear Algebra Appl.* 114/115 (1989) 455–499.

[89] R. Connelly, Rigidity and energy, *Invent. Math.* 66 (1982) 11–33.

[90] S. A. Cook, A hierarchy for nondeterministic time complexity, *J. Comput. System Sci.* 7 (1973) 343–353.

[91] W. J. Cook, W. H. Cunningham, W. R. Pulleyblank and A. Schrijver, *Combinatorial Optimization*, Wiley-Interscience Series in Discrete Mathematics and Optimization, John Wiley & Sons, Inc., New York, 1998.

[92] W. Cook and S. Dash, On the matrix-cut rank of polyhedra, *Math. Oper. Res.* 26 (2001) 19–30.

[93] W. Cook, A. M. H. Gerards, A. Schrijver and É. Tardos, Sensitivity theorems in integer programming, *Math. Program.* 34 (1986) 251–264.

[94] G. Cornuéjols, Valid inequalities for mixed integer linear programs, *Math. Program.* 112 (2008) 3–44.

[95] G. Cornuéjols, *Combinatorial Optimization, Packing and Covering*, SIAM, 2001.

[96] G. Cornuéjols and Y. Li, Elementary closures for integer programs, *Oper. Res. Lett.* 28 (2001) 1–8.

[97] G. Cornuéjols and R. Tütüncü, *Optimization Methods in Finance*, Cambridge University Press, Cambridge, 2007.

[98] P. Crescenzi, V. Kann, R. Silvestri and L. Trevisan, Structure in approximation classes, *SIAM J. Comput.* 28 (1999) 1759–1782.

[99] R. E. Curto and L. A. Fialkow, Recursiveness, positivity, and truncated moment problems, *Houston J. Math.* 17 (1991) 603–635.

[100] G. B. Dantzig, *Linear Programming and Extensions*, Princeton University Press, Princeton, NJ, USA, 1963.

[101] E. de Klerk and D. V. Pasechnik, Approximation of the stability number of a graph via copositive programming, *SIAM J. Optim.* 12 (2002) 875–892.

[102] E. de Klerk, D. V. Pasechnik and A. Schrijver, Reduction of symmetric semidefinite programs using the regular $*$-representation, *Math. Program.* 109 (2007) 613–624.

[103] E. de Klerk, D. V. Pasechnik and J. P. Warners, On approximate graph coloring and MAX-$k$-CUT algorithms based on the $\vartheta$ function, *J. Comb. Optim.* 8 (2004) 267–294.

[104] C. Delorme and S. Poljak, The performance of an eigenvalue bound on the max-cut problem in some classes of graphs, *Disc. Math.* 111 (1993) 145–156.

[105] C. N. Delzell, A continuous, constructive solution to Hilbert's 17th problem, *Invent. Math.* 76 (1984) 365–384.
[106] M. Deza and M. Laurent, *Geometry of Cuts and Metrics*, Springer-Verlag, 1997.
[107] J.-P. Doignon, Convexity in cristallographical lattices, *Journal of Geometry* 3 (1973) 71–85.
[108] M. Doob, An interrelation between line graphs, eigenvalues and matroids, *J. Combin. Theory B* 15 (1973) 40–50.
[109] R.J. Duffin, Infinite programs, In A.W. Tucker, editor, *Linear Equalities and Related Systems*, pages 157–170. Princeton University Press, Princeton, NJ, 1956.
[110] J. Edmonds, Systems of distinct representatives and linear algebra, *Journal of Research of the National Bureau of Standards B* 71 (1967) 241–245.
[111] J. Edmonds, Maximum matching and a polyhedron with 0,1-vertices. *Journal of Research of the National Bureau of Standards B* 69 (1965) 125–130.
[112] J. Edmonds and R. Giles, Total dual integrality of linear inequality systems, *Progress in combinatorial optimization* (Waterloo, Ont., 1982), Academic Press, Toronto, ON, 1984, pp. 117–129.
[113] M. Einsiedler and S. Tuncel, When does a polynomial ideal contain a positive polynomial?, *J. Pure Appl. Algebra* 164 (2001) 149–152.
[114] F. Eisenbrand, On the membership problem for the elementary closure of a polyhedron, *Combinatorica* 19 (1999) 297–300.
[115] F. Eisenbrand and A. S. Schulz, Bounds on the Chvátal rank of polytopes in the 0/1 cube, *Combinatorica* 23 (2003) 245–261.
[116] P. Erdős, On bipartite subgraphs of graphs (in Hungarian), *Matematikai Lapok* 18 (1967) 283–288.
[117] M. S. Escalante, M. S. Montelar and G. L. Nasini, Minimal $N_+$-rank graphs: Progress on Lipták and Tunçel's conjecture, *Oper. Res. Lett.* 34 (2006) 639–646.
[118] M. S. Escalante, G. L. Nasini and M. C. Varaldo, On the commutativity of antiblocker diagrams under lift-and-project operators, *Discrete Appl. Math.* 154 (2006) 1845–1853.
[119] R. Fagin, Generalized first-order spectra and polynomial-time recognizable sets, Complexity of computation, *Proc. SIAM-AMS Sympos. Appl. Math.*, New York, 1973), pp. 43–73.
[120] J. Faraut and A. Korányi, *Analysis on Symmetric Cones*, Oxford University Press, NY, USA, 1994.
[121] L. Faybusovich, On Nesterov's approach to semi-infinite programming, *Acta Appl. Math.* 74 (2002) 195–215.
[122] U. Feige, Randomized graph products, chromatic numbers, and the Lovász $\vartheta$-function, *Combinatorica* 17 (1997) 79–90.
[123] U. Feige and M.X. Goemans, Approximating the value of two prover proof systems, with applications to MAX 2SAT and MAX DICUT, *Proceedings of the Third Israel Symposium on Theory of Computing and Systems*, Tel Aviv, Israel, pp. 182–189, 1995.
[124] U. Feige and R. Krauthgamer, The probable value of the Lovász-Schrijver relaxations for maximum independent set, *SIAM J. Comput.* 32 (2003) 345–370.
[125] W. Firey, Approximating convex bodies by algebraic ones, *Arch Math.* 25 (1974) 424–425.
[126] M. L. Flegel, and C. Kanzow, On the Guignard constraint qualification for mathematical programs with equilibrium constraints, *Optimization* 54 (2005) 517–534.
[127] R. M. Freund, Complexity of convex optimization using geometry-based measures and a reference point, *Math. Program.* 99 (2004) 197–221.
[128] R. M. Freund, Complexity of an algorithm for finding approximate solution of a semi-definite program with no regularity assumption, *Working Paper*, Operations Research Center, MIT, MA, USA, 1994.
[129] R. M. Freund and J. Vera, Equivalence of convex problem geometry and computational complexity in the separation oracle model, *manuscript* 2009.
[130] R. M. Freund and J. Vera, Condition-based complexity of convex optimization in conic linear form via the ellipsoid algorithm, *SIAM J. Optim.* 10 (2000) 155–176.
[131] S. Friedland and R. Loewy, Subspaces of symmetric matrices containing matrices with multiple first eigenvalue, *Pacific J. Math.* 62 (1976) 389–399.
[132] A. Frieeze and M. Jerrum, Improved approximation algorithms for MAX $k$-CUT and MAX BISECTION, *Algorithmica* 18 (1997) 67–81.
[133] T. Fujie and M. Kojima, Semidefinite relaxation for nonconvex programs, *J. Global Optim.* 10 (1997) 367–380.

[134] D. R. Fulkerson, Blocking and anti-blocking pairs of polyhedra, *Math. Prog.* 1 (1971) 168–194.

[135] D. R. Fulkerson, Anti-blocking polyhedra, *J. Combin. Theory B* 18 (1972) 50–71.

[136] D. Gale and V. Klee, Continuous convex sets, *Math. Scand.* 7 (1959) 379–391.

[137] M. R. Garey and D. S. Johnson, *Computers and Intractability A Guide to the Theory of NP-Completeness*, W. H. Freeman and Company, NY, USA, 1979.

[138] K. Gatermann and P. A. Parrilo, Symmetry groups, semidefinite programs, and sums of squares, *J. Pure Appl. Algebra* 192 (2004) 95–128.

[139] A. M. H. Gerards, G. Maróti and A. Schrijver, Note on: N. E. Aguilera, M. S. Escalante, G. L. Nasini, "A generalization of the perfect graph theorem under the disjunctive index," *Math. Oper Res.* 28 (2003) 884–885.

[140] M. X. Goemans, Semidefinite programming in combinatorial optimization, *Math. Program.* 79 (1997) 143–161.

[141] M. X. Goemans, Worst-case comparison of valid inequalities for the TSP, *Math. Program.* 69 (1995) 335–349.

[142] M. X. Goemans and F. Rendl, Semidefinite programs and association schemes, *Computing* 63 (1999) 331–340.

[143] M. X. Goemans and L. Tunçel, When does the positive semidefiniteness constraint help in lifting procedures, *Math. Oper. Res.* 26 (2001) 796–815.

[144] M. X. Goemans and D. P. Williamson, Approximation algorithms for MAX-3-CUT and other problems via complex semidefinite programming, *J. Comput. System Sci.* 8 (2004) 442–470.

[145] M. X. Goemans and D. P. Williamson, Improved approximation algorithms for maximum cut and satisfiability problems using semidefinite programming, *J. ACM* 42 (1995) 1115–1145.

[146] M. X. Goemans and D. P. Williamson, A new 3/4 approximation algorithm for max sat, In G. Rinaldi and L. Wolsey, eds. *Proceedings of the Third Integer Programming and Combinatorial Optimization Conference*, 1993, 313–321.

[147] D. Goldfarb and M. J. Todd, Linear Programming, In Handbooks in Operations Research and Management Science, Vol. 1, Optimization, North-Holland, Amesterdam, 1989.

[148] G. H. Golub and C. F. Van Loan, *Matrix Computations*, Third edition, Johns Hopkins Studies in the Mathematical Sciences, Johns Hopkins University Press, Baltimore, MD, USA, 1996.

[149] C. C. Gonzaga, Path-following methods for linear programming, *SIAM Review* 34 (1992) 167-224.

[150] D. Grigoriev, E. A. Hirsch and D. V. Pasechnik, Complexity of semi-algebraic proofs, *Moscow Math. Journal* 2 (2002) 647–679.

[151] P. Gritzmann and Klee, Separation by hyperplanes in finite-dimensional vector spaces over Archimedean ordered fields, *J. Convex Anal.* 5 (1998) 279–301.

[152] M. Grötschel and M. Henk, The representation of polyhedra by polynomial inequalities, *Disc. Comput. Geom.* 29 (2003) 485–504.

[153] M. Grötschel, L. Lovász and A. Schrijver, *Geometric Algorithms and Combinatorial Optimization* Springer, New York, 1988.

[154] M, Grötschel, L. Lovász and A. Schrijver, Relaxations of vertex packing, *J. Combin Theory B* 40 (1986) 330–343.

[155] M, Grötschel, L. Lovász and A. Schrijver, Polynomial algorithms for perfect graphs, *Annals of Discrete Mathematics* 21 (1984) 325–356.

[156] M. Guignard, Generalized Kuhn-Tucker conditions for mathematical programming problems in a Banach space, *SIAM J. Control* 7 (1969) 232–241.

[157] L. Gurvits, Hyperbolic polynomials approach to Van der Waerden/Schrijver-Valiant like conjectures: sharper bounds, simpler proofs and algorithmic applications, *STOC'06: Proceedings of the 38th Annual ACM Symposium on Theory of Computing*, pp. 417–426, ACM, New York, 2006.

[158] G. Gutoski, *Quantum strategies and local operations*, Ph.D. Thesis, David Cheriton School of Computer Science, University of Waterloo, Ontario, Canada, 2009.

[159] G. Gutoski and J. Watrous, Toward a general theory of quantum games, *STOC'07—Proceedings of the 39th Annual ACM Symposium on Theory of Computing*, ACM, New York, USA, 2007, pp. 565–574.

[160] O. Güler, Hyperbolic polynomials and interior point methods for convex programming, *Math. Oper. Res.* 22 (1997) 350–377.
[161] O. Güler, A. J. Hoffman and U. G. Rothblum, Approximations to solutions to systems of linear inequalities, *SIAM J. Matrix Anal. Appl.* 16 (1995) 688–696.
[162] D. Handelman, Representing polynomials by positive linear functions on compact convex polyhedra, *Pac. J. Math.* 132 (1988) 35–62.
[163] J. Håstad, Some optimal inapproximability results, *J. ACM* 48 (2001) 798–859.
[164] C. Helmberg, F. Rendl, R. Vanderbei and H. Wolkowicz, An interior-point method for semidefinite programming, *SIAM J. Optim.* 6 (1996) 342–361.
[165] F. Hausdorff, Summationmethoden und Momentfolgen I, *Math. Z.* 9 (1921) 74–109.
[166] H. Hatami, A. Magen and V. Markakis, Integrality gaps of semidefinite programs for Vertex Cover and relations to $\ell_1$ embeddability of negative type metrics, *manuscript*, 2006.
[167] J. W. Helton and J. Nie, Semidefinite representation of convex sets, *Math. Program.* 122 (2010) 21–64.
[168] J. W. Helton and J. Nie, Sufficient and necessary conditions for semidefinite representability of convex hulls and sets, *SIAM J. Optim.* 20 (2009) 759–791.
[169] J. W. Helton and V. Vinnikov, Linear matrix inequality representation of sets, *Technical Report*, 2002.
[170] D. Henrion and J.-B. Lasserre, GloptiPoly: global optimization over polynomials with Matlab and SeDuMi, *ACM Trans. Math. Software* 29 (2003) 165–194.
[171] Ch. Hermite, *Intermėd des Math.* 1 (1894) 65.
[172] N. J. Higham, *Functions of Matrices. Theory and Computation*, Society for Industrial and Applied Mathematics (SIAM), Philadelphia, PA, USA, 2008.
[173] R. D. Hills and S. R. Waters, On the cone of positive semi definite matrices, *Linear Algebra Appl.* 90 (1987) 81–88.
[174] J.-B. Hiriart-Urruty and C. Lemaréchal, *Convex analysis and minimization algorithms. I. Fundamentals*, Grundlehren der Mathematischen Wissenschaften [Fundamental Principles of Mathematical Sciences], Springer-Verlag, Berlin, 1993.
[175] J.-B. Hiriart-Urruty and C. Lemaréchal, *Convex analysis and minimization algorithms. II. Advanced theory and bundle methods*, Grundlehren der Mathematischen Wissenschaften [Fundamental Principles of Mathematical Sciences], Springer-Verlag, Berlin, 1993.
[176] J. C.-K. Ho and L. Tunçel, Reconciliation of various complexity and condition measures for linear programming problems and a generalization of Tardos' theorem, *FOUNDATIONS OF COMPUTATIONAL MATHEMATICS, Proceedings of the Smalefest*, World Scientific, 2002, pp. 93–147.
[177] A. J. Hoffman, On approximate solutions of systems of linear inequalities. *J. Res. Nat. Bur. Stand.* 49 (1952) 263–265.
[178] S.-P. Hong and L. Tunçel, Unification of lower-bound analyses of the lift-and-project rank of combinatorial optimization polyhedra, *Discrete Appl. Math.* 156 (2008) 25–41.
[179] R.A. Horn and C.R. Johnson, *Topics in matrix analysis*, Cambridge University Press, Cambridge, 1994. Corrected reprint of the 1991 original.
[180] A. Hurwitz, A. (1891) Üeber die angenäherte Darstellung der Irrationalzahlen durch rationale Brüche, *Mathematische Annalen* 39 (1891) 279–284.
[181] S. Iwata, L. Fleischer and Fujishige, A combinatorial strongly polynomial algorithm for minimizing submodular functions, *J. ACM* 48 (2001) 761–777.
[182] V. Jeyakumar, G. M. Lee andG. Y. Li, Alternative theorems for quadratic inequality systems and global quadratic optimization, *SIAM J. Optim.* 20 (2009) 983–1001.
[183] D. Jibetean and E. de Klerk, Global optimization of rational functions: a semidefinite programming approach, *Math. Program.* 106 (2006) 93–109.
[184] D. Jibetean and M. Laurent, Semidefinite approximations for global unconstrained polynomial optimization, *SIAM J. Optim.* 16 (2005) 490–514.
[185] W. B. Johnson and J. Lindenstrauss, Extensions of Lipschitz mappings into a Hilbert space, Conference in modern analysis and probability (New Haven, Conn., 1982), 189–206, *Contemp. Math.* 26, Amer. Math. Soc., Providence, RI, 1984.
[186] F. Juhász, The asymptotic behaviour of Lovász' $\theta$ function for random graphs, *Combinatorica* (1982) 153–155.
[187] G. Kalai, Linear programming, the simplex algorithm and simple polytopes, *Math. Program.* 79 (1997) 217–233.

[188] D. Karger, R. Motwani and M. Sudan, Approximate graph coloring by semidefinite programming, *J. ACM* 45 (1998) 246–265.

[189] S. E. Karisch and F. Rendl, Semidefinite programming and graph equipartition, In P. M. Pardalos and H. Wolkowicz, eds., *Topics in Semidefinite and Interior-Point Methods, Fields Institute Communications*, vol. 18, American Math. Soc. 1998, pp. 77-95.

[190] H. Karloff, How good is the Goemans-Williamson max cut algorithm?, *SIAM J. Comput.* 29 (2000) 336–350.

[191] H. Karloff and U. Zwick, 7/8-approximation algorithm for MAX 3SAT?, In *Proc. 38th FOCS* (1997) 406–415.

[192] N. Karmarkar, A new polynomial time algorithm for linear programming, *Combinatorica* 4 (1984) 373–395.

[193] L. G. Khachiyan, A polynomial algorithm in linear programming, *Soviet Math. Dokl.* 20 (1979) 191-194.

[194] A. Khintchine, Neuer Beweis und Verallgemeinerung eines Hurwitzschen Satzes, *Math. Ann.* 111 (1935) 631–637.

[195] A. Kitaev, Quantum computations: algorithms and error correction, *Russian Mathematical Surveys* 52 (1997) 1191–1249.

[196] V. Klee, Maximal separation theorems for convex sets, *Trans. Amer. Math. Soc.* 134 (1968) 133–147.

[197] V. L. Jr. Klee, Strict separation of convex sets, *Proc. Amer. Math. Soc.* 7 (1956) 735–737.

[198] V. L. Jr. Klee, Convex sets in linear spaces. II, *Duke Math. J.* 18 (1951) 875–883.

[199] V. L. Jr. Klee, On certain intersection properties of convex sets, *Canadian J. Math.* 3 (1951) 272–275.

[200] V. Klee, E. Maluta and C. Zanco, Basic properties of evenly convex sets, *J. Convex Anal.* 14 (2007) 137–148.

[201] V. Klee and G. J. Minty, How good is the simplex algorithm?, in: Inequalities, III (Proc. Third Sympos., Univ. California, Los Angeles, Calif., 1969; dedicated to the memory of Theodore S. Motzkin), Academic Press, NY, USA, 1972, pp. 159–175.

[202] J. Kleinberg and M. X. Goemans, The Lovász theta function and a semidefinite programming relaxation of vertex cover, *SIAM J. Discrete Math.* 11 (1998) 196–204.

[203] D. E. Knuth, The sandwich theorem, *Electron. J. Combin.* 1 (1994), Article 1, approx. 48 pp. (electronic).

[204] M. Kojima, S. Kim and H. Waki, Sparsity in sums of squares of polynomials, *Math. Program.* 103 (2005) 45–62.

[205] M. Kojima, S. Mizuno and A. Yoshise, A polynomial time algorithm for a class of linear complementarity problems, *Math. Program.* 44 (1989) 1-26.

[206] M. Kojima and M. Muramatsu, An extension of sums of squares relaxations to polynomial optimization problems over symmetric cones, *Math. Program.* 110 (2007) 315–336.

[207] M. Kojima, M. Shida and S. Shindoh, Search directions in the SDP and the monotone SDLCP: generalization and inexact computation, *Math. Program.* 85 (1999) 51–80.

[208] M. Kojima, S. Shindoh and S. Hara, Interior-point methods for the monotone linear complementarity problem in symmetric matrices, *SIAM J. Optim.* 7 (1997) 86–125.

[209] M. Kojima and A. Takeda, Complexity analysis of successive convex relaxation methods for nonconvex sets, *Math. Oper. Res.* 26 (2001) 519–542.

[210] M. Kojima and L. Tunçel, Cones of matrices and successive convex relaxations of nonconvex sets, *SIAM J. Optim.* 10 (2000) 750–778.

[211] M. Kojima and L. Tunçel, Discretization and localization in successive convex relaxation methods for nonconvex quadratic optimization problems, *Math. Program.* 89 (2000) 79–111.

[212] M. Kojima and L. Tunçel, On the finite convergence of successive SDP relaxation methods, *European Journal of Operational Research* 143 (2002) 325-341.

[213] M. Kojima and L. Tunçel, Some fundamental properties of successive convex relaxation methods on LCP and related problems, *J. Global Optim.* 24 (2002) 333–348.

[214] A. Kotlov, L. Lovász and S. Vempala, The Colin de Verdière number and sphere representations of a graph, *Combinatorica* 17 (1997) 483–521.

[215] K. Kretschmer, Programming in paired spaces, *Canad. J. Math.* 13 (1961) 221–238.

[216] S. Kuhlmann and M. Marshall, Positivity, sums of squares and the multi-dimensional moment problem, *Trans. Amer. Math. Soc.* 354 (2002) 4285–4301.

[217] J. B. Lasserre, Convex sets with semidefinite representation, *Math. Program.* 120 (2009) 457–477.

[218] J. B. Lasserre, Convexity in semialgebraic geometry and polynomial optimization, *SIAM J. Optim.* 19 (2008) 1995–2014.

[219] J. B. Lasserre, Convergent SDP-relaxations in polynomial optimization with sparsity, *SIAM J. Optim.* 17 (2006) 822–843.

[220] J. B. Lasserre, Polynomials nonnegative on a grid and discrete optimization, *Trans. Amer. Math. Soc.* 354 (2002) 631–649.

[221] J. B. Lasserre, An explicit exact SDP relaxation for nonlinear 0-1 programs, in *Lecture Notes in Computer Science* (K. Aardal and A. M. H. Gerards, eds.) 2001, pp. 293–303.

[222] M. Laurent, Semidefinite representations for finite varieties, *Math. Prog.* 109 (2007) 1–26.

[223] M. Laurent, Lower bound for the number of iterations in semidefinite hierarchies for the cut polytope, *Math. Oper. Res.* 28 (2003) 871–883.

[224] M. Laurent, A comparison of the Sherali-Adams, Lovász-Schrijver, and Lasserre relaxations for 0-1 programming, *Math. Oper. Res.* 28 (2003) 470–496.

[225] M. Laurent, Tight linear and semidefinite relaxations for max-cut based on the Lovász-Schrijver lift-and-project technique, *SIAM J. Opt.* 12 (2002) 345–375.

[226] T. Leighton and S. Rao, Multicommodity max-flow min-cut theorems and their use in designing approximation algorithms, *J. ACM* 46 (1999) 787–832. (Preliminary version in *STOC* 1988.)

[227] A. S. Lewis and M. L. Overton, Eigenvalue optimization, *Acta Numerica* 7 (1996) 149-190.

[228] A. S. Lewis, P. A. Parrilo and M. V. Ramana, The Lax conjecture is true, *Proc. Amer. Math. Soc.* 133 (2005) 2495–2499.

[229] M. H. Lim, Linear transformations on symmetric matrices, *Linear and Multilinear Algebra* 7 (1979) 47–57.

[230] N. Linial and A. Magen, Least-distortion Euclidean embeddings of graphs: products of cycles and expanders, *J. Combin. Theory B* 79 (2000) 157–171.

[231] N. Linial, A. Magen and A. Naor, Girth and Euclidean distortion, *Geom. Funct. Anal.* 12 (2002) 380–394.

[232] L. Lipták, *Critical Facets of the Stable Set Polytope*, Ph.D. Thesis, Yale University, 1999.

[233] L. Lipták and L. Lovász, Critical facets of the stable set polytope, *Combinatorica* 21 (2001) 61–88.

[234] L. Lipták and L. Tunçel, Lift-and-project ranks and antiblocker duality, *Oper. Res. Lett.* 33 (2005) 35–41.

[235] L. Lipták and L. Tunçel, Stable set problem and the lift-and-project ranks of graphs, *Math. Program.* 98 (2003) 319–353.

[236] R. Loewy and H. Schneider, Positive operators on the $n$-dimensional ice cream cone, *J. Math. Anal. Appl.* 49 (1975) 375–392.

[237] L. Lovász, Geometric representations of graphs, manuscript, December 2009.

[238] L. Lovász, Semidefinite programs and combinatorial optimization, Recent advances in algorithms and combinatorics, CMS Books Math./Ouvrages Math. SMC, 11, Springer, New York, 2003, pp. 137–194.

[239] L. Lovász, Steinitz representations of polyhedra and the Colin de Verdière number, *J. Combin. Theory B* 82 (2001) 223–236.

[240] L. Lovász, Integer sequences and semidefinite programming, *Publ. Math. Debrecen* 56 (2000) 475–479.

[241] L. Lovász, Combinatorial optimization: some problems and trends, *DIMACS Tech. Report* 92-53, NJ, USA, 1992.

[242] L. Lovász, *An Algorithmic Theory of Numbers, Graphs and Convexity*, CBMS-NSF Regional Conference Series in Applied Mathematics 50, Society for Industrial and Applied Mathematics (SIAM), Philadelphia, PA, USA, 1986.

[243] L. Lovász, On the Shannon capacity of a graph, *IEEE Transactions on Information Theory* 25 (1979) 1–7.

[244] L. Lovász, Normal hypergraphs and the weak perfect graph conjecture, *Discrete Math.* 2 (1972) 253–267.

[245] L. Lovász and A. Schrijver, Cones of matrices and set-functions and 0-1 optimization, *SIAM J. Optim.* 1 (1991) 166–190.

[246] K. Löwner, Über monotone matrixfunktionen, *Math. Z.* 38 (1934) 177–216.

[247] Z.-Q. Luo, N. D. Sidiropoulos, P. Tseng and S. Zhang, Approximation bounds for quadratic optimization with homogeneous quadratic constraints, *SIAM J. Optim.* 18 (2007) 1–28.

[248] Z.-Q. Luo, J. Sturm and S. Zhang, Conic convex programming and self-dual embedding, *Optim. Methods Softw.* 14 (2000) 169–218.

[249] A. W. Marshall and I. Olkin, *Inequalities: theory of majorization and its applications*, Mathematics in Science and Engineering, Academic Press, Inc. [Harcourt Brace Jovanovich, Publishers], New York-London, 1979.

[250] M. Marshall, Representations of non-negative polynomials, degree bounds and applications to optimization, *Canad. J. Math.* 61 (2009) 205–221.

[251] M. Marshall, *Positive polynomials and sums of squares*, Mathematical Surveys and Monographs, 146. American Mathematical Society, Providence, RI, USA, 2008.

[252] C. Mathieu and A. Sinclair, Sherali-Adams relaxations of the matching polytope, *STOC'09: Proceedings of the 41st Annual ACM Symposium on Theory of Computing*, pp. 293–302, ACM, New York, 2009.

[253] J. Matoušek, On variants of the Johnson-Lindenstrauss lemma, *Random Structures Algorithms* 33 (2008) 142–156.

[254] A. Megretski, Relaxations of quadratic programs in operator theory and system analysis, *Oper. Theory Adv. Appl.* 129 (2001) 365–392.

[255] M. Mesbahi and G. P. Papavassilopoulos, A cone programming approach to the bilinear matrix inequality problem and its geometry, *Math. Program.* 77 (1997) 247–272.

[256] J. E. Mitchell, Restarting after branching in the SDP approach to MAX-CUT and similar combinatorial optimization problems, *J. Combinatorial Optimization* 5 (2001) 151–166.

[257] R. D. C. Monteiro, Primal-dual path following algorithms for semidefinite programming, *SIAM J. Optim.* 7 (1997) 663–678.

[258] R. D. C. Monteiro and I. Adler, Interior path following primal-dual algorithms, Part I: Linear programming, *Math. Program.* 44 (1989) 27-42.

[259] R. D. C. Monteiro and I. Adler, Interior path following primal-dual algorithms, Part II: Convex quadratic programming, *Math. Program.* 44 (1989) 43-66.

[260] R. D. C. Monteiro, I. Adler and M. G. C. Resende, A polynomial time primal-dual affine scaling algorithm for linear and convex quadratic programming and its power series extension, *Math. Oper. Res.* 15 (1990) 191–214.

[261] R. D. C. Monteiro and Y. Zhang, A unified analysis for a class of long-step primal-dual path-following interior-point algorithms for semidefinite programming, *Math. Program.* 81 (1998) 281–299.

[262] M. Muramatsu, A unified class of directly solvable semidefinite programming problems, *Ann. Oper. Res.* 133 (2005) 85–97.

[263] K. Murota, *Discrete Convex Analysis*, SIAM Monographs on Discrete Mathematics and Applications, SIAM, Philadelphia, PA, 2003.

[264] G. L. Nemhauser and L. A. Wolsey, *Integer and Combinatorial Optimization*, Wiley-Interscience Series in Discrete Mathematics and Optimization, John Wiley & Sons, Inc., New York, 1988.

[265] A. Nemirovski, Sums of random symmetric matrices and quadratic optimization under orthogonality constraints, *Math. Program.* 109 (2007) 283–317.

[266] A. Nemirovski, Proximal method with rate of convergence $O(1/t)$ for variational inequalities with Lipschitz continuous monotone operators and smooth convex-concave saddle point problems, *SIAM J. Optim.* 15 (2004) 229–251.

[267] A. Nemirovski and D. B. Yudin, *Problem Complexity and Method Efficiency in Optimization*, Wiley-Interscience Series in Discrete Mathematics, John Wiley & Sons, Inc., New York, 1983.

[268] Yu. Nesterov, Unconstrained convex minimization in relative scale, *Math. Oper. Res.* 34 (2009) 180–193.

[269] Yu. Nesterov, Rounding of convex sets and efficient gradient methods for linear programming problems, *Optim. Methods Softw.* 23 (2008) 109–128.

[270] Yu. Nesterov, Dual extrapolation and its applications for solving variational inequalities and related problems, *Math. Prog.* 109 (2007) 319–344.

[271] Yu. Nesterov, Smooth minimization of nonsmooth functions, *Math. Prog.* 103 (2005) 127–152.

[272] Yu. Nesterov, *Introductory Lectures on Convex Optimization. A basic course*, Applied Optimization, 87, Kluwer Academic Publishers, Boston, MA, 2004.
[273] Yu. Nesterov, Squared functional systems and optimization problems, High performance optimization, *Appl. Optim.* 33, Kluwer Acad. Publ., Dordrecht, 2000, pp. 405–440.
[274] Yu. Nesterov, Semidefinite relaxation and nonconvex quadratic optimization, *Optimization Methods and Software* 9 (1998) 141–160.
[275] Yu. E. Nesterov and A. S. Nemirovskii, *Interior-Point Polynomial Algorithms in Convex Programming*, SIAM, Philadelphia, PA, USA, 1994.
[276] Nesterov, Yu. E. and M. J. Todd, Self-scaled barriers and interior-point methods for convex programming, *Math. Oper. Res.* 22 (1997) 1–46.
[277] Yu. E. Nesterov and M. J. Todd, Primal-dual interior-point methods for self-scaled cones, *SIAM J. Optim.* 8 (1998) 324–364.
[278] J. Nie, K. Ranestad and B. Sturmfels, The algebraic degree of semidefinite programming, *Math. Program.* 122 (2010) 379–405.
[279] J. Nie and B. Sturmfels, Matrix cubes parameterized by eigenvalues, *SIAM J. Matrix Anal. Appl.* 31 (2009) 755–766.
[280] C. H. Papadimitriou, *Computational Complexity*, Addison-Wesley Publishing Company, CA, USA, 1994.
[281] C. H. Papadimitriou and M. Yannakakis, Optimization, approximation, and complexity classes, *J. Comput. System Sci.* 43 (1991) 425–440.
[282] P. A. Parrilo, Semidefinite programming relaxations for semialgebraic problems, *Math. Program.* 96 (2003) 293–320.
[283] P. A. Parrilo, A. Robertson and D. Saracino, On the asymptotic minimum number of monochromatic 3-term arithmetic progressions, *J. Combin. Theory A* 115 (2008) 185–192.
[284] P. A. Parrilo and B. Sturmfels, Minimizing polynomial functions, Algorithmic and quantitative real algebraic geometry (Piscataway, NJ, 2001), pp. 83–99, DIMACS Ser. Discrete Math. Theoret. Comput. Sci., 60, Amer. Math. Soc., Providence, RI, 2003.
[285] D. V. Pasechnik, Bipartite sandwiches: semidefinite relaxations for maximum biclique, 6th Twente Workshop on Graphs and Combinatorial Optimization (Enschede, 1999), 5 pp. (electronic), *Electron. Notes Discrete Math.*, 3, Elsevier, Amsterdam, 1999.
[286] G. Pataki, Cone LP's and semidefinite programs: Geometry and a simplex type method, *Proc. Fifth IPCO Conference*, W. H. Cunningham, S. T. McCormick and M. Queyranne (eds.), Springer-Verlag, pp. 161–174, 1996.
[287] G. Pataki, On the rank of extreme matrices in semidefinite programs and the multiplicity of optimal eigenvalues, *Math. Oper. Res.* 23 (1998) 339–358.
[288] G. Pataki and L. Tunçel, On the generic properties of convex optimization problems in conic form, *Math. Program.* 89 (2001) 449–457.
[289] J. Peña, J. Vera and L. F. Zuluaga, Computing the stability number of a graph via linear and semidefinite programming, *SIAM J. Optim.* 18 (2007) 87–105.
[290] J.-P. Penot, Softness, sleekness and regularity properties in nonsmooth analysis, *Nonlinear Anal.* 68 (2008) 2750–2768.
[291] I. Pólik and T. Terlaky, A survey of the S-lemma, *SIAM Review* 49 (2007) 371–418.
[292] S. Poljak and F. Rendl, Solving the max-cut problem using eigenvalues. Partitioning and decomposition in combinatorial optimization, *Discrete Appl. Math.* 62 (1995) 249–278.
[293] S. Poljak, F. Rendl and H. Wolkowicz, A recipe for semidefinite relaxation for (0,1)-quadratic programming, *J. Global Optim.* 7 (1995) 51–73.
[294] L. Porkolab and L. Khachiyan, On the complexity of semidefinite programs, *J. Global Optim.* 10 (1997) 351–365.
[295] V. Powers and B. Reznick, A new bound for Pólya's theorem with applications to polynomials positive on polyhedra, *J. Pure Appl. Algebra* 164 (2001) 221–229.
[296] V. Powers and B. Reznick, Polynomials that are positive on an interval, *Trans. Amer. Math. Soc.* 352 (2000) 4677–4692.
[297] A. Prestel and C. N. Delzell, *Positive Polynomials. From Hilbert's 17th problem to real algebra*, Springer Monographs in Mathematics. Springer-Verlag, Berlin, 2001.
[298] M. Putinar, Positive polynomials on compact semi-algebraic sets, *Indiana Univ. Math. J.* 42 (1993) 969–984.
[299] M. V. Ramana, An exact duality theory for semidefinite programming and its complexity implications, *Math. Program.* 77 (1997) 129–162.

[300] M. V. Ramana, L. Tunçel and H. Wolkowicz, Strong duality for semidefinite programming, *SIAM J. Optim.* 7 (1997) 641–662.

[301] J. Renegar, Hyperbolic programs, and their derivative relaxations, *Found. Comput. Math.* 6 (2006) 59–79.

[302] J. Renegar, *A Mathematical View of Interior-Point Methods in Convex Optimization*, MPS/SIAM Series on Optimization, Philadelphia, PA, USA, 2001.

[303] J. Renegar, A polynomial-time algorithm based on Newton's method for linear programming, *Math. Program.* 40 (1988) 59-93.

[304] B. Reznick, Some concrete aspects of Hilbert's 17th problem, Real algebraic geometry and ordered structures (Baton Rouge, LA, 1996), pp. 251–272, Contemp. Math., 253, Amer. Math. Soc., Providence, RI, 2000.

[305] R. T. Rockafellar, *Convex Analysis*, Princeton University Press, Princeton, New Jersey, 1970.

[306] R. T. Rockafellar and R. J. B. Wets, *Variational Analysis*, Springer-Verlag, Berlin, 1998.

[307] C. Roos, T. Terlaky and J.-P. Vial, *Interior Point Methods for Linear Optimization*, Second edition, Springer, New York, 2006.

[308] K. F. Roth, Remark concerning integer sequences, *Acta Arithmetica* 9 (1964) 257–260.

[309] W. Rudin, *Functional Analysis*, second edition, Mc-Graw Inc., NY, USA, 1991.

[310] J. M. Ruiz, On Hilbert's 17th problem and real Nullstellensatz for global analytic functions, *Math. Z.* 190 (1985) 447–454.

[311] M. G. Safonov, K. G. Goh and J. H. Ly, Control system synthesis via bilinear matrix inequalities, In *Proceedings of the 1994 American Control Conference*, Baltimore, MD, 1994.

[312] B. D. Saunders and H. Schneider, Cones, graphs and optimal scalings of matrices, *Linear and Multilinear Algebra* 8 (1979/80) 121–135.

[313] C. Scheiderer, Stability index of real varieties, *Invent. Math.* 97 (1989) 467–483.

[314] K. Schmüdgen, The $K$-moment problem for compact semi-algebraic sets, *Math. Ann.* 289 (1991) 203-206.

[315] R. Schneider, *Convex bodies: the Brunn-Minkowski theory*, Encyclopedia of Mathematics, Cambridge University Press, 1993.

[316] G. Schoenebeck, Linear level Lasserre lower bounds for certain $k$-CSPs, *FOCS* 2008.

[317] A. Schrijver, *Combinatorial Optimization, Polyhedra and Efficiency*, Springer-Verlag, 2003.

[318] A. Schrijver, A combinatorial algorithm minimizing submodular functions in strongly polynomial time, *J. Combin. Theory B* 80 (2000) 346–355.

[319] A. Schrijver, *Theory of Linear and Integer Programming*, John Wiley and Sons, 1986.

[320] M. Schweighofer, Global optimization of polynomials using gradient tentacles and sums of squares, *SIAM J. Optim.* 17 (2006) 920–942.

[321] M. Schweighofer, On the complexity of Schmüdgen's Positivstellensatz, *J. Complexity* 20 (2004) 529–543.

[322] P. D. Seymour, Decomposition of regular matroids, *J. Combin. Theory B* 8 (1980) 305–359.

[323] A. Shapiro, First and second order analysis of nonlinear semidefinite programs, *Math. Program.* 77 (1997) 301–320.

[324] H. D. Sherali and W. P. Adams, *A Reformulation-Linearization Technique for Solving Discrete and Continuous Nonconvex Problems*, Nonconvex Optimization and its Applications, 31. Kluwer Academic Publishers, Dordrecht, 1999.

[325] H. D. Sherali and W. P. Adams, A hierarchy of relaxations between the continuous and convex hull representations for zero-one programming problems, *SIAM J. Disc. Math.* 3 (1990) 411–430.

[326] H. D. Sherali and A. Alameddine, A new reformulation-linearization technique for bilinear programming problems, *J. Global Optim.* 2 (1992) 379–410.

[327] H. D. Sherali and Y. Lee, Tighter representations for set partitioning problems, *Discrete Appl. Math.* 68 (1996) 153–167.

[328] H. D. Sherali and C. H. Tuncbilek, A reformulation-convexification approach for solving nonconvex quadratic programming problems, *J. Global Optim.* 7 (1995) 1–31.

[329] J. W. J. Sikora, *Applications of Semidefinite Programming in Quantum Cryptography,* M. Math. Thesis, Dept. of Combinatorics and Optimization, University of Waterloo, Canada, 2007.

[330] N. Z. Shor, Dual quadratic estimates in polynomial and boolean programming, *Annals of Operations Research* 25 (1990) 163–168.

[331] N. Z. Shor, Quadratic optimization problems, *Soviet Journal of Computer and Systems Sciences* 25 (1987) 1–11.

[332] N. Z. Shor, *Metody minimizatsii nedifferentsiruemykh funktsii i ikh prilozheniya* (Russian), [Minimization methods for nondifferentiable functions and their applications] "Naukova Dumka", Kiev, 1979.

[333] A. M.-C. So and Y. Ye, A semidefinite programming approach to tensegrity theory and realizability of graphs, *manuscript*, 2005.

[334] A. M.-C. So, Y. Ye and J. Zhang, A unified theorem on SDP rank reduction, *Math. Oper. Res.* 33 (2008) 910–920.

[335] O. Stein, On constraint qualifications in nonsmooth optimization, *J. Optim. Theory Appl.* 121 (2004) 647–671.

[336] G. Stengle, A nullstellensatz and a positivstellensatz in semialgebraic geometry, *Math. Ann.* 207 (1974) 87–97.

[337] T. Stephen and L. Tunçel, On a representation of the matching polytope via semidefinite liftings, *Math. Oper. Res.* 24 (1999) 1–7.

[338] J. F. Sturm, Error bounds for linear matrix inequalities, *SIAM J. Optim.* 10 (2000) 1228–1248.

[339] J. Sturm and S. Zhang, Symmetric primal-dual path-following algorithms for semidefinite programming, *Appl. Numer. Math.* 29 (1999) 301–315.

[340] M. Talagrand, Embedding subspaces of $L_p$ in $l_p^N$, Geometric aspects of functional analysis (Israel, 1992–1994), pp. 311–325, *Oper. Theory Adv. Appl.*, 77, Birkhaüser, Basel, 1995.

[341] É. Tardos, A strongly polynomial algorithm to solve combinatorial linear programs, *Operations Research* 34 (1986) 250–256.

[342] É. Tardos, A strongly polynomial mimimum cost circulation algorithm, *Combinatorica* 5 (1985) 247–255.

[343] M. Tawarmalani and N. V. Sahinidis, *Convexification and Global Optimization in Continuous and Mixed-Integer Nonlinear Programming*, Kluwer Academic Publishers, Dordrecht, The Netherlands, 2002.

[344] M. J. Todd, Semidefinite optimization, *Acta Numer.* 10 (2001) 515–560.

[345] M. J. Todd, Potential-reduction methods in mathematical programming, *Math. Program.* 76 (1997) 3–45.

[346] L. Tunçel, Generalization of primal-dual interior-point methods to convex optimization problems in conic form, *Foundations of Computational Mathematics* 1 (2001) 229–254.

[347] L. Tunçel, On the Slater condition for the SDP relaxations of nonconvex sets, *Oper. Res. Lett.* 29 (2001) 181–186.

[348] L. Tunçel, Potential reduction and primal-dual methods, *Handbook of Semidefinite Programming: Theory, Algorithms and Applications*, H. Wolkowicz, R. Saigal and L. Vandenberghe (eds.), Kluwer Academic Publishers, Boston, MA, USA, 2000, pp. 235–265.

[349] L. Tunçel and H. Wolkowicz, Strong duality and minimal representations for cone optimization, *Research Report* 2008–07, Department of Combinatorics and Optimization, Faculty of Mathematics, University of Waterloo, Waterloo, Ontario, Canada, August 2008.

[350] L. Tunçel and H. Wolkowicz, Strengthened existence and uniqueness conditions for search directions in semidefinite programming, *Linear Algebra Appl.* 400 (2005) 31–60.

[351] L. Tunçel and S. Xu, Complexity analyses of discretized successive convex relaxation methods, *Research Report* 99–37, Department of Combinatorics and Optimization, University of Waterloo, Waterloo, Ontario, Canada, 1999.

[352] W. T. Tutte, Lectures on Matroids, *J. Res. Nat. Bur. Standards Sect. B* 69B (1965) 1–47.

[353] L. Vandenberghe and S. Boyd, A primal-dual potential reduction method for problems involving matrix inequalities, *Math. Program.* 69 (1995) 205–236.

[354] L. Vandenberghe and S. Boyd, Semidefinite programming, *SIAM Review* 38 (1996) 49–95.

[355] L. Vandenberghe, S. Boyd and S.-P. Wu, Determinant maximization with linear matrix inequality constraints, *SIAM J. Matrix Anal. Appl.* 19 (1998) 499–533.

[356] R. J. Vanderbei, *Linear programming. Foundations and Extensions*, Third edition, International Series in Operations Research & Management Science, 114 Springer, New York, USA, 2008.

[357] R. J. Vanderbei and B. Yang, The simplest semidefinite programs are trivial, *Math. Oper. Res.* 20 (1995) 590–596.

[358] C. F. Van Loan, The ubiquitous Kronecker product, *J. Comput. Appl. Math.* 123 (2000) 85–100.

[359] S. A. Vavasis, *Nonlinear Optimization. Complexity Issues*, Oxford University Press, New York, 1991.

[360] S. A. Vavasis and Y. Ye, A primal-dual interior point method whose running time depends only on the constraint matrix. *Mathematical Programming A* 74 (1996) 79–120.

[361] V. Vinnikov, Self-adjoint determinantal representations of real plane curves, *Mathematische Annalen* 296 (1993) 453–479.

[362] H. Waki, S. Kim, M. Kojima and M. Muramatsu, Sums of squares and semidefinite program relaxations for polynomial optimization problems with structured sparsity, *SIAM J. Optim.* 17 (2006) 218–242.

[363] R. Webster, *Convexity*, Oxford University Press, New York, USA, 1994.

[364] H. Wolkowicz, Explicit solutions for interval semidefinite linear programs, *Linear Algebra Appl.* 236 (1996) 95–104.

[365] H. Wolkowicz, Some applications of optimization in matrix theory, *Linear Algebra Appl.*, 40 (1981) 101–118.

[366] H. Wolkowicz, R. Saigal and L. Vandenberghe, editors. *Handbook of Semidefinite Programming: Theory, Algorithms, and Applications*, Kluwer Academic Publishers, Boston, MA, 2000.

[367] H. Wolkowicz and Q. Zhao, Semidefinite programming relaxations for the graph partitioning problem, *Discrete Appl. Math.* 96/97 (1999) 461–479.

[368] L. A. Wolsey, Further facet generating procedures for vertex packing polytopes, *Math. Prog.* 11 (1976) 158–163.

[369] S. J. Wright, *Primal-Dual Interior-Point Methods*, SIAM, Philadelphia, 1997.

[370] V. A. Yakubovich, S-procedure in nonlinear control theory, *Vestnik Leningrad. Univ.* 1 (1971) 62–77.

[371] M. Yannakakis, On the approximation of maximum satisfiability, In *Proceedings of the Third ACM-SIAM Symposium on Discrete Algorithms*, 1992, pp. 1–9.

[372] Y. Ye, Approximating quadratic programming with bound and quadratic constraints, *Math. Program.* 84 (1999) 219–226.

[373] Y. Ye, *Interior-Point Algorithms: Theory and Analysis*, John Wiley and Sons, NY, 1997.

[374] D. B. Yudin and A. S. Nemirovskii, Informational complexity and effective methods for the solution of convex extremal problems (Russian), *E'konom. i Mat. Metody* 12 (1976) 357–369.

[375] Y. Zhang, On extending some primal-dual interior-point algorithms from linear programming to semidefinite programming, *SIAM J. Optim.* 8 (1998) 365–386.

[376] Q. Zhao, S. E. Karisch, F. Rendl and H. Wolkowicz, Semidefinite programming relaxations for the quadratic assignment problem, *J. Combinatorial Optimization* 2 (1998) 71–109.

[377] G. Ziegler, *Lectures on Polytopes*, Springer-Verlag, Berlin, 1995.

# Index

# Titles in This Series

27 **Levent Tunçel,** Polyhedral and semidefinite programming methods in combinatorial optimization, 2010

26 **Clifton Cunningham and Monica Nevins, Editors,** Ottawa lectures on admissible representations of reductive $p$-adic groups, 2009

25 **Eric T. Sawyer,** Function theory: Interpolation and corona problems, 2009

24 **Kenji Ueno,** Conformal field theory with gauge symmetry, 2008

23 **Marcelo Aguiar and Swapneel Mahajan,** Coxeter groups and Hopf algebras, 2006

22 **Christian Meyer,** Modular Calabi-Yau threefolds, 2005

21 **Arne Ledet,** Brauer type embedding problems, 2005

20 **James W. Cogdell, Henry H. Kim, and M. Ram Murty,** Lectures on automorphic $L$-functions, 2004

19 **Jeremy P. Spinrad,** Efficient graph representations, 2003

18 **Olavi Nevanlinna,** Meromorphic functions and linear algebra, 2003

17 **Vitaly I. Voloshin,** Coloring mixed hypergraphs: theory, algorithms and applications, 2002

16 **Neal Madras,** Lectures on Monte Carlo Methods, 2002

15 **Bradd Hart and Matthew Valeriote, Editors,** Lectures on algebraic model theory, 2002

14 **Frank den Hollander,** Large deviations, 2000

13 **B. V. Rajarama Bhat, George A. Elliott, and Peter A. Fillmore, Editors,** Lectures in operator theory, 2000

12 **Salma Kuhlmann,** Ordered exponential fields, 2000

11 **Tibor Krisztin, Hans-Otto Walther, and Jianhong Wu,** Shape, smoothness and invariant stratification of an attracting set for delayed monotone positive feedback, 1999

10 **Jiří Patera, Editor,** Quasicrystals and discrete geometry, 1998

9 **Paul Selick,** Introduction to homotopy theory, 1997

8 **Terry A. Loring,** Lifting solutions to perturbing problems in $C^*$-algebras, 1997

7 **S. O. Kochman,** Bordism, stable homotopy and Adams spectral sequences, 1996

6 **Kenneth R. Davidson,** C*-Algebras by example, 1996

5 **A. Weiss,** Multiplicative Galois module structure, 1996

4 **Gérard Besson, Joachim Lohkamp, Pierre Pansu, and Peter Petersen Miroslav Lovric, Maung Min-Oo, and McKenzie Y.-K. Wang, Editors,** Riemannian geometry, 1996

3 **Albrecht Böttcher, Aad Dijksma and Heinz Langer, Michael A. Dritschel and James Rovnyak, and M. A. Kaashoek Peter Lancaster, Editor,** Lectures on operator theory and its applications, 1996

2 **Victor P. Snaith,** Galois module structure, 1994

1 **Stephen Wiggins,** Global dynamics, phase space transport, orbits homoclinic to resonances, and applications, 1993